THE SCIENTIFIC NAMES OF THE BRITISH LEPIDOPTERA

THE SCIENTIFIC NAMES OF THE BRITISH LEPIDOPTERA

Their History and Meaning

A. MAITLAND EMMET

1991

Harley Books (B. H. & A. Harley Ltd.)
Martins, Great Horkesley,
Colchester, Essex CO6 4AH, England

Text set in Linotron 202 Palatino by
Rowland Phototypesetting Ltd., and
printed in Great Britain by
St Edmundsbury Press Ltd.,
Bury St Edmunds, Suffolk
Bound by Hunter & Foulis Ltd., Edinburgh

The Scientific Names of the British Lepidoptera – their History and Meaning

British Library Cataloguing-in-Publication Data

Emmet, A. M.
The Scientific names of the British lepidoptera.
1. Lepidoptera. Great Britain
I. Title
595.780941

ISBN 0 946589 28 3 (Hardback edition)
ISBN 0 946589 35 6 (Paperback edition)

Contents

Foreword

The study of the Lepidoptera attracts a wide variety of persons, their common characteristic perhaps being the possession of an enlarged curiosity: a description applied to Charles Darwin who entered biology via entomology. Thus although the major interest of most lepidopterists will be some or all of the many aspects of the biology of their favourite insects, it will be an unusual person indeed whose curiosity is not also aroused by their scientific names.

The purpose of scientific nomenclature is to give each species a unique label, so that all the information may be correctly collated. If we were embarking on this task today, I fear some numerical scheme, particularly suited to computers, would be adopted. Indeed, as Colonel Emmet points out in this work, Linnaeus himself utilized a simple form of this approach with the plume moths, but the great Swede and those others who laid the foundations of our present system have given us something just as versatile and precise as a numerical system, but infinitely more interesting. Just how interesting is apparent from the briefest glance at any of the following pages in which, with infinite care and a remarkable blend of entomological and classical scholarship, Colonel Emmet has provided a thorough guide to the history and meanings of the names of the British Lepidoptera. The great majority of the authors of scientific names have attempted to make the name apt and, hence, frequently provide some descriptive information. Many of the specific names of our butterflies indicate their larval food plants, e.g. 1546 *Gonepteryx rhamni* or 1559 *Satyrium pruni*, though some associations are misleading, e.g. 1555 *Callophrys rubi*. Less prosaically, who pursuing a hairstreak on the edge of a woodland clearing will not concur with Colonel Emmet's conclusion that *Satyrium* refers to the spritely flight, like the dances of the mythical satyrs, or when the Dingy Skipper one has been hunting for rises up from ground at one's feet not think that Linnaeus must have had similar experiences when he named it *tages*, after the boy in the Etruscan myth who suddenly rose up from a furrow.

In spite of all his endeavours, the author has been unable to resolve a few names, and these are listed in Appendix 3. I feel sure that this will provide a challenge to those with a classical bent and a code breaking (or crossword) caste of mind, but I do not think they will find it easy to succeed where Colonel Emmet has abandoned the chase. Other names mislead, but then so do the popular names: Kentish Glory, a glorious moth certainly and a particular favourite of the late Duke of Newcastle (to whom this work is so appropriately dedicated), but no longer one with a S.E. England distribution. The scientific name 1644 *Endromis versicolora*, as derived by Colonel Emmet, does at least tell us something reliable about its appearance.

Whether one has a good grasp of Latin and Greek or whether, as more common today, the merest acquaintance, the derivations and explanations provided in this work will intrigue one in the study and their recollection add spice to the pursuit of the name-bearer in the field. Once again the entomological world is greatly indebted to Colonel Emmet for providing a masterly and informative work.

Professor Sir Richard Southwood FRS
Vice-Chancellor, University of Oxford
September, 1990

To
Edward Charles Pelham-Clinton
10th Duke of Newcastle (1920–88)
('Teddy')

In gratitude for the interest he took
in the preparation of this book
and the help that he gave

Introduction

Nomenclature, classification and conventions

The nomenclature adopted in this work is that of Kloet & Hincks (1972), as emended and supplemented by Bradley & Fletcher (1974; 1979; 1983; 1986), Emmet (1987a) and Agassiz (1987).

In the Systematic Section specific names are preceded by the Log Book numbers allocated by Bradley & Fletcher (1979; 1983; 1986). Generic and suprageneric names were not allocated Log Book numbers and are therefore cited under the number of the first species to follow. Where suprageneric names are explained under a generic name (in one case a specific name), the relevant Log Book number is added in parenthesis. In cross references the Log Book number is normally given, but may be omitted if there is no need for the reader to refer to the meaning of the name. Author's names and dates of publication are given in the main entry for each genus and species but are not repeated in cross references unless of importance in the context. Where a series of species is cited, Log Book numbers only may be given. Where reference is made to another species, the generic name or, if in the same genus, the generic initial is given, but if the reference is to the name as such and not to the insect, no generic name or initial is used.

Synonyms are not normally included unless listed by Bradley & Fletcher (1986); where a name has only recently been introduced, the superseded name is necessary for recognition. However *nomina dubia* (see p. 10) which have been accepted on the Continent but not in Britain and are therefore added in synonymy by Bradley & Fletcher are not included in this work. Synonyms not listed by Bradley & Fletcher are sometimes included here if they are important, e.g. for the explanation of a suprageneric name.

The order in which the species are presented is that of Bradley & Fletcher (1986) which differs in certain respects from that of their earlier publications. This means that the Log Book numbers do not always follow in numerical sequence; accordingly both page and Log Book numbers are given in the index and a special key (p. 288) is given for the Phycitinae where the rearrangement is most extensive.

Certain names may be given in Roman or italic type according to the context in which they appear; thus, Hesperia is the family set up by Fabricius to accommodate all the blue and skipper butterflies, *Hesperia* Fabricius, 1793, is the genus with its modern restricted application.

I have followed the convention of lexicons and dictionaries in giving the first person singular of the present indicative of a classical verb, but the infinitive in its English translation, e.g. φιλέω (phileō), *amo*, to love; the literal translation would be 'I love'. A few verbs are defective and have no present indicative; in these cases I give the infinitive, e.g. φαγεῖν (phagein), to eat.

After a Greek word written in Greek characters, I give its transliteration in Roman letters, e.g. κύκλος (kuklos), but if this word were to be incorporated into a name, it would take the form 'cyclus'. It is therefore necessary to give the rules for the Latinization of Greek words.

αι (ai) becomes 'ae', e.g. 686 *Exaeretia* from ἐξαίρετος (exairetos)

γκ (gk), γχ (gkh) become 'nc', 'nch', e.g. 1115 *Ancylis* from ἀγκυλίς (agkulis); 761 *Rhynchopacha* from ῥύγχος (rhugkhos) and παχύς (pakhus)

ϰ (k) becomes 'c', e.g. 1160 *Acroclita* from ἄϰρον (akron) and ϰλιτύς (klitus)
χ (kh) becomes 'ch', e.g. 1531 *Ochlodes* from ὀχλώδης (okhlodes)
οι (oi) becomes 'oe', e.g. 651 *Oecophora* from οἶϰος (oikos) and φορέω (phoreo).
-ος (-os) becomes -us', e.g. 1571 *argus* from Ἄργος (Argos)
ου (ou) becomes 'u', e.g. 1995 *Cerura* from ϰέρας (keras) and οὐρά (oura)
υ (u) becomes 'y', e.g. 1240 *Cydia* from ϰῦδος (kudos)
The diphthongs αυ (au) and ου (ou) do not alter

Greek and Latin are inflected languages, i.e. the termination of a word is altered to express case, number, gender, person, tense, etc. The nominative case of a noun does not always show the stem, i.e. the base of the word to which the terminations are added. Where the stem needs to be shown, the genitive case is added after the nominative, e.g. 992 *rurinana* from *rus, ruris*; 234 *Trichophaga* from θρίξ, τριχός (thrix, trikhos) and φαγεῖν (phagein).

The formation of some words changes when they are used in composition; thus εἶδος (eidos), form, often becomes -ωδ- (-od-) and is shown thus, εἶδος, ωδ- (eidos, od-), the hyphen placed conventionally after the word, e.g. 1001 *Lozotaeniodes*.

Paradoxically, for the purpose of this book old editions of lexicons and dictionaries are better than new ones. Names bestowed by Linnaeus and Hübner should, if possible, be explained after reference to the dictionaries they themselves used. Dictionaries from the 18th century are seldom available in libraries as they prefer to keep up to date. My own dictionaries, inherited from my grandfather, date from the middle of the 19th century, *A Greek-English Lexicon* (Liddell & Scott, Edn 6, 1869), *A Latin-English Dictionary* (Smith, of about the same date, but the title page is missing) and (from my father) *A smaller classical Dictionary* (Smith, Edn 17, 1877), the last differing little from the first edition of 1852. I have on occasion consulted the edition of Liddell & Scott, revised and augmented by Jones (1968), which is the most authoritative lexicon available, the Latin-English dictionary of Lewis & Short and a wide range of classical dictionaries. Though older dictionaries might have been better, I feel that those in my possession represent fairly faithfully the information available to the entomologists of the 18th and 19th centuries when most of our insects received their names.

Terms and abbreviations

adj., adjective
alpha copulative, α, the letter alpha, added as a prefix to express similarity
alpha intensive, α, the letter alpha, added as a prefix to strengthen the meaning
alpha privative, α, the letter alpha, added as a prefix to a Greek word to remove or negate its meaning, as in 'asexual'
b., born
c., *circa* (Lat.), about
cf., *confer* (Lat.), compare
d., died
dim. diminutive
e.g., *exempli gratia* (Lat.), for example
et al., *et alii* (Lat.), and others, used to abbreviate a list of authors
fem., feminine
fl., *floruit* (Lat.), he flourished (followed by a date)
gen., genitive
Gr., Greek
I.C.Z.N., The International Commission on Zoological Nomenclature
i.e., *id est* (Lat.), that is
i.h.o., in honour of
Lat., Latin
masc., masculine
MBGBI, *The moths and butterflies of Great Britain and Ireland*
n., noun
nec (Lat.), and not, but not
neut., neuter
nom., nominative
nomen dubium (Lat.), a doubtful name, one not ascribable with certainty to any one species
nomen nudum (Lat.), a 'naked' name, one unaccompanied by a description to indicate the species to which it was intended to apply

p., pp. (pl.), *pagina, paginae* (Lat.), page, pages
part., participle
phonetic insertion, a letter added, e.g. after an alpha privative, to aid pronunciation
pl., plural
q.v., qq.v. (pl.), *quod vide, quae vide* (Lat.), 'which see', which you should consult
sensu (Lat.), in the opinion (of)
sensu auctt., sensu auctorum (Lat.), in the opinion of authors
sic (Lat.), thus; used to draw attention to a misspelling in a quotation
sing., singular
sp., spp. (pl.), species (sing.), species (pl.) (Lat.)
subsp., subspp. (pl), subspecies (sing.), subspecies (pl.) (Lat.)
teste according to the evidence (of)
usu (Lat.), as used by
viz., *videlicet* (Lat.), namely (vi- + z, late Lat. symbol for contraction)

Acknowledgements

I wish to express my thanks to Basil Harley, who suggested the subject of this book and gave extensive editorial assistance; to the dedicatee, the late Duke of Newcastle (E. C. Pelham-Clinton), for various suggestions and for help in tracing the identity of the entomologists in whose honour species have been named; to Mrs Joan Heath and Canon D. J. L. Agassiz for the loan of books; to the Royal Irish Academy for information on the meaning of the generic names bestowed by B. P. Beirne; to the staff of the Entomological Library, British Museum (Natural History) and to that of the Rare Books Room in the Library of the University of Cambridge, where most of my research has been conducted, for providing me with the literature I needed; and to the British Museum (Natural History) and particularly to Miss Pamela Gilbert, Deputy Head of Library Services, for providing the illustrations. Plate VII depicting Jacob Hübner is reproduced from an illustration in Hemming's *Hübner*, 1937, of a miniature in the possession of Naturwissenschaftliche Gesellschaft zu Augsburg.

A. M. E.

Plate I Carl von Linné (Linnaeus), 1707–1778

A History of the Scientific Nomenclature of Lepidoptera

A book for the curious

When a specimen of 1273/1286 *Dichrorampha sedatana* Busck emerged recently, I was pleased to have a reared specimen of a moth I had hitherto encountered only as an adult, but even more so to know that I had before me the 'staid' member of the 'two-coloured bill-hook' genus. A scientist is expected to treat his subject with detachment and objectivity, but the bestowal of names is an aspect of his work in which he can indulge his fantasy and introduce poetic licence or anthropomorphism without reproof. Guenée must have enjoyed the creative activity he exercised in devising a generic name that was imaginative and mysterious, yet at the same time scientifically informative. Busck lacked the Gallic panache, but proffered a sensible, workaday specific name.

Scientific names have much in common with crossword puzzles. The nomenclator is the setter; he searches for a name that is neat and appropriate and if he can mystify his fellow entomologists, he will derive sadistic pleasure in so doing. His successors are the solvers, seeking the answer to the riddle he has set. Some clues are cryptic, others matter-of-fact; some poetic, others pedestrian. There is no need to seek the explanations of scientific names, just as there is no need to do crossword puzzles. There are, however, many who derive pleasure from both pursuits. This book is written for the reader who is curious about the names themselves, the reasons why they were given and, in the case of the older supraspecific names, the history of their application.

Literature devoted to the explanation of scientific names

When Linnaeus (1758) inaugurated the binominal system of nomenclature, he did not explain the meaning of the names he gave except in rare cases like that of 1053 *hastiana*. His immediate successors followed the pattern set by their master. Fabricius, in particular, obviously enjoyed posing riddles. By the early 19th century curiosity over the meaning of scientific names began to assert itself and this led some authors like Ochsenheimer to give the derivations of those he bestowed. Others like Hübner, who probably coined more generic and specific names than any other lepidopterist, maintained the Linnaean tradition of silence. To know in part stimulates the wish to know in full and the field was open to the entomological exegetist.

The first author to devote a paper to the meaning of scientific names was Sodoffsky (1837). He confined his research to the generic names. Not only did he endeavour to explain their meaning, but he also emended them as he saw fit. Some of his changes were harmless as when he inserted the aitches omitted by French authors (e.g. *Hyponomeuta* for *Yponomeuta*), but others altered the sense to what he supposed it should have been; for instance, *Hepialus* puzzled him so he changed it to *Hepiolus* taken from a Greek word for a moth. This was a dangerous practice later quashed by the rules of the I.C.Z.N. which state that the original spelling must stand even if the author himself later sought to correct it.

The first British book wholly about scientific names was *An accentuated list of the British Lepidoptera* (1858) by members of the Entomological Societies of Oxford and Cambridge, cited in this work as Pickard *et al.*, since the Rev. H. A. Pickard, the Oxford president, is the first of the eleven names cited from the two Councils. He was, indeed, one of the principal authors, the other being J. W. Dunning, a vice-president of the Cambridge Society (Cowan, 1971). Eminent names on the editorial panel are J. O. Westwood, the first Hope Professor of Entomology at Oxford, and Professor C. C. Babington of Cambridge, who traced the true locality of the black hairstreak to Monks Wood when Seaman, the dealer who discovered it, had tried to put collectors off the scent by saying that the butterflies were from Yorkshire. Besides explaining the meaning of scientific names, Pickard *et al.* gave advice on pronunciation and included interesting accounts of many of the principal nomenclators. The book must have enjoyed a wide circulation since today, well over a century after its first publication, it is by no means a rarity in second-hand bookshops.

A few of the authors of text-books on the Lepidoptera give brief etymological notes on the names, stating the Greek or Latin words from which they believe them to be derived, but generally without further detail. Important among these is Spuler (1903–10). Such embellishment was a sideline conducted without full research and errors are only to be expected.

In relatively recent years a second book has appeared in Britain which is wholly devoted to explaining the meaning of the scientific names of our Lepidoptera. This is the *Key to the names of British butterflies and moths* by R. D. Macleod. It is still in print and enjoys a good sale. Why then, it may be asked, is yet another book needed?

In 1959, when the *Key* was published, there was no up-to-date standard check list of our Lepidoptera. There was a range of sources for Macleod to choose from and he was eclectic. He took some names from a list produced by the Amateur Entomologist's Society (Cooper & O'Farrell, 1946), some from *An indexed check-list of the British Lepidoptera* Edn 1 (Heslop, 1947), some (Tortricidae) from a paper in the *Entomologist's Gazette* (Bradley & Martin, 1956), some from the typescript of the new edition of South's *The Moths of the British Isles* which was published two years after the *Key*, and yet others on the verbal advice of the officers of the British Museum (Natural History) who were working on the Lepidoptera section of the second edition of Kloet & Hincks (1972). Clearly he went to a great deal of trouble, but in the circumstances it is not surprising that many of the names he uses differ from those current today. In a sample of four pages from the *Key*, two of Macrolepidoptera and two of Microlepidoptera chosen at random except that I avoided those with long runs in a single genus, I found that 10 per cent of the specific names and 47 per cent of the generic names had been changed. Had I taken into account variation in spelling, the percentage would have been greater. Moreover, many species have been added to the British list in the last thirty years. Since Macleod gave no index, species are hard to find when they are listed in genera which are now unfamiliar. Clearly a new list is needed to cover these changes.

Macleod had already written successful books on the scientific names of birds, plants and mammals. He tells us that the Lepidoptera confronted him with more names than there were in his three previous volumes put together, and in order to produce a matching volume he had to 'practise the utmost conciseness possible'. He therefore omitted authors' names and dates, this accounting for the anachronisms noted in the Systematic Section below. Being a general naturalist rather than an entomologist, he frequently sought help at the British Museum (Natural History), where he is remembered with regard by some of the senior officers. Some of his errors may have resulted from his adopting off-the-cuff suggestions made by experienced entomologists. Although about 15 per cent of Macleod's explanations are faulty (see Appendix 4), his book is a very useful source. However he was somewhat inventive when at a loss and it has been considered necessary to draw attention throughout this work to his misinterpretations and fictitious derivations which might otherwise get wider currency.

Names of Lepidoptera before Linnaeus

Although most common species of mammals, birds, fish and plants were given names in each language from early times, the same was not true of insects. There were names for classes of insects such as bees, wasps and flies, but none for the members of those classes unless they made some special impact on mankind. Aristotle and Pliny the Elder had included insects in their studies, but thereafter they were almost wholly ignored until the 16th century. The first book about insects to be published in Britain was *Theatrum Insectorum* by Thomas Mouffet*. This was compiled from the notes of several authors in the 1580s but not published until 1634 when it aroused considerable interest, so much so that the original Latin text was soon followed by an English translation. The insects were described using a strange mixture of fact and myth, and some were crudely figured. Names were almost wholly lacking; however, two were given to lepidopterous larvae and, since they were adopted by Linnaeus, they are in use today, viz. 1992 *porcellus* and 1995 *vinula*. It is worth mentioning by way of digression that Mouffet refers to hairy caterpillars as woolly bears. The next author was Christopher Merrett whose *Pinax* (1666), a general work on British natural history including Lepidoptera, used no names and therefore does not concern us here.

James Petiver, a general naturalist and the father of British entomology, was the first to use vernacular names for the Lepidoptera. He formed a collection of natural history specimens which was destined to become, via Sir Hans Sloane, the basis of the British Museum. In 1695 he began to publish illustrated lists of his specimens with brief supporting text, partly in Latin and partly in English. His first entry is the brimstone butterfly, under that very name. Details of his other English names for butterflies may be found in MBGBI 7 (1), chapter 1. Here we are more concerned with his contribution towards the evolution of scientific names. Most of his Latin names were in effect brief descriptions and were not binomial. Thus we have *Papilio testudinarius major*, *P. testudinarius minor* and *P. testudinarius alis lacertis* for the large tortoiseshell, the small tortoiseshell and the comma respectively; the third name may be translated as 'the tortoiseshell butterfly with jagged wings'. He called the clouded yellow *Papilio croceus*, and if the rules allowed he could be cited as author of the name in current use. Some of his other names were fanciful, like *P. oculus pavonis* for his 'peacock's eye' and *P. Bella Donna* for his 'painted lady'. Probably the English names came first and the Latin ones were simply translations, but one cannot be certain.

With regard to other pre-Linnaean English entomological writers, John Ray (1710) relied mainly on his friend Petiver for his information and names, while Albin (1720) and Wilkes (1741–42; 1747–49) improved the English names without bothering about the Latin ones.

Linnaeus

In his earlier writings such as *Fauna Suecica* (Edn 1, 1746), Linnaeus' (Pl. I) use of Latin names was co-ordinate with, or a development of, that of Petiver; in fact he used some of Petiver's names such as *Oculus pavonis*, *Bella Donna* and *Ammiralis*, the (red) admiral. It was later that he formulated what is known as the binomial system of nomenclature whereby every plant and animal had two names, the first denoting the group to which it belonged and the second peculiar to the creature itself. This plan first found written expression in *Systema Naturae* (Edn 10, 1758) and scientific nomenclature is deemed to have originated in that work.

* Also spelt Moufet, Moffett, Moffet and Mofet; the spelling adopted here is that used by Linnaeus. In the nursery rhyme, his daughter is spelt 'Muffet' to rhyme with 'tuffet'.

The genus as conceived today is not Linnaean. His supraspecific taxa were more like our families or superfamilies and the genus containing only a few or even a single species was a later development. I shall therefore start with the specific names, leaving the evolution of the genus until later.

The specific names of Linnaeus

At the end of the part of *Systema Naturae* (Edn 10) which dealt with insects, Linnaeus wrote, *'ea quae scimus sunt pars minima eorum, quae ignoramus'* (the ones we know form only a fraction of the many of which we have no knowledge). He had no illusions over the problem of finding enough names. He had to start where Adam left off: 'And out of the ground the Lord God formed every beast of the field, and every fowl of the air; and brought them unto Adam to see what he would call them; and whatsoever Adam called every living creature, that was the name thereof' (Genesis 2:19). The God-Adam partnership had jibbed at insects. If you had to invent an almost infinite number of Latin or Latinized names starting from scratch, how would you set about it?

I shall work my way through the pages of *Systema Naturae* that cover the Lepidoptera (458 (Pl. II) –542, 822–23), noting each type of name as it appears, but first a general statement on the formation of the names is needed. Linnaeus' supraspecific names are all nouns, e.g. *Papilio*. His specific names, however, are of three kinds:

1. Nouns in apposition, e.g. *Papilio Machaon*, the butterfly Machaon.
2. Nouns in the genitive case, e.g. *Sphinx Convolvuli*, the hawk-moth of the *Convolvulus* genus.
3. Adjectives in agreement, e.g. *Sphinx fuciformis*, the hawk-moth formed like a drone.

Linnaeus spelt all nouns with a capital letter and all adjectives with a small letter unless formed from a noun that had to have a capital such as the name of a person or a genus of plants, e.g. *lubricipeda*, *Alniaria*. Today all specific names are spelt with the lower case, but in this section I am quoting from Linnaeus and shall therefore follow his usage.

Linnaeus found a treasure-house of names in the Greek and Roman literature which formed the basis of contemporary education. For the swallowtails he turned to Homer and especially the *Iliad*. He applied the names of Trojan heroes to those that had the thorax marked with red, starting with Priamus, King of Troy. The names of Greek heroes (Achivi) were given to those that lacked this red, headed appropriately enough by Helena, 'the face that launched a thousand ships', and her rightful husband Menelaus. 1539 *Machaon* and 1540 *Podalirius* are in this group.

After the Equites or Knights came the Heliconii which took their names from the Muses and Graces that dwelt on Mt Helicon. The first name is 1536 *Apollo*, their patron god, and this is the only one on the British list, the *Aglaja* of this group not being the fritillary of this name.

The third section are the Danai. Danaus (see 1630) conveniently had fifty daughters, a splendid source for names. Linnaeus divided the Danai into two groups, those that were white (Candidi), and those gaily coloured (Festivi). Here we encounter a new type of name, for eight of the eighteen Danai candidi are named from their foodplant. Among them are 1548 *Crataegi*, 1549 *Brassicae*, 1550 *Rapae*, 1551 *Napi*, 1541 *Sinapis*, 1553 *Cardamines* and 1546 *Rhamni*. Interspersed are the daughters of Danaus like 1552 *Daplidice* and 1543 *Hyale*.

The Nymphales follow, named predominantly after nymphs like 1620 *Galathea*, but some from other mythological beings like 1616 *Maera*, who was a dog. There are two new brands of name, 1594 *polychloros*, the first to be spelt without a capital, taken from earlier non-classical literature, and 1598 *C.album* [sic], the first name descriptive of a character in

III. LEPIDOPTERA.

Alæ IV, *imbricatæ ſquamis.*
Os *Lingua involuta ſpirali.*
Corpus *piloſum.*

203. PAPILIO. *Antennæ* apicem verſus craſſiores, ſæpius clavato - capitatæ.
Alæ (ſedentis) erectæ, ſurſumque conniventes (volatu diurno).

* EQUITES *Trojani.*

Priamus. 1. P. E. alis denticulatis tomentoſis ſupra viridibus : inſtitis atris; poſticis maculis ſex nigris.
Vincent. muſ. 10. Papilio amboinenſis viridi & nigro-holoſericeus inſignis.
Muſ. petrop. 664. Papilio ſ. Atlas amboinenſis, alis ſuperioribus holoſerice nigris : inſtitis viridibus.
Habitat in Amboina.
Papilionum omnium Princeps longe auguſtiſſimus, totus holoſericeus, ut dubitem pulchrius quidquam a natura in inſectis productum.

Cor-

Papiliones *dividuntur* in VI phalanges :
a. Equites *Alis primoribus ab angulo poſtico ad apicem longioribus, quam ad baſin; his ſæpe Antennæ filiformes.*
— — Trojani *ad Pectus maculis ſanguineis, (ſæpius nigri).* 1 - 17.
— — Achivi *Pectore incruento, ocello ad angulum ani :*
— — *Alis abſque faſciis.* 18, 19.
— — *Alis faſciatis,* 20 - 40.
b. Heliconii *Alis anguſtis integerrimis ſtriatis : primoribus oblongis; poſticis breviſſimis.* 41 - 55.
c. Danai *Alis integerrimis.*
Candidi *Alis albidis* 56 - 74.
Feſtivi *Alis variegatis* 75 - 87.
d. Nymphales *Alis denticulatis :*
Gemmati *Alis ocellatis :*
Ocellis in Alis omnibus 88 - 103.
in Alis primoribus 104 - 106.
in Alis poſticis 107 - 110.
Phalerati *Alis cæcis abſque ocellis* 111 - 144.
e. Plebeji *parvi : Larva ſæpius contracta :*
Rurales *Alis maculis obſcurioribus* 145 - 161.
Urbicolæ *Alis ſæpius maculis pellucidis* 162 - 168.
f. Barbari *corollarii loco adjecti, ad ordines non relati.* 169 - 192.

Plate II Page 458 from Linnaeus' *Systema Naturae* (Edn 10), 1758; Papiliones – the Butterflies

the wing pattern. A little later we find a group of three non-British species named in sequence from their similarity to each other, *similis, assimilis* and *dissimilis*. In the 12th edition of *Systema Naturae* (1767) Linnaeus was to use the same device for species on our list (2316 *affinis* and 2317 *diffinis*).

After the Nymphales, we have the Plebeji or Commoners (the smaller butterflies), and here we get a variation of the last type of name in that the model and the matching name are both nouns: 1571 *Argus* is followed by 1580 *Argiolus*, '*praecedenti similis, sed minor*' (like the last, but smaller).

The butterflies are followed by the Sphinges or hawk-moths. Many of their names follow earlier patterns, but new ones also emerge. In 1980 *ocellata* we have an adjectival name descriptive of the wing pattern; it differs from *C.album* which was a noun in apposition. Our vernacular names, the eyed hawk-moth and the comma, echo this distinction. Another innovation is the 'comparative' name which compares its bearer with some other person, creature or thing, e.g. 1983 *fuciformis*, the hawk-moth like a drone. In *pectinicornis* (non-British) we have the first name descriptive of a structural character, in this case of the antenna which is pectinated in its basal half. On later pages such names appear for British species, e.g. 2477 *proboscidalis*.

The Phalaenae*, the rest of the moths, come next (Pl. III). One of the first is 1643 *pavonia*, an adjectival metonymous name. Then there is 1642 *quercifolia*, descriptive of adult behaviour in that the moth rests in a posture imitative of a bunch of leaves, and 1640 *potatoria*, from the behaviour of the larva which likes drinking dew. The name 1634 *Neustria* has been supposed, probably wrongly, to have been taken from a district in France; 2300 *maura* is certainly named from Mauretania, the type locality, and 1037 *Holmiana* from Stockholm. Linnaeus also turned to human occupations and behaviour; 1635 *castrensis*, the camper, is both metonymous and descriptive of larval behaviour.

New ground is broken with names like 2057 *Caja*, – *virgo* and 2058 *villica*. Caja, the feminine form of Caius, is a Roman lady's name which, like our Mary, belongs to nobody in particular; *virgo* refers to a woman's condition and *villica* to her occupation (housekeeper). Such names are fanciful and need not have any entomological application, though they invite speculation (see 2026 *antiqua*).

In the Geometrae Linnaeus introduced a new convention, that of giving a common adjectival termination to the names of related species or those having a similar structural character. In this case he gave the ending *-aria* to species that had pectinate antennae and *-ata* to those with simple antennae. He also introduced several new categories of name. 1376 *Hortulata*, a pyralid treated as a geometrid by Linnaeus, is named from its habitat, '*habitat in* Urtica, *hortis pomonae*' (it lives on nettle in orchards). Another name refers to uncertainty over classification; of 1790 *dubitata* (*dubius*, doubtful), Linnaeus wrote '*statura* Geometrae, *magnitudo* Noctuae, *facies* Tineae' (the build of a geometrid, the size of a noctuid and the look of a tineid). A third innovation in the Geometrae is exemplified by 1799 *brumata*, named from winter, the season of the moth's appearance. By this time Linnaeus was beginning to let his imagination run away with him, as in 1839 *succenturiata*, q.v.

*From φάλαινα (φάλλαινα) (phalaina, phallaina), a word used by Aristotle for a moth. It also signified a whale or devouring monster and so may be derived from the destructive properties of the clothes-moth. However, scholars have supposed that it is akin to φαλλός (phallos), the *phallus* or erect penis, in view of the Greek association between Lepidoptera and semen which was supposed to attract or even generate them; see Davies & Kathirithamby, 1986, pp. 108–109 and also figs 23, 24. Another explanation is that φάλαινα is derived from the same root as φάος (phaos), light, and refers to the attraction of moths to a lamp (*ibid.*, p. 108).

ſuperioribus) intrantibus regionem exteriorem alæ, & vix manifeſto margine nigro cinctæ: colore demum alarum pallidius flaveſcente nec ferrugineo. Quomodo hæc a priori orta, dies docebit.

Cecropia. 3. P. *Bombyx* elinguis, alis patulis ſubfalcatis griſeis: faſcia fulva, ſuperioribus ocello ſubfeneſtrato ferrugineo. *M. L. U.*
Catesb. car. 2. *p.* 86. *t.* 86.
Habitat in America *ſeptentrionali.*

Paphia. 4. P. *Bombyx* elinguis flava, alis patulis falcatis concoloribus ocello feneſtratis. *M. L. U.*
Pet. gaz. t. 29. *f.* 3. *Catesb. car.* 2. *p.* 91. *t.* 91?
Habitat in Guinea.

Luna. 5. P. *Bombyx* elinguis, alis patulis caudatis flavo-virentibus concoloribus, ocello diſci lunato.
Catesb. car. 2. *p.* 84. *t.* 84. *Pet. gaz. t.* 14. *f.* 5.
Habitat in America *ſeptentrionali. Kalm.*

pavonia. 6. P. *Bombyx* elinguis, alis patulis rotundatis griſeo nebuloſis ſubfaſciatis: ocello nictitante ſubfeneſtrato.

α *Fn.*

Phalænæ dividendæ, quo facilius inquirantur,
Primariæ *Alis incumbenti depreſſis*:
1. - BOMBYCES *Antennis Pectinatis*:
- - Elingues *abſque lingua manifeſte ſpirali.*
- - - læves *dorſo, nec criſtatæ*:
- - - - *Alis* patulis 1 — 7.
- - - - *Alis* reverſis 8 — 21.
- - - - *Alis* deflexis 22 — 34.
- - - criſtatæ *dorſo faſciculis exaſperato.* 35 — 40.
- - Spirilingues *Lingua involuto-ſpirali*:
- - - læves, *Alis* patulis 41 — 44.
Alis deflexis 45 — 52.
- - - criſtatæ *dorſo* 53 — 58.
2. - NOCTUÆ *Antennis ſetaceis, nec pectinatis.*
- - Elingues 59 — 63.
- - Spirilingues; læves *dorſo* 64 — 85.
criſtatæ *dorſo* 86 — 126.
3. GEOMETRÆ *Alis patentibus horizontalibus quieſcentes*:
Pectinicornes: *alis poſticis* angulatis *ſ. dentatis* 127 — 133.
alis poſticis rotundatis *integris* 134 — 154.
Seticornes: *alis* angulatis 155 — 161.
alis rotundatis 162 — 201.
4. TORTRICES *Alis obtuſiſſimis ut fere retuſis*, planiuſculis 202 — 225.
5. PYRALIDES *Alis conniventibus in figuram deltoideam forficatam* 226 — 233.
6. TINEÆ *Alis convolutis fere in cylindrum*, fronte prominula 234 — 298.
7. ALUCITÆ *Alis digitatis fiſſis ad baſin* 299 — 304.

Plate III Page 496 from Linnaeus' *Systema Naturae* (Edn 10), 1758; Phalaenae – the Moths

Geometra is followed by Tortrix, with the family termination *-ana*. Here Linnaeus initiated the practice of naming species after people, in this instance his pupils, 1053 *Hastiana*, q.v., being a moving example. Pyralis, termination *-alis*, contains only eight species and no innovations, but in Tinea, termination *-ella*, Linnaeus began to commemorate naturalists of the past; the Englishmen Mouffet, Petiver, Ray and Wilkes are so honoured.

The final family is Alucita, the plumes. Linnaeus included only six species, giving them names that numbered them serially (*mono-*, *di-*, *tri-*, *tetra-*, *penta-* and *hexa-*), followed by the termination *-dactyla* from the Greek word for finger, signifying the lobes into which the wings are divided. All except for 1288 *hexadactyla* have the same number of lobes, so it is better to regard, for example, *tetradactyla* as meaning 'plume no. 4' rather than 'the plume with four lobes'.

Thus in the year 1758 did Linnaeus initiate the naming of the Lepidoptera. He listed only 542 species, but used such an extensive pattern of names that in future years the wit of man was hard put to it to devise new forms. Hardly a name has been bestowed since that is not modelled on one that is found in *Systema Naturae*, Edition 10. There follows a summary of the name-types used by Linnaeus, arranged more or less in the order in which they first appear in his pages. Examples are given from the British list, now using the lower case, with the Log Book numbers for easy reference.

1. From a name taken from classical literature (1590 *atalanta*)
2. From the larval foodplant (1593 *urticae*)
3. From a name used for the insect in earlier literature (162 *cossus*; 1995 *vinula*)
4. From a character in the wing pattern (2441 *gamma*)
5. From comparison with another species (1580 *argiolus*)
6. From a structural character of the adult (2477 *proboscidalis*) or larva (2297 *pyramidea*)
7. From the behaviour of the adult (1641 *ilicifolia*) or the larva (1640 *potatoria*)
8. From resemblance in appearance or behaviour to some other person, creature or thing (2033 *monacha*; 1635 *castrensis*)
9. From fancy (2452 *nupta*; 1839 *succenturiata*)
10. From the habitat (1300 *pratella*)
11. From a problem in classification (1790 *dubitata*)
12. From the time of appearance (1799 *brumata*)
13. From a contemporary entomologist or friend (263 *clerkella*) or a distinguished naturalist now deceased (1273 *petiverella*)
14. From a place name, sometimes the type locality (1037 *holmiana*; 2300 *maura*)
15. From position in a series (1524 *monodactyla*)
16. From aesthetic appreciation (2054 *pulchella*)

We have seen that Linnaeus not only initiated the sources from which specific names should be derived, but also devised a system of family terminations to indicate affinity. Since these terminations form an integral part of more than half the names in the British list and had a meaning for Linnaeus which is not readily apparent today, our next task is to consider them in greater detail. They affect only the Phalaenae, i.e. moths other than hawk-moths.

Linnaeus divided these into seven families on the basis partly of antennal structure but mainly on the way in which the wings were disposed when the moth was at rest. The families are defined in *Systema Naturae* (Edn 10), p. 496 (Pl. III), here freely translated and slightly abridged.

The first groups with the wings folded over the resting insect (*alis depressis*)

1. Bombyces – with pectinate antennae
2. Noctuae – with bristly but not pectinate antennae
3. Geometrae – resting with wings spread horizontally
   Pectinicornes – with pectinate antennae [termination *-aria*]
   Seticornes – with bristly antennae [termination *-ata*]
4. Tortrices – with wings so very blunt as to seem almost truncate and indented; rather flat and small species (*'planiusculae'*) [termination *-ana*]
5. Pyralides – with wings closing (literally 'winking') into a scissored (i.e. overlapping) deltoid [termination *-alis*, perhaps influenced by the name Pyralis]
6. Tineae – with wings rolled almost into the shape of a cylinder [termination *-ella*]
7. Alucitae – with wings cleft into fingers in a basal direction [termination *-dactyla*]

Subdivisions within these families are omitted here because they do not influence the formation of specific names. They were based on a more detailed analysis of wing-folding technique and the presence or absence of a coiled haustellum, dorsal crests and angulations on the terminal margins of the wings.

Had this system of classification proved fully satisfactory, little more need have been said. We now know, however, that noctuids can have pectinate antennae or rest with their wings disposed as a deltoid, that pyrales can wrap their wings round their bodies in cylindrical fashion and that geometrids can overlap their wings like the blades of a pair of scissors. This leads to some interesting anomalies in nomenclature which are here briefly considered.

Because Bombyces were supposed to have pectinate antennae and Noctuae not to do so, some species were described in the 'wrong' family; for example, 1994 *Phalera bucephala* was placed in Noctua and 2176 *Cerapteryx graminis* in Bombyx. This did not affect the name. However, when noctuids were assigned to Pyralis, as were 2085 *Agrotis vestigialis*, 2178 *Tholera decimalis* and most of the hypenines, the termination *-alis* which they accordingly received is historically informative, the first two having been disqualified from their membership of Noctua because of their pectinate antennae and the hypenines because of the configuration of their wings in the resting position. Most of the Drepanidae spread their wings and have pectinate antennae and so were assigned to the Pectinicornes group of the Geometrae (termination *-aria*). The smaller noctuids such as 2413 *Deltote bankiana*, the Chloephorinae and the Sarrothripinae with squared wing-tips were classified in Tortrix, whilst others like 2414 *Emmelia trabealis* joined the hypenines in Pyralis.

The Geometrae and Pyrales are broad-winged moths and rest with their wings extended over the substrate. The Linnaean difference between them was that in the Geometrae the dorsal margins of the forewings might reach inwards to the abdomen but always left it exposed; in the Pyrales they touched each other or even overlapped in scissor fashion, so concealing the abdomen. In some species the posture they adopt when newly alighted differs from the one they assume when fully dormant. Furthermore, many species were named from dead specimens whose true resting position was bound to be largely a matter of conjecture. This is the explanation for the 'wrong' termination in the names of many members of the two families. Names such as 1672 *fimbrialis*, 1788 *cervinalis* and 1944 *punctinalis* indicate that the dorsal margins of their forewings meet and conceal the abdomen in the resting position, or at any rate did so in the specimens Scopoli had before him when he bestowed the names. Likewise Linnaeus treated 1345 *nymphaeata* and 1376 *hortulata* as geometrids because they rest with the abdomen exposed. 1864 *Chesias legatella*, which 'scissors' its wings in repose and has forewings too narrow to appear deltoid, was named as a tineid, extraordinary to us but fully in compliance with the Linnaean canons.

In the Geometrae, pectinate antennae grade into those that are setose, posing a problem to the nomenclator, and if the type specimen was a female, the termination *-ata* might well supplant the rightful *-aria*. There is hardly a species that lacks a synonym involving the other termination, often a spelling variant of the same name, e.g. *grossulariata* and *grossulariaria*.

Not all tortrices are *'planiusculae'* (see above), some having tectiform or involute wings in their resting position. Accordingly, they received the tineid termination, e.g. 1086 *salicella* and 1214 *resinella*. Indeed, there are many species that wrap their wings in repose and were described in Tinea, wrongly by modern criteria. The Crambinae and Galleriinae are in this category and relatively early in taxonomic history Fabricius (1798) removed them from Tinea and set up new families to accommodate them (see p. 29). The Tineoidea do not all adopt the convolute posture; the Depressariinae, for example, rest with their wings folded flat over the abdomen. They were consequently classified as tortrices and the specific names bestowed on them by Linnaeus and Fabricius end in *-ana*. Fabricius set up the name *depressana* (683) because it rests *'alis depressis'* in 1775; in 1798, with a fuller understanding of taxonomy, he wrote *depressella*. The future International Commission for Zoological Nomenclature (I.C.Z.N.) rules were to allow him to alter the classification but not the termination of his original name.

The attraction of a common family termination is considerable and Haworth (1802) went so far as to propose that the names of all Bombyces should end in *'-us'* and of all Noctuae in *'-ina'*. He accordingly set about restyling them wholesale, so that *vinula* became *'vinulus'* and *pronuba 'pronubina'*. His experiment was short-lived.

Specific names after Linnaeus

Once the pattern had been set, the task of naming the Lepidoptera proceeded apace. Then, as now, there were three main types of entomologist, the systematist, the collector and the artist, but the roles overlapped. One of the first in the field was an artist, Clerck, a pupil and friend of Linnaeus who had already named a species in his honour. In his *Icones Insectorum rariorum* (1759), Clerck figured the Lepidoptera with exceptional fidelity. A number of new names appear and, although there is no printed text, the quality of draughtsmanship is so good that there is seldom any doubt over the identity of the insect. Clerck's new names were probably suggested or at least approved by Linnaeus and several of them are repeated and even explained in the 12th edition of *Systema Naturae* (1767).

The compilation of local lists is mainly a collector's pastime. The first of significance as far as new scientific names were concerned was Linnaeus' own *Fauna Suecica* (Edn 2, 1761). The new nomenclature was applied to the species that had been included in the first edition and there were a number of additions. The year 1763 saw the publication of Scopoli's *Entomologia Carniolica*, one of the most attractive of early entomological works, which dealt with the insects of that part of Yugoslavia that borders the northern shores of the Adriatic Sea. In the following year O. F. Müller's *Fauna Insectorum Fridrichsdalina* appeared, being a list for the Copenhagen district. Richer in new names was Hufnagel's list for the Berlin district which appeared in the *Berlinische Magazin* during 1766 and 1767. The greatest 'local list' ever written was that for the Vienna district by Denis & Schiffermüller (1775), a substantial book that was to become the standard European text-book for many years to come. Linnaeus had attempted to cover the whole realm of nature and insects and the Lepidoptera formed only a part of his study, but his successors who specialized on a single Order and single region could pursue their researches in far greater depth and so were in a better position to extend the bounds of entomological knowledge.

For a time the University of Upsala (now Uppsala) in Sweden, where Linnaeus was professor, was the centre of biological and botanical research in Europe. The pupils of Linnaeus, their training done, travelled the world collecting and recording. Possibly the greatest of them was Fabricius, a Dane (Pl. IV), who came to England as a young man, resided here for several years and throughout his life was a frequent visitor. Then, as now, Britain was rich in collections, both of native specimens and of those brought back from overseas by the scientists that accompanied Captain Cook and other navigators on their voyages. Fabricius studied these collections, naming the species and arranging the material according to Linnaean principles. At the same time, he enjoyed field work and some of the British species he named (e.g. 1011 *conwagana*) commemorate the companions with whom he collected. He was first and foremost a systematist and after his death Haworth was to describe him as 'that greatest of entomologists', a judgement that few would dispute.

To return to the illustrator, the closing decades of the 18th century saw possibly the greatest of them all starting on his career. Nowadays we do not expect an entomological artist to be a leading taxonomist, but that is what Hübner (Pl. VII) became. Once his reputation was established, many collectors sent him their specimens to figure and name. An artist perforce studies the insects he figures in great detail and is therefore well able to recognize the specific differences between them. We shall return to Hübner when we come to consider the evolution of genera.

English lepidopterists do not feature prominently as nomenclators in the years between 1758 and the end of the century. Moses Harris did not adopt the Linnaean system of nomenclature in *The Aurelian* (1766), though he did so in the 2nd edition nine years later, published after the arrival of Fabricius. When he found that some of our butterflies had not yet been named, he did not attempt to supply the deficiency. However Lewin (1795) introduced several new names which he ascribed to Fabricius, who had probably suggested them to him. Forster was the first 'Briton' to give new scientific names to Lepidoptera, the inverted commas because he was more of a cosmopolitan than a national of any one country. Born of British stock in Poland, he later settled in England where his abilities earned him an honorary degree at Oxford University and the appointment as chief scientist on one of Captain Cook's expeditions. While in England he wrote *A catalogue of English insects* (1770) and *Novae Species Insectorum* (1771). Later he returned to Europe to take the Chair of Natural History at Halle University. Pickard *et al.* record that he could speak seventeen languages. The first major British nomenclator was A. H. Haworth (Pl. VI) in his *Lepidoptera Britannica* of 1803–28.

As the number of unnamed European butterflies and larger moths declined, attention began to be focused more upon the Microlepidoptera, the study of which called for a degree of specialization. The middle years of the 19th century saw great activity in this field, with Stainton (Pl. VIII) in Britain and Zeller and Frey on the Continent being the protagonists.

Modern entomologists still follow the pattern and principles established by Linnaeus in naming new species, but tend to be less poetic and imaginative.

Generic names

Once the binomial system had been presented to the world, a flood of new specific names followed. Generic names have quite a different story. The genus as we know it today was not a Linnaean invention and did not emerge until about half a century after the publication of *Systema Naturae*. I shall suggest tentatively that Schrank paved the way

Plate IV Johann Christian Fabricius, 1745–1808

for the genus in the early 1800s and that it came into being in a process of gradual evolution*.

Linnaeus divided the Lepidoptera into three main groups, the Papiliones or butterflies, the Sphinges or hawk-moths and the Phalaenae or all the other moths. He subdivided the Phalaenae into seven sections for easier study (*quo facilius inquirantur*) (Pl. III). These correspond with some of our present superfamilies and are as follows: the Bombyces, Noctuae, Geometrae, Tortrices, Pyrales or Pyralides, Tineae and Alucitae. Some of these groupings were further divided on characters such as the presence or absence of the haustellum, of antennal pectinations or of dorsal crests on the abdomen. Names, however, were given only to the subdivisions of the butterflies (Pl. II). These were Equites (Knights), the swallowtails; Heliconii (Muses), mainly tropical species with entire (untoothed) wings and the hindwing short in relation to the forewing; Danai (children of King Danaus, who had fifty daughters), mainly Pieridae and Satyrinae; Nymphales (Nymphs), mainly Nymphalinae; Plebeji (Commoners), the smaller butterflies, mainly Lycaenidae and Hesperiidae; and Barbari (Barbarians, foreigners), species added to the butterflies 'as a gift or perquisite' (*corolarii loco*), but which could not be assigned to any of the former groupings. Further unnamed divisions based on colour, the presence or absence of eye-spots, etc., were of the nature of distinctive characters in a dichotomous key and lacked taxonomic importance.

Linnaeus, therefore, did not invent the genus: his groupings were at family or superfamily level. The earliest additions to his system bore similar rank: they were collateral with, not subdivisions of, the Linnaean categories. Names such as *Zygaena* Fabricius, 1775, now have generic status but in their inception they belonged to families and were then written in Roman type. I shall trace the evolution of the genus by examining the early supraspecific additions to the British list in chronological order; although at first they were suprageneric, I shall give the Log Book numbers they bear as genera in modern usage.

The first name to be added to the Linnaean system was 1510 Pterophorus of Geoffroy, 1762. However, Geoffroy did not use it in combination with a specific name and under rules made many years later it was deemed invalid for generic use; Schäffer (1766) first used it 'correctly', so he is now regarded as the author. It was proposed as a substitute for the Alucita of Linnaeus and as a family name; the Law of Priority not yet having been invented, entomologists had no qualms over its adoption. Linnaeus had named all his 'plumes' from the Greek and Geoffroy naturally used the same language for Pterophorus. It is possible that the post-Linnaean tradition of forming generic names from Greek rather than from Latin words arose from this accident.

The next author of new supraspecific names was Fabricius in his *Systema entomologica* of 1775. He made no new groupings himself but supplied titles for three of the previously unnamed subgroups of Linnaeus. Two of them were for sections of the Sphinges. One of these consisted of the bee hawk-moths, the humming-bird hawk-moths and the clearwings. Several had names like 1983 *fuciformis*, having the form of a drone, so Fabricius called the group 370 Sesia from the Greek σής (sēs), a moth, and the combination *Sesia*

*The terms used by the early entomologists for supraspecific taxa differ from those in use today and may cause confusion. The main divisions into which Linnaeus and his successors divided the Lepidoptera were called 'genera', the word then corresponding with our family or superfamily. When subdivision began, the new lower taxa were called 'families', i.e. genera contained families. Ochsenheimer in *Die Schmetterlinge von Europa* **4** (1816) divided the 'genus' Hipparchia of 77 species into seven families (then unnamed); Hipparchia now corresponds with our subfamily Satyrinae and the families have become the genera into which the subfamily is divided, with certain modifications. I have failed to identify the author who reversed the usage and the date on which he did so.

fuciformis could be translated 'the moth formed like a drone'. Linnaeus had called another section of Sphinx *'adscitae habitu & larva diversae'* (adopted or associated species differing in appearance and larva); these formed a mixed bag but included the burnets. Fabricius wanted a name analagous to Sphinx but which conserved or at any rate recalled the sense of *'adscitae'* or linked species. He chose 166 Zygaena, from the Greek ζύγαινα (zugaina), the hammer-headed shark, because of its pun with ζυγόν (zugon), a yoke. A third small group which Linnaeus had called *Noctuae elingues* (noctuids without haustellum) Fabricius placed in 14 Hepialus, from the Greek ἠπίαλος (hēpialos), a fever. Again, this was a diverse assemblage, but it included the 'swifts' and one of these, 16 *hecta*, prompted the name. Fabricius intended these names to rank equal with Sphinx and Noctua and accordingly I have used Roman type; later in their history, after the number of species they embraced had been greatly reduced, they became more homogeneous in their composition and acquired the characteristics of present-day genera.

Linnaeus died in 1778, twenty years after the publication of his masterpiece. His system of classification had suffered no modification other than the bestowal of names on some of his subordinate groupings. Two supraspecific names date from 1783. One came into being almost by accident. Linnaeus had placed 163 *statices* in his Sphinges adscitae (see above) and Retzius not unnaturally referred to it as *Adscita statices* and was later deemed to have established a new genus. The other was given for a scientific reason and is historically important as marking the first extension of the Linnaean system. Fabricius had placed 174 *asella* in Hepialus; Knoch saw that it was a misfit and erected the first completely new family, Heterogenea, q.v., to accommodate it, framing the name so as to draw attention to the innovation. He was ahead of his time and the name was not immediately accepted.

Ten more years elapsed before Fabricius (1793) provided the first new suprageneric name for butterflies, still, however, within the Linnaean framework. Linnaeus had grouped the smaller species, i.e. the blues and skippers, together under the name Plebeji parvi; Fabricius deemed them worthy of family status and established Hesperia to receive them. It may come as a shock to some readers to learn that the blues were once in Hesperia and Fabricius (1807) would have kept it that way if Latreille (1804) had not already restricted the name to the skippers (see p. 31). *Hesperia* now is a genus containing only one British species: how are the mighty fallen!

A year later Fabricius made a revisionary change. Linnaeus had placed 162 *cossus* amongst his *'Bombyces elingues, alis depressis, dorso cristato'* (Bombyces without haustellum, with wings folded over the body and with a dorsal crest, where it kept company with species like 2003 *ziczac* and 2028 *pudibunda*. He had put 161 *pyrina* in Noctua. Fabricius saw that a new family was needed for these species, so he renamed the goat moth *ligniperda* and appropriated its former specific name for the family. He was probably motivated by piety towards Linnaeus' memory: if he must deviate from his old master, he would do so using a name Linnaeus himself had used. He may also have been aware that the Roman concept of *cossus* contained more than one species. Then there were no I.C.Z.N. rules to prohibit such manipulation.

In 1796 Latreille (Pl. V) summed up the current classification of the Lepidoptera in his *Précis des caractères génériques des insectes*, increasing the nine Linnaean 'genera' to twenty-one as follows:– Papilio, Linnaeus; Hesperia, Fabricius; Sphinx, Linnaeus; Sesia, Fabricius; Zygaena, Fabricius; Bombyx, Linnaeus; Hepialus, Fabricius; Cossus, Fabricius; Noctua, Linnaeus; Phalaena, Linnaeus; Pyralis, Linnaeus; Hyblaea, Fabricius (oriental and not our concern here); Aglossa, mihi; Ypsolopha, mihi; Tinea, Linnaeus; Yponomeuta, mihi; Oecophora, mihi; Adela, mihi; Alucita, Fabricius; Orneodes, mihi; Pterophorus, Geoffroy. There is no mention of Heterogenea, Knoch. *'Mihi'* (to me) is the

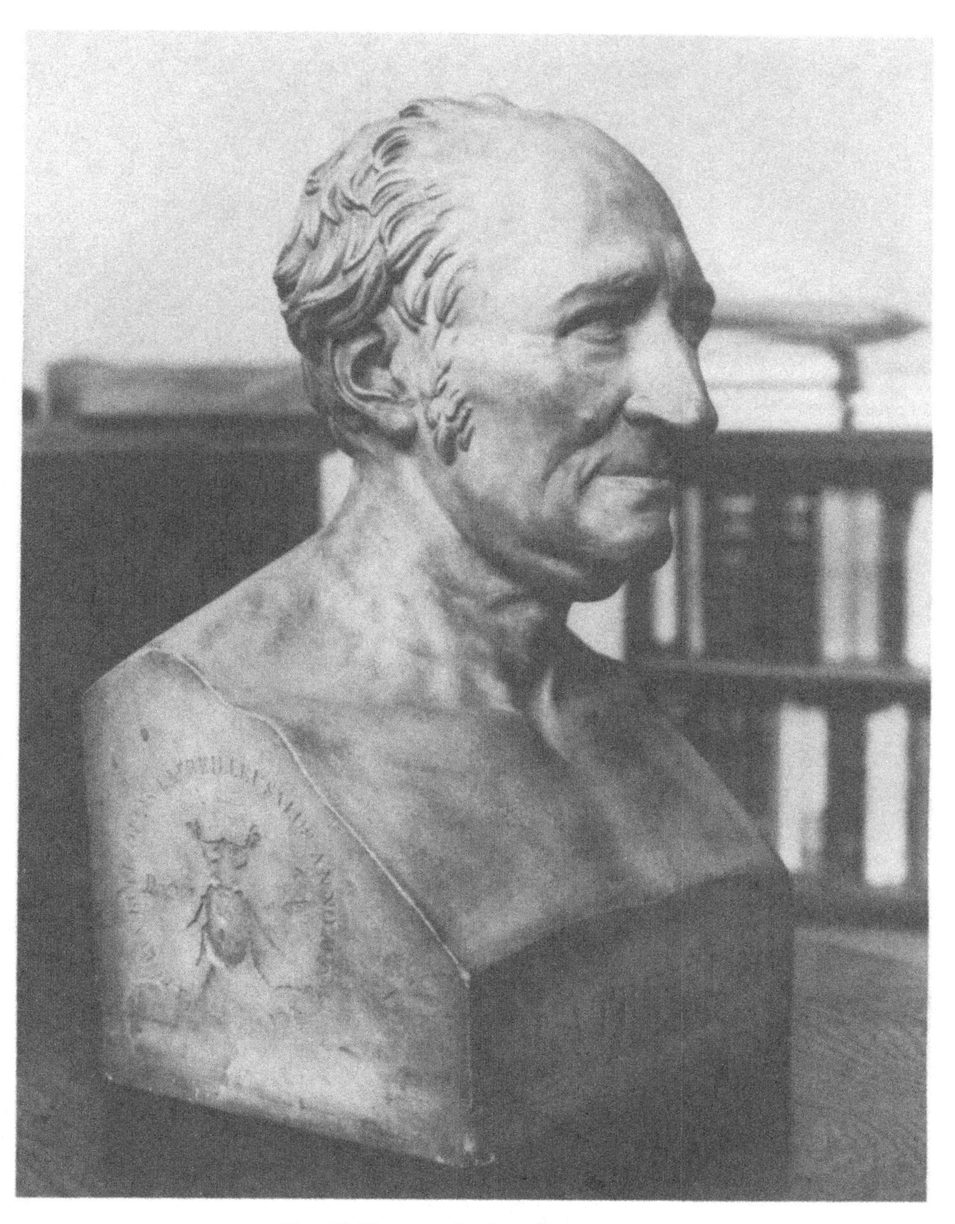

Plate V Pierre André Latreille, 1762–1833

Plate VI Adrian Hardy Haworth, 1767–1833

conventional term used by an author to indicate that a name is new and is to be ascribed to himself. Latreille introduced six new families, viz. 1420 Aglossa, 451 Ypsolopha, 424 Yponomeuta, 651 Oecophora, 149 Adela and 1288 Orneodes, qq.v. All have now been reduced to generic status and the last to synonymy, though two provided the source for a current family name and one for a subfamily.

Some of the family names ascribed by Latreille to Linnaeus had been redeployed with an application out of keeping with their author's intention, changes first made by Fabricius (1775). Phalaena was originally the collective name for all the moths other than the hawk-moths, but was now applied by Fabricius and Latreille to a grouping which comprised the Linnaean Geometrae and Pyrales (part of the Pyralidae + the Hypeninae), the latter in a subdivision headed '*Alis forficatis*', with 'scissored' wings (see 1356 *forficalis*). The Linnaean Tortrices were now assigned to Pyralis, together with the Depressariinae, Choreutidae and Chloephorinae which were considered congeneric; hence it came about that 387 *Prochoreutis sehestediana* (Fabricius, 1777) and 388 *P. myllerana* (Fabricius, 1794) were originally described in Pyralis (see MBGBI 2, pp. 394, 395), though as tortrices in the meaning of the current terminology. Tinea, Linnaeus, was divided between Tinea and Alucita, Fabricius. Linnaeus had intended Alucita for the plumes and the twenty-plume moth, but these had now been assigned to Pterophorus, Geoffroy, and Orneodes, Latreille, leaving Alucita available for other use. Fabricius had accordingly commandeered it for a new family comprising the Nemapogoninae and certain Plutellinae and Oecophorinae. Constructive progress was being made in taxonomy, but chaos threatened with the arbitrary application of new meanings to the old names; Hübner was largely responsible for restoring the Linnaean usage. Later, the designation of type species stabilized the nomenclature.

In Latreille's classification there was still no intermediary taxon between the species and the 'genus' (our family or superfamily).

Illiger (1798) also published a classification of the Lepidoptera. He gave the families named by Linnaeus, Fabricius and Geoffroy but omitted those of Latreille. He added one of his own, 144 Nemophora, with the same application as Adela, Latreille. He attributed the name to Hoffmannsegg and it has been incorrectly cited as *Nemophora* Illiger & Hoffmannsegg although according to the I.C.Z.N. rules Illiger was the sole author (see p. 49). Both Latreille and Illiger were of the opinion that the 'long-horns' merited a family of their own ahd the name proposed by each author survives in generic usage. In the same year Fabricius added 1294 Crambus, 1425 Galleria and 2051 Lithosia (qq.v.) with parity of rank to the new family names in Latreille's list.

We have now reached the year 1800, forty-two years after the publication of *Systema Naturae*. Including those bestowed by Linnaeus, there were twenty-five supraspecific names available for the species occurring in Britain, all with equal rank; the concept of taxa ranking midway between the family and species did not yet exist. Yet there was a growing belief among taxonomists that smaller groupings were needed. The first 19th century additions were of minor importance: Kluk (1801–02) has been deemed to have created three new butterfly genera when he used the Linnaean names *Nymphalis*, *Plebejus* and *Danaus* in a way acceptable to future legislators on nomenclature; compare *Adscita* Retzius above (p. 26).

The initiator of change was Franciscus von Paula von Schrank, a professor of theology who transmogrified himself into a professor of botany and wrote also about insects – a polymath of the old school. His *Fauna Boica* was produced in parts and it is the second volume containing the Lepidoptera (Part 1, 1801; Part 2, 1802) which concerns us here, the innovation coming in Part 2. Part 1 covered the butterflies. Fabricius had given two families, Papilio and Hesperia, but Schrank proposed five, Erynnis (the Hesperiidae),

Pieris (the Papilionidae + the Pieridae), Maniola (the Satyrinae + the Apaturinae), Papilio (the Limenitinae + the Nymphalinae + the Argynninae) and Cupido (the Lycaenidae). Whereas Fabricius had included the blues and skippers in a single family, Schrank emphasised their disparity by placing them at either end of his sequence of families. Also in Volume 2, Part 1 he added Psyche as a new family in the moths.

All these were families of rank equal to those already in existence. The change came in Part 2, where the new families were expressed as subdivisions of the older families. For example, he divided the comprehensive Linnaean family Bombyx into seven sections, Saturnia, Bombyx, Laria (roughly our Lymantriidae), Arctia, Lasiocampa, Cerura and Drepana. In modern taxonomic terms, these were families and the Linnaean Bombyx had been elevated to the rank of superfamily. Though many of Schrank's names were later adapted for generic usage, he did not invent the genus. He did, however, pave the way for the hierarchical system of classification which is the one that is adopted today. New names bestowed by Schrank which have not yet been mentioned include Stigmella, Nemapogon, Plutella, Nymphula, Pyrausta, Agrotera, Scopula, Setina, Hadena, Cucullia, Catocala and Hypena.

Two schools of thought on classification were now emerging. The older school favoured increasing the number of families and giving them all equal status: the newer preferred subdivision. The two methods are placed side by side in an anonymous article in the sixth volume of Illiger's *Magazin für Insektenkunde* (1807) which dealt with the classification of the two Linnean 'superfamilies', Papilio and Sphinx. First the author gives what was to be Fabricius' last word on the subject. He divided the butterflies into forty-one families, seventeen of which occur in Britain, and Sphinx into eight families, six of them British. All these families bore equal rank and there was no subdivision. After Fabricius' scheme we have that of Latreille which had been published in more detail three years earlier in Buffon's *Nouvelle Dictionnaire d'Histoire naturelle* (Latreille, 1804). Whereas in 1796 (see p. 26) Latreille had observed the same method as Fabricius, here he gives three taxonomic grades corresponding to our present superfamily, family and subfamily. He used only eight families for Papilio and five for Sphinx, but these had subsections, not all of which were named. The article ends with a brief reference to Schrank, giving his classification in bare outline; the author evidently felt that his work did not merit comparison with the other two, an opinion not shared by posterity.

The article stands at the parting of ways and is of great historical importance. I shall therefore give the system adopted by each author, but in abbreviated form; for example, I shall omit or abridge the diagnoses they give based on structural features such as legs, palps and antennae. I shall follow the original order of presentation and begin with Fabricius, although Latreille's names are in fact senior. In Fabricius' list I shall give only those families that are represented in Britain, placing before each name the serial number allocated by Fabricius and after it the species included, mainly in vernacular names, and the total number of species world-wide he assigned to the family.

PAPILIONIDES

3. Papilio (swallowtails), 125 species
8. Euploea (milkweed butterflies), 32 species
9. Apatura (emperors), 14 species
10. Limenitis (white admirals), 14 species
11. Cynthia (painted ladies, pansies (*Precis*), etc.), 95 species
12. Vanessa (other vanessids), 30 species
14. Hipparchia (satyrines), 119 species
19. Argynnis (the larger fritillaries), 41 species

22. Doritis (= Parnassius Latreille) (apollos), 4 species
23. Pontia (whites and orange-tips), 94 species
24. Colias (clouded yellows and brimstones), 35 species
29. Melitaea (small fritillaries, including the Duke of Burgundy), 15 species
31. Hesperia (blues, including the long-tailed blue (see below)), 108 species
32. Lycaena (coppers and all other blues in the British list (see below)), 150 species
35. Thecla (hairstreaks), 8 species
39. Thymela ('black' skippers), 131 species
41. Pamphila (orange skippers), 34 species.

When Fabricius first established Hesperia in 1793 (p. 26), he applied it collectively to the blues and skippers; now that he had separated them, he wanted to use the name for some of the blues, but could not be permitted to do so for a reason explained after Latreille's list below. The distinction between Hesperia and Lycaena was based on the structure of the labial palpus.

Latreille's classification now follows; it has a more modern guise. I give all his families and it is necessary to include a proportion of his diagnoses.

LEPIDOPTERA

Diurni

Family 1: Papilionides

I. Antennae approximated at base; apical club straight; all four wings elevated in the resting position
   A. Four functional legs
      1. Nymphalis
         a) *Nymphales proprie dicti* (true nymphalids) (vanessids and emperors)
         b) *Perlati* (latinized from Fr. *perle*, fritillaries, including the Duke of Burgundy, with pearly markings)
         c) Satyri (satyrines)
      2. Heliconius
      3. Danais (milkweed butterflies, including *Danaus plexippus*)
   B. Six functional legs
      4. Papilio (swallowtails)
      5. Parnassius (apollos)
      6. Pieris (whites, clouded yellows, brimstones and orange-tips)
      7. Polyommatus (hairstreaks, coppers and blues)

II. Antennae well separated at base, terminating in a hook; two of the wings held almost horizontal in the resting position
      8. Hesperia (skippers)

The reason why Fabricius could not use Hesperia, of which he himself was the author, for a family of blues is now clear; Latreille, whose names have priority, had already appropriated it for the skippers.

Fabricius had forty-one butterfly families, Latreille only eight, but his scheme invited further subdivision. There is still no genus, but its potential is there. Superimpose Fabricius' classification upon that of Latreille and observe the result. In Nymphalis you would find *Apatura, Limenitis, Cynthia, Vanessa, Argynnis* and *Melitaea*; these are no longer families but genera and are accordingly printed in italics.

We now turn to Sphinx which was also included in the article in *Magazin für*

Insektenkunde and we find much the same. Their Sphinx was still based on the Linnaean 'superfamily' and comprises more than the hawk-moths.

Of Fabricius' eight families, the six that occur in Britain are as follows; I still give Fabricius' serial number and the number of species.

42. Laothoe (hawk-moths with reduced haustellum such as *ocellata* and *tiliae*), 21 species
43. Sphinx (hawk-moths with functional haustellum such as *ligustri*), 74 species
44. Sesia (bee hawk-moths and humming-bird hawk-moths), 18 species
45. Aegeria (clearwings), 1 [*sic*] species; three examples are cited by name
47. Zygaena (burnets), 17 species
49. Procris (foresters), 9 species

Latreille gave only five families but divided one of them into two subfamilies.

9. Sphinx (hawk-moths with functional haustellum)
   (a) Without anal tuft (e.g. *atropos*)
   (b) With anal tuft (e.g. *stellatarum* and *fuciformis*)
10. Smerinthus (hawk-moths without functional haustellum (e.g. *ocellata*))
11. Sesia (clearwings)
12. Zygaena (burnets and foresters)
13. Stygia (non-British and now placed as a genus of Cossidae)

Sesia has the same history as Hesperia; although he was author of the name, Fabricius could not use it for the bee hawk-moth half of the original family because Latreille had already assigned it to the clearwing half.

Meanwhile, important developments were taking place elsewhere. I have already said that the artist looks with particular application at the insects he depicts (p. 23). Hübner at first accepted the Linnaean classification without demur but as his experience broadened he began to wish to draw attention to the narrower affinities he was observing. Accordingly, between 1804 and 1806 he wrote a pamphlet, consisting of a single sheet of paper printed on both sides, which he called his *Tentamen* and circulated it privately among his entomological friends. In it he set out his views on classification. Those who wish to study them in detail are recommended to consult Hemming (1937, **1**: 8–9, 14, 592–600). Suffice it here to say that he evolved the concept of genera ranking between the species and family and each consisting of some 1–10 species which evinced some form of similarity. He gave 107 examples but, as the *Tentamen* was published privately, these names were not valid. Some of them, however, were subsequently used by Ochsenheimer who is therefore credited with their authorship, while others were incorporated by Hübner himself in his later work. A number of years elapsed before he had the confidence to give full rein to his ideas. He waited to see whether they would be accepted by his contemporaries and then, between 1819 and 1826, he flooded taxonomy with new generic names, many of which are still in use today.

What was now needed was an entomologist of intellectual ability who could synthesize the current theories of classification. Ochsenheimer had been a recipient of Hübner's *Tentamen* and was influenced by it. After an early academic career, he became a playwright and then an actor, rising to the top of his profession. Outside the theatre, his all-consuming interest was entomology and, having been encouraged by Laspeyres to become an author, he embarked on *Die Schmetterlinge von Europa* which was to become a standard work on the Continent. The first volume appeared in 1807 but Ochsenheimer lived to see only the first four published. After his death in 1822, Treitschke, a professional colleague and a close friend who shared his interest in entomology, continued the work to the tenth and final volume of 1835. This work incorporated the

Hübnerian principle of dividing the families into units of relatively few species. Ochsenheimer's own work embraced the families from Papilio to Noctua of the Linnaean classification. In the volume of 1816 he introduced over thirty generic names still in use, including five taken from Hübner's *Tentamen*. Linnaeus had met the challenge of finding suitable sources for a very large number of specific names: Ochsenheimer faced a similar problem with generic names, albeit on a smaller scale. Over half of his are descriptive of a character of the adult or its behaviour, like 1642 *Gastropacha* from the stout abdomen. Six are drawn from classical personages or objects mentioned by classical writers and eight are 'geographical'. Linnaeus had made use of personal names from mythology and it was logical to extend this practice to place names, though Ochsenheimer did not restrict himself to those mentioned by classical authors. Three of Fabricius' enigmatic names of 1807, 1542 *Colias*, 1552 *Pontia* and 1611 *Melitaea* may in part be geographical, as may Leach's problematic name 2077 *Nola* of 1815, but it was Ochsenheimer who established the genre. After Ochsenheimer's death, Treitschke continued to make use of this source and other entomologists like Stephens (1007 *Capua*; 1110 *Bactra*) did so from time to time; however, it was not favoured by Hübner and never achieved widespread popularity.

By now the genus as a taxon between the species and the family was firmly established. Genera set up in the first half of the 19th century were usually based on 'look-alike' characters such as colour or the pattern of the forewing and in consequence their composition lacked stability. An event which had an important influence on classification and so on the composition of genera was the publication in 1859 of Charles Darwin's *Origin of Species*. Once the idea of evolution had been accepted, entomologists began to look for 'primitive' and 'advanced' characters and many new combinations ensued. A second factor was the development of the study of venation, the genitalia and other structural features which gave a clearer picture of natural affinities. With the wholesale transference of species from one genus to another, some sort of stability had to be found. This was achieved by the establishment of a 'type species'. One of the original members of the genus was so designated and thereafter, as long as the genus remained in use, it had to contain that species; others that were found to be related were moved into its company. In many cases the type species is the first of those included in the original genus, but the generic name may be based on a character shown by another member since transferred. 2152 *Sideridis* in its original composition included 2262 *Agrochola circellaris*, then known as *ferruginea* [Denis & Schiffermüller]. *Sideridis* is a rough Greek translation of the Latin *ferruginea*, but the latter was not selected as the type species. So it comes about that a genus with its name descriptive of the 'rusty brown' coloration of a cuculline is now applied to a hadenine that is not rusty brown. 2306 *Phlogophora* now contains only *meticulosa*; the name is a Greek translation of 2305 *lucipara* which Treitschke had placed in the genus but has now been shifted to *Euplexia*. The meaning of the name *Phlogophora* has no relevance to *meticulosa*. Generic names can often be understood only in the context of the species originally included in the genus.

As entomologists became more scientific, so did the generic names they proposed. Many are based on characters of the genitalia or other structures that are not readily visible. Such names are appropriate and informative. A counter-development is the meaningless neologism, a made-up word of no signification, entomological or otherwise. This expedient is not new: I suspect that 2421 *Bena* Billberg, 1820, is an early example. Guenée admitted that some of his generic names were 'sans étymologie'. Walker, who worked through the collections at the British Museum (Natural History) over a hundred years ago, was partial to this sort of name. I dread them because one can never be quite sure whether there is not, after all, some hidden meaning; the fruitless research is time-consuming and the outcome frustration.

Plate VII Jacob Hübner, 1761–1826

Because of their diversity, generic names are harder to categorize than specific names, but most types are included in the list which follows. As with the specific names on p. 20, examples are given from the British list.

1. From a name found in classical mythology (1569 *Cupido;* 1979 *Mimas*)
2. From a deceased naturalist (751 *Aristotelia*)
3. From a contemporary entomologist, friend or kinsman (182 *Bankesia*)
4. From the name of a place, generally ancient, with no relevance to the insects in the genus (1652 *Thyatira*)
5. From a place with entomological connections (2376 *Sedina*)
6. From the habitat (1303 *Agriphila*)
7. From a character in the wing pattern of a species in the genus (2315 *Dicycla*)
8. From a structural character of the adults (757 *Recurvaria*)
9. From the behaviour of the adults (385 *Anthophila*)
10. From a character of the larvae (1636 *Lasiocampa*)
11. From the behaviour of the larvae (489 *Coleophora;* 2179 *Panolis*)
12. From the larval foodplant (777 *Bryotropha*)
13. From the season of appearance of the adults (1661 *Archiearis*)
14. From an object or creature having a fancied similarity (2053 *Coscinia;* 2415 *Acontia*)
15. From a name used by Linnaeus, but not by him in the accepted generic sense (1571 *Plebejus*)
16. From the Greek translation of the Latin-derived name of one of the species in the genus (2277 *Moma;* 2306 *Phlogophora*)
17. From aesthetic considerations (2068 *Callimorpha*)
18. From a problem in classification (590 *Perittia*)
19. From a made-up word apparently without meaning (643 *Dafa*)
20. From an anagram of an existing name, this being especially favoured when a replacement name has to be found for one that is preoccupied. For example, out of 55 replacement names proposed by Nye (1975) in the Noctuoidea, 40 are anagrams of the invalid homonyms, e.g. *Goonallica* Nye, 1975, for *Callogonia* Hampson, 1908. The British list is virtually free of such anagrams, but 1582 Riodininae, q.v., is an example.
21. From affinity with another genus, shown by adding a prefix or suffix to the older generic name. Prefixes commonly used include *Apo-* (1486 *Apomyelois*); *Bi-, Bis-* (624 *Biselachista*); *Eu-* (731 *Eulamprotes*); *Neo-* (797 *Neofaculta*); *Para-* (440 *Paraswammerdamia*); *Peri-* (1005 *Periclepsis*); *Pro-* (387 *Prochoreutis*); *Pseudo-* (436 *Pseudoswammerdamia*); *Syn-* (1568 *Syntarucus*). Common suffixes are *-ella* (2009 *Ptilodontella*); *-idia* (932 *Phalonidia*); *-morpha* (1217 *Eucosmomorpha*); *-odes* (765 *Teleiodes*); *-oides* (199 *Psychoides*); *-opsis* (776 *Teleiopsis*). Alternatively, another word may be added (2430 *Ctenoplusia;* 2432 *Trichoplusia*).

Suprageneric names

Butterflies and moths are known collectively as Lepidoptera from λεπίς, λεπίδος (lepis, lepidos), a scale and πτερόν (pteron), a wing, since the order is characterized by the presence of scales on the wings of the adults. Largely unscientific divisions of the Lepidoptera are Rhopalocera from ῥόπαλον (rhopalon), a club and κέρας (keras), a horn, or Diurni from *diurnus*, occurring by day, for the butterflies which have knobbed antennae and fly by day; and Heterocera from ἕτερος (heteros), other, and κέρας, or Nocturni from *nocturnus*, occurring by night, for the moths which have differently formed antennae and fly mostly by night. Another convenient but non-scientific grouping is into Macrolepidoptera from μακρός (makros), large and Microlepidoptera from μικρός (mikros), small, for the families containing the larger and smaller species respectively.

The categories used in the scientific classification of insects are set out below, together with an example of each to show the conventional termination.

Order	Lepidoptera
Superfamily	Papilionoidea
Family	Papilionidae
Subfamily	Papilioninae
Tribe	Pyralini
Genus	*Papilio* Linnaeus, 1758
Subgenus	*Argynnis* (*Fabriciana* Reuss, 1920)
Species	*machaon* Linnaeus, 1758
Subspecies	*machaon britannicus* Seitz, 1907

Tribal and subgeneric names are seldom used by British authors and are omitted in this work; the species group, named from one of the species included in it, tends to be preferred to the subgenus. Suprageneric names are not explained if the generic name from which they are formed appears below; for example, Papilionoidea, Papilionidae and Papilioninae are all deemed to be explained under 1539 *Papilio* (Limacodidae is explained under the specific name 173 *limacodes*); however, where the name is derived from a genus not represented in the British list (e.g. 2425 Pantheinae) the meaning is given. Generic synonyms are sometimes included to show the source of a suprageneric name. Where necessary, a Log Book number is given to indicate where the explanation may be found.

The scientific name of an insect as used today

The full scientific name of an insect has four elements, the generic name + the specific name + the author of the specific name + the date on which the name was first published in conformity with the rules laid down by the International Commission for Zoological Nomenclature.

a) *Generic name*. The generic name is a noun and is spelt with a capital letter. As already stated (p. 25), it is usually formed from one or more Greek words but in Latinized form. As a Latinized noun, it has a gender (*Crambus*, masculine; *Saturnia*, feminine; *Caryocolum*, neuter), but the gender of many is not readily apparent. When cited without a species, it should be followed by the author's name; the date should also be added when relevant or in a formal context. Unlike a suprageneric name, it is always written in italics.

b) *Specific name*. There are three forms of specific names (see also p. 16):

(i) A noun in apposition, e.g. *Papilio machaon* Linnaeus, 1758
(ii) A noun in the genitive case, e.g. *Lasiocampa quercus* (Linnaeus, 1758)
(iii) An adjective, e.g. *Crambus pratella* (Linnaeus, 1758)

The third example should strictly be written '*Crambus pratellus*' since an adjective must agree with the noun it qualifies; however, the modern practice, followed by Kloet & Hincks (1972) and therefore in this work, is to treat the generic name as genderless and to retain the original gender and spelling of the specific name. This contravenes I.C.Z.N. Article 30, but most modern taxonomists have 'small Latin and less Greek' and in consequence are unable to operate the rule.

Linnaeus himself followed the rules of gender. *Papilio*, a butterfly, is masculine and accordingly the few adjectival names he bestowed on butterflies are also masculine, e.g. 1567 *boeticus*. Later authors such as Verity who used the feminine termination for subspecies of butterflies with masculine names, like 1580 *argiolus britanna*, are clearly at fault. Likewise the names of the 'phalanges' or regiments into which Linnaeus divided the butterflies, together with their subdivisions, are masculine; see Pl. II, where *urbicolae* in the penultimate line, although belonging to the predominantly feminine first declension, is a masculine word like *nauta*, a sailor. In digression, it is noteworthy that Petiver (1695–1703a; 1702b–1706) started correctly by treating *Papilio* as masculine, but later as feminine, wherein he was followed by his friend Ray (1710). Aesthetic considerations seem to require creatures of beauty and elegance such as butterflies to be assigned to the fairer sex.

Sphinx, Bombyx and Phalaena, the names Linnaeus used for the higher divisions of the moths, are all feminine and reference to Pl. III shows that his lower categories like Tortrix were also spelled in the feminine form. Most of his specific names for moths are adjectival and these without exception are feminine.

Had later authors been prepared to follow the grammatically correct lead given by Linnaeus and to make all adjectival butterfly names masculine and moth names feminine, the scientific nomenclature of the Lepidoptera would have followed a smoother course; however, the very first new supraspecific name for moths, Pterophorus Schäffer, 1766 (really Geoffroy, 1762) was masculine and the first for butterflies, Hesperia Fabricius, 1793, was feminine. Thereafter authors set up feminine genera to accommodate masculine specific names, masculine genera for those that were feminine, and yet others that were neuter. The simple and workable pattern bequeathed by Linnaeus had been torn into shreds.

c) *Author's name.* The author of a name is the writer who first uses it together with a description or figure in a properly published book or paper. If there is no description, it is a '*nomen nudum*' and invalid; a name that appears in a manuscript or private publication is likewise invalid. The first person who subsequently uses the name in accordance with the rules laid down by the I.C.Z.N. will be deemed to be the author. 1163 *Zeiraphera ratzeburgiana* Ratzeburg, 1840 is an example; the name was proposed by Saxesen but Ratzeburg himself was the first author to cite the name correctly and in consequence is open to the unjustified reproach of having named a species in his own honour. When an entomologist first describes a species, he may well give it a name suggested by one of his friends and acknowledge the source; nevertheless he and not the friend will be the rightful author. Many of the names of Microlepidoptera in Stainton's works were proposed by Douglas but stand in Stainton's name.

Brackets are used for the author's name and date when the species is no longer in the genus in which it was first described. In the examples given in paragraph (*b*) above, brackets are used in (ii) and (iii) because Linnaeus described *quercus* in *Bombyx* and *pratella* in *Tinea*; they are not used in (i) because he described *machaon* in *Papilio*. Square brackets are used for an author's name when the work was published anonymously and the authorship established at a later date. This applies especially to Denis and Schiffermüller.

d) *Date.* The date is important so that the Law of Priority may be operated (see next section). If the precise month and day cannot be established, the date is deemed to be the 31st of December; this applies to 2485 *turfosalis* Wocke, 1850. The name *humidalis* Doubleday, 1850, has priority because the exact date of its publication is known.

Square brackets should be used for a date that has been established by research for an originally undated publication. This applies to many of Hübner's names. I had originally included these brackets in the systematic section, but later deleted them as I considered (perhaps wrongly) that they were a distraction. Square brackets may also be used for

Plate VIII Henry Tibbats Stainton, 1822–1892

serial publications that are running behind schedule. Stainton's *Entomologist's Annual* used to appear in the December prior to the date on the cover, perhaps so that it could serve as a Christmas present. If a new species was described, say, in the *Annual* for 1860, the first reference should be cited thus: '*Entomologist's Annual* 1860, [1859]'.

Why scientific names sometimes need to be changed

Collectors get very annoyed when a name they have used all their lives is relegated to synonymy and they find themselves obliged to use one that is unfamiliar. The taxonomist who makes the change is not the malicious demon they imagine, but is following simple and necessary rules. I propose to give the main reasons for such changes and to explain the conventional terms used in check lists to indicate that they have taken place.

a) The Law of Priority. Under this rule, a species must be called by the earliest name bestowed upon it. For instance, 1331 *Acentria nivea* (Olivier, 1791) is now *A. ephemerella* ([Denis & Schiffermüller], 1775). Some early names are classed as *nomina dubia*; continental taxonomists are readier than their English counterparts to jump to conclusions and introduce *nomina dubia* into the nomenclature. I am strongly on the British side. There is one way in which a change that ought to be made in accordance with the I.C.Z.N. rules can be avoided. That is to appeal to the Commission itself. For instance, everyone is familiar with the name 373 *Synanthedon tipuliformis* (Clerck, 1759), but now that name is deemed to be a junior synonym of *salmachus* Linnaeus, 1758. This is not at all obvious from Linnaeus' own description or that of Petiver which he quotes, but what matters is that the decision was made. Now this moth is a pest species and appears under the name *tipuliformis* in many horticultural and agricultural handbooks. To change the name would confuse non-entomological fruit-farmers and accordingly the I.C.Z.N. ruled that the junior name should be conserved. In check lists, synonyms follow the valid name in order of seniority and are indented.

b) Homonymy. The same specific name may be used for two species if they are in different genera, but not if they are in the same genus; the genus for this purpose is the one in which the species was first described. It is permissible to have *Quercusia quercus* (Linnaeus, 1758) and *Lasiocampa quercus* (Linnaeus, 1758) because Linnaeus described the first species in Papilio (Plebejus) and the second in Phalaena (Bombyx). It is not permissible to have *Cucullia umbratica* (Linnaeus, 1758) and *Rusina umbratica* (Goeze, 1781) because both authors described their species in *Noctua*. Goeze's name is therefore described as a junior homonym and cannot be used. The *Rusina umbratica* of the earlier editions of South's *The Moths of the British Isles* is now known as *R. ferruginea* (Esper, 1781), the next available name in order of seniority, and the old name is shown in synonymy thus: *umbratica* (Goeze, 1781) *nec* (Linnaeus, 1758).

c) Misidentification. It not infrequently happens that authors apply a name to the wrong species through misidentification and the error may persist for many years. For example, in the earlier editions of South the blood-vein is known as *amata* Linnaeus. The true *amata* of Linnaeus is not that species at all, but a form of *Cyclophora punctaria* (Linnaeus, 1758). So *amata* became a junior synonym of *punctaria* which Linnaeus named first, and a new name, *griseata* (Petersen, 1902), had to be given to the blood-vein. The synonymy is written thus:

griseata (Petersen, 1902)
    amata sensu auctt.

'Sensu auctt.' is the abbreviation for *sensu auctorum*, the Latin for 'in the opinion of the authors', the double 't' for 'auctt.' being the conventional way of indicating the plural (cf.

'pp.' for *paginae*, pages). Alternatively, one can write '*sensu* South, 1908' if South was the only author, or the relevant author, to make the misidentification.

d) *Incompatability with the type specimen.* In modern entomology, when a new species is named the author designates a type specimen and, if there is any later dispute over the identity of the species to which the name refers, it is settled by reference to the type specimen. Early entomologists did not do this, so a lectotype is designated, if possible from the nomenclator's own collection and the specimen or specimens he had before him when he made the description and bestowed the name. It occasionally happens that a specimen bearing a given name in the author's collection turns out to be a different species from the one he had described under that name. The rule is that in these cases the specimen takes precedence over the description. For example, the moth now known as 688 *Agonopterix heracliana* (Linnaeus, 1758) is the one so labelled in Linnaeus' own collection housed at the Linnean Society in London, but the species and life history described under this name by Linnaeus in *Systema Naturae* is without any doubt the moth now called 672 *Depressaria pastinacella* (Duponchel, 1838). As a result, the name *heracliana* was transferred from the species that had borne it for 200 years and applied to one that had been known as 688 *applana* Fabricius, 1777, for almost as long. This is also the reason why *Coleophora vestianella* (Linnaeus, 1758), the larva of which feeds on the seeds of *Atriplex*, bears a name intended for a clothes-moth. In 1862 Heinemann made a good description of the early stages of an undescribed nepticulid which he called *distinguenda*, but the species so labelled in his collection is *Stigmella glutinosae* (Stainton, 1858). As a result of Heinemann's carelessness in muddling the adults that emerged from his cages, *distinguenda* is reduced to synonymy and a new name had to be found for the species he had been the first to describe. It is a pity that there is no conventional term to be used in check lists to identify changes made for this reason. It is also a pity that some taxonomists do not know the virtue of a blind eye.

Systematic Section

ZEUGLOPTERA

ζεύγλη (zeuglē), a yoke; πτερόν (pteron), a feather, πτερά (ptera) (pl.), a wing: from the mechanism for linking the forewing and hindwing in flight.

MICROPTERIGOIDEA (1)

MICROPTERIGIDAE (1)

Micropterix Hübner, 1825 – μικρός (mikros), little; πτέρυξ (pterux), a wing: from the small size of the adult.

1 **tunbergella** (Fabricius, 1794) – i.h.o. K. P. Thunberg (1743–1828), a successor of Linnaeus in the Chair of Botany at Uppsala University.

2 **mansuetella** Zeller, 1844 – *mansuetus*, tame: from the docile behaviour of the adult female when feeding on pollen.

3 **aureatella** (Scopoli, 1763) – *aureatus*, golden: from the submetallic markings on the forewing.

4 **aruncella** (Scopoli, 1763) – *Spiraea aruncus*, a plant on the pollen of which the adults had been found feeding.

5 **calthella** (Linnaeus, 1761) – *Caltha palustris*, marsh marigold, on the pollen of which the adult had been found feeding.

DACNONYPHA

δάκνω (daknō), to bite; νύφη, νύμφη (nuphē, numphē), an immature insect stage, here the pupa: from the mandibles used to cut open the cocoon.

ERIOCRANIOIDEA (6)

ERIOCRANIIDAE (6)

Eriocrania Zeller, 1851 – ἔριον (erion), wool; κρανίον (kranion), the upper part of the head: from the hair-scales on the vertex of the adult head.

6 **subpurpurella** (Haworth, 1828) – *sub-*, somewhat; *purpureus*, purple: from the purplish strigulae on the forewing.

7 **chrysolepidella** Zeller, 1851 – χρυσός (khrusos), gold; λεπίς, λεπίδος (lepis, lepidos), a scale: from the golden ground colour of the forewing.

8 **unimaculella** (Zetterstedt, 1840) – *unus*, one; *macula*, a spot: from the conspicuous pretornal white spot on the forewing.

9 **sparrmannella** (Bosc, 1791) – i.h.o. A. Sparrmann (1748–1820), a Swedish entomologist.

10 **salopiella** (Stainton, 1854) – Salop, Shropshire: the county in which the type locality is situated.

11 **haworthi** Bradley, 1966 – i.h.o. A. H. Haworth (1767–1833), the British entomologist. The name is a replacement for *purpurella* (Haworth, 1828) which was in use for nearly 150 years before it was found to be preoccupied.

12 **sangii** (Wood, 1891) – i.h.o. J. Sang (1828–87), a leading lepidopterist who operated in the north-east of England.

13 **semipurpurella** (Stephens, 1835) – *semi-*, half; *purpureus*, purple: from the coloration of the forewing.

EXOPORIA

ἔξ, ἔξο- (ex, exo-), out, outwards; πῶρος (pōros), a node, a projection: from the jugum or coupling mechanism projecting from the forewing.

HEPIALOIDEA (14)

HEPIALIDAE (14)

Hepialus Fabricius, 1775 – ἠπίαλος (hēpialos), a fever: 'from the fitful, alternating flight of these insects' (Pickard *et al.*). Fabricius bestowed this name to accommodate the 'Noctuae *elingues*' (Noctuae without haustellum) of *Systema Naturae* (pp. 496 (Pl. III), 508, 822). Linnaeus had listed six species, three of them 'swifts'; Fabricius excluded the other three but added others such as the 173 Limacodidae. The name was suggested by 16 *hecta*, q.v., of which he wrote later *'vespere in aere fluctitat motu pendulo at solitarius'*, in the evening it hovers in the air with a pendulous motion, but alone (i.e. not in a swarm, like gnats) (Fabricius, 1793). Sodoffsky, Spuler and others suggested emendation to *Hepiolus*, from ἠπίολος, a name used by Aristotle for a moth, and cite 370 *Sesia* as an analogy, but as Fabricius consistently spelt the name *Hepialus* in all his subsequent works, that must be what he intended. However, a play on words is possible (cf. 166 *Zygaena*, etc.).

14 **humuli** (Linnaeus, 1758) – *Humulus*, the hop genus: Linnaeus wrongly supposed the larva to feed on the roots.

subsp. **thulensis** Newman, 1865 – *'Ultima Thulē'*, an island six days' sailing north of Orkney, discovered by Pythias in the 4th century B.C. and commonly supposed to be Shetland: the subspecies occurring in Shetland.

15 **sylvina** (Linnaeus, 1761) – *sylvinus*, alternative spelling of *sylvanus* (*silvanus*), belonging to a wood: from the habitat, although the moth is not restricted to woodland.

16 **hecta** (Linnaeus, 1758) – ἑκτικός (hectikos), hectic, feverish: either from the moth's flight (see generic name), or from the flushed forewing, or from both.

17 **lupulinus** (Linnaeus, 1758) – *Humulus lupulinus*, hop: Linnaeus bestowed the name to indicate affinity with 14 *H. humuli*; he had no knowledge of the foodplant.

18 **fusconebulosa** (DeGeer, 1778) – *fuscus*, dark; *nebulosus*, cloudy: from the dark variegation on the forewing of certain forms.

subsp. **shetlandicus** Viette, 1958 – the subspecies occurring in the Shetland Islands.

MONOTRYSIA

μόνος (monos), single, solitary; τρῦπα, τρυσ- (trūpa, trūs-), a hole: from the female genitalia which have only a single orifice (cf. Ditrysia, p. 51).

NEPTICULOIDEA (50)

NEPTICULIDAE (50)

The systematic arrangement is that of Bradley & Fletcher (1986) but the log book numbers of Bradley & Fletcher (1979) are retained.

Bohemannia Stainton, 1859 – i.h.o. C. H. Boheman (1796–1868), the Swedish entomologist who named the next species.

19 **quadrimaculella** (Boheman, 1851) – *quadri-*, four; *macula*, a spot: from the four golden spots, two on each forewing.

33 **auriciliella** (Joannis, 1908) – *aurum*, gold; *cilia*, eyelashes, the fringe of an insect's wing: from the forewing cilia.

= **bradfordi** (Emmet, 1974) – i.h.o. E. S. Bradford, contemporary British entomologist, who took a specimen of what was then supposed to be an unnamed species.

40 **pulverosella** (Stainton, 1849) – *pulverosus*, dusty: from the dark irroration on the forewing.

Etainia Beirne, 1945 – Professor B. P. Beirne (see 47 *beirnei*) turned to Gaelic mythology and the legends of his native Ireland for the names of the new genera he erected in the Nepticulidae (B.P. Beirne, pers. comm.). Étaín, originally a sun-goddess, was also the name of several beautiful, noble or saintly women in legend or early history (Ó Corráin & Maguire, 1981).

20 **decentella** (Herrich-Schäffer, 1855) – *decens*, comely: from the attractive black and white pattern of the forewing.

21 **sericopeza** (Zeller, 1839) – σηρικόπεζα (sērikopeza), silken-footed (Pickard *et al.*): 'from moth's glistening legs' (Macleod); the reference may, however, be to the smooth, silky speed with which the moth runs.

22 **louisella** (Sircom, 1849) – probably i.h.o. a relative or friend; Sircom gives no explanation.

= **sphendamni** (Hering, 1937) – σφένδαμνος (sphendamnos), maple: the larval foodplant.

Ectoedemia Busck, 1907 – ἐκτός (ektos), outside; οἴδημα (oidēma), a swelling, a tumour: from the larval feeding habits of the type species. The genus was erected to accommodate the North American nepticulid *E. populella* Busck, 1907, a close relative of 23 *E. argyropeza*; the larva feeds in a leaf petiole of *Populus* spp., causing a gall to develop.

41 **atrifrontella** (Stainton, 1851) – *ater*, black; *frons*, *frontis*, the forehead: from the black frons of the imago.

23 **argyropeza** (Zeller, 1839) – ἀργυρόπεζα (arguropeza), silver-footed: 'from colour of moth's legs' (Macleod); the legs are creamy white but hardly silvery. The Greek word is used as an epithet of Aphrodite, the goddess of love, and Zeller may have intended the name to be laudatory in a general sense.

24 **turbidella** (Zeller, 1848) – *turbidus*, confused: from the ill-defined pattern of the forewing.

25 **intimella** (Zeller, 1848) – *intimus*, innermost: from the life history of the larva which in its early instars feeds concealed in the midrib of a leaf of sallow (*Salix* spp.).

26 **agrimoniae** (Frey, 1858) – *Agrimonia eupatoria*, agrimony: the larval foodplant.

27 **spinosella** (Joannis, 1907) – *Prunus spinosa*, blackthorn: the larval foodplant.

28 **angulifasciella** (Stainton, 1849) – *angulus*, an angle; *fascia*, a band: from the angled fascia on the forewing.

29 **atricollis** (Stainton, 1857) – *ater*, black; *collum*, the neck: from the conspicuous black prothoracic plate of the larva.

30 **arcuatella** (Herrich-Schäffer, 1855) – *arcuatus*, curved: from the shape of the fascia on the forewing.

31 **rubivora** (Wocke, 1860) – *Rubus*, the bramble genus; *voro*, to devour: from the larval foodplant.

32 **erythrogenella** (Joannis, 1907) – ἐρυθρός (eruthros), red; γεν- (gen-) found only in compounds, begetting: from the reddish discoloration in the leaves of bramble (*Rubus fruticosus*) caused by the larval mine.

33 see below 19

34 **occultella** (Linnaeus, 1767) – *occultus*, concealed: not only is the larva a leaf-miner, but it also rests hidden, when not feeding, under a black central blotch formed on the upper epidermis of the leaf.

= **mediofasciella** (Haworth, 1828) – *medius*, middle; *fascia*, a band: from the fascia on the forewing which is placed more centrally than it is in the *atricollis* group. The name was misapplied to the next species.

= **argentipedella** (Zeller, 1839) – *argentum*, silver; *pes*, *pedis*, the foot: a Latin translation of 23 *argyropeza*, q.v.

35 **minimella** (Zetterstedt, 1839) – *minimus*, smallest, very small: from the small size of the moth, though, for a nepticulid, it is fairly large.

= **mediofasciella** sensu auctt. – see last species.

= **woolhopiella** (Stainton, 1887) – Woolhope, a village in Herefordshire where Wood took the first British specimens.

36 **quinquella** (Bedell, 1848) – *quinque*, five: from the silvery spots on the forewings, three on each, but when the wings are folded the dorsal spots merge, giving the appearance of only five spots.

37 **albifasciella** (Heinemann, 1871) – *albus*, white; *fascia*, a band: from the white fascia on the forewing.

38 **subbimaculella** (Haworth, 1828) – *sub-*, somewhat; *bi-*, two; *macula*, a spot: from the forewing fascia which is broken into two spots. Haworth could not use *bimaculella* because the name was preoccupied; see 902 *lathamella*.

39 **heringi** (Toll, 1935) – i.h.o. E. M. Hering (1893–1967), the distinguished German entomologist and authority on leaf-mining insects.

= **quercifoliae** (Toll, 1836) – *Quercus*, the oak genus; *folium*, a leaf: from the larval foodplant. The name is malformed; Toll evidently supposed that *folia*, neut. pl., leaves, was the nom. sing. of a first declension noun and formed '*foliae*' as its genitive case as in 118 *acetosae* which is orthographically correct; cf. 494a *prunifoliae* and 498 *alnifoliae* in which the same blunder is made. Linnaeus (1641 *ilicifolia*, 1642 *quercifolia*) and Stainton (501 *siccifolia*) were better Latinists.

40 see below 19

41 see below 22

Fomoria Beirne, 1945 – see 20 *Etainia*. The Fomhoire or Fomoir were a race of demonic and malevolent beings who ruled Ireland on several occasions, exacting heavy tribute from the people. They were eventually defeated in battle and expelled for ever (Mac Cana, 1983).

42 **septembrella** (Stainton, 1849) – from September, the month in which the first moths were reared, though May and August are more usual months for emergence.

43 **weaveri** (Stainton, 1855) – i.h.o. R. Weaver (19th century), a British entomologist who lived near Birmingham. According to Allan (1980: 77, 126), he was a shoemaker and a professional collector.

Fedalmia Beirne, 1945 – see 20 *Etainia*. Fedelm or Feidelm was the name of several royal, beautiful or martial women in Irish legend, and of no fewer than six saints (Ó Corráin & Maguire, 1981: Ms Maguire's first name is Fidelma).

44 **headleyella** (Stainton, 1854) – Headley Down, Surrey is the type locality.

Trifurcula Zeller, 1848 – *trifurcus*, three-pronged: there are three veins emanating from the median vein on the forewing; in *Stigmella* there are only two.

45 **subnitidella** (Duponchel, 1843) – *sub-*, somewhat; *nitidus*, shining: from the faint gloss on the forewing.

= **griseella** Wolff, 1957 – *griseus* (late Lat.), grey: from the colour of the forewing.

46 **immundella** (Zeller, 1839) – *immundus*, dirty, unadorned: from the dull grey coloration of the forewing.

46a **squamatella** Stainton, 1849 – *squamatus*, scaled: from the coarse scaling of the forewing which gives a roughened appearance.

47 **beirnei** Puplesis, 1984 – i.h.o. B. P. Beirne, the contemporary Irish/English/Canadian entomologist who first figured the genitalia.

= **pallidella** sensu auctt. – *pallidus*, pale: from the colour of the forewing. The true *pallidella* Zeller is an eastern European species not found in Britain.

48 **cryptella** (Stainton, 1856) – κρυπτός (kruptos), hidden: from the larval habit of mining in the leaves of bird's-foot trefoil (*Lotus* spp.).

49 **eurema** (Tutt, 1899) – εὕρημα (heurēma), a discovery, a windfall: from Durrant's unexpected discovery that it was distinct from 48 *cryptella*.

Stigmella Schrank, 1802 – στίγμα (stigma), a brand, a small dot: possibly from the small size but more probably from the conspicuous, sometimes metallic, fascia on the forewing of many of the species; cf. 288 *stigmatella*. Since slaves were branded, the word 'stigma' in English sometimes signifies a mark of disgrace, but not here; Schrank called the moths 'Edelmotte', noble moths. The name in its first conception had family rather than generic status; see p. 30.

= **Nepticula** Heyden, 1843 – dim. of *neptis*, a granddaughter, potentially the smallest member of the family: from the very small size of the moths.

50 **aurella** (Fabricius, 1775) – *aurum*, gold: from the golden metallic fascia on the forewing.

51 **fragariella** (Heyden, 1862) – *Fragaria vesca*, wild strawberry: a foodplant.

52 **dulcella** (Heinemann, 1862) – *dulcis*, sweet: 'a sweet little moth'. Nos 51 and 52 are probably only forms of 50 *S. aurella*.

53 **splendidissimella** (Herrich-Schäffer, 1855) – *splendidissimus*, very brilliant: from the glossy forewing and metallic fascia.

54 **auromarginella** (Richardson, 1890) – *aurum*, gold; *margo*, *marginis*, a border: from the golden metallic scales on the termen of the forewing.

55 **aeneofasciella** (Herrich-Schäffer, 1855) – *aëneus*, brazen; *fascia*, a band: from the metallic brassy fascia on the forewing.

56 **dryadella** (Hofmann, 1868) – *Dryas octopetala*, mountain avens: the larval foodplant.

57 **filipendulae** (Wocke, 1871) – *Filipendula vulgaris*, dropwort: the larval foodplant.

58 **ulmariae** (Wocke, 1879) – *Filipendula ulmaria*, meadowsweet: the larval foodplant.

59 **poterii** (Stainton, 1857) – *Poterium sanguisorba*, salad burnet: the larval foodplant.

60 **tengstroemi** (Nolken, 1871) – i.h.o. J. M. J. von Tengström (1821–90), a Swedish entomological author.

61 **serella** (Stainton, 1888) – σήρ (sēr), an oriental, a Chinaman, a silkworm: from the silky texture of the forewing; cf. Latin *sericus*, silky.

62 **tormentillella** (Herrich-Schäffer, 1860) – *Potentilla erecta*, formerly known as *P. tormentilla*, tormentil: the larval foodplant. Placed on the British list through misidentification.

63 **marginicolella** (Stainton, 1853) – *margo, marginis*, border; *colo*, to inhabit: from the larval mine which frequently follows the margin of a leaf of elm.

64 **continuella** (Stainton, 1856) – *continuus*, unbroken: from the frass in the larval mine which extends over the whole breadth of the mine. Stainton had described this mine before he had bred the moth.

65 **speciosa** (Frey, 1858) – *speciosus*, beautiful: in approbation.

66 **sorbi** (Stainton, 1861) – *Sorbus aucuparia*, rowan: the larval foodplant.

67 **plagicolella** (Stainton, 1854) – *plaga*, flat, open ground; *colo*, to inhabit: the larva feeds on blackthorn, as often as not occurring in such situations. The suggestion of Macleod that it refers to the flat cocoon has no foundation.

68 **salicis** (Stainton, 1854) – *Salix*, sallow: the larval foodplant.

69 **auritella** (Skåla, 1932) – *Salix aurita*, eared sallow: the larval foodplant. Probably conspecific with the last species.

70 **obliquella** (Heinemann, 1862) – *obliquus*, slanting: with reference to the fascia on the forewing.

= **vimineticola** sensu auctt. – *viminetum*, an osier bed; *colo*, to inhabit: placed on the British list through misidentification, *S. vimineticola* (Frey, 1856) being a species confined to central and southern Europe.

71 **zelleriella** (Snellen, 1875) – i.h.o. P. C. Zeller (1808–83), the distinguished German entomologist.

= **repentiella** (Wolff, 1955) – *Salix repens*, dwarf sallow, creeping willow: the larval foodplant.

72 **myrtillella** (Stainton, 1857) – *Vaccinium myrtillus*, bilberry: the larval foodplant.

73 **trimaculella** (Haworth, 1828) – *tri-*, three; *macula*, a spot: from the markings on the forewing.

74 **assimilella** (Zeller, 1848) – *assimilis*, like: from its resemblance to 23 *Ectoedemia argyropeza* which it follows and with which it is compared in Zeller's paper.

75 **floslactella** (Haworth, 1828) – *flos*, flower; *lac, lactis*, milk: from the cream-coloured or yellowish ground colour of the forewing, variably overlaid with fuscous scales.

76 **carpinella** (Heinemann, 1862) – *Carpinus betula*, hornbeam: the larval foodplant.

77 **tityrella** (Stainton, 1854) – Tityrus, a mythical shepherd who sang songs whilst sitting under a beech-tree and tending his sheep, '*Tityre, tu patulae recubans sub tegmine fagi*' (Tityrus, you who are sitting under the shade of a spreading beech-tree) (Virgil, *Ecologue* 1:1): beech is the larval foodplant.

78 **pomella** (Vaughan, 1858) – *poma*, a fruit, especially an apple: apple (*Malus* spp.) is the larval foodplant.

79 **perpygmaeella** (Doubleday, 1859) – *per-*, intensive prefix, very; *pygmaeus*, a pigmy: from the small size of the moth. It had been named *pygmaeella* by Haworth (1828), but as this name was preoccupied Doubleday proposed a replacement name close to the original.

80 **ulmivora** (Fologne, 1860) – *Ulmus*, elm; *voro*, to devour: from the larval foodplant.

81 **hemargyrella** (Kollar, 1832) – ἡμι- (hēmi-), half; ἄργυρος (arguros), silver: from the broad silver fascia on the forewing.

82 **paradoxa** (Frey, 1858) – παράδοξος (paradoxos), contrary to expectation: from the larval mine which was quite unlike any other that Frey had described.

83 **atricapitella** (Haworth, 1828) – *ater*, black; *caput, capitis*, head: from the black head of the adult.

84 **ruficapitella** (Haworth, 1828) – *rufus*, red; *caput, capitis*, head: from the red head of the female (that of the male is black).

85 **suberivora** (Stainton, 1869) – *Quercus suber*, cork-oak; *voro*, to devour: from the foodplant in the Mediterranean region; in Britain it feeds on holm-oak (*Q. ilex*).

86 **roborella** (Johansson, 1971) – *Quercus robur*, gen. *roboris*, pedunculate oak: one of the oak species utilized by the larva.

87 **svenssoni** (Johansson, 1971) – i.h.o. I. Svensson, the distinguished contemporary Swedish entomologist.

88 **samiatella** (Zeller, 1839) – *samiatus*, polished with Samian stone: from the gloss on the forewing.

89 **basiguttella** (Heinemann, 1862) – *basis*, base; *gutta*, a spot: from the whitish spot at the base of the forewing.

90 **tiliae** (Frey, 1857) – *Tilia cordata*, small-leaved lime: the larval foodplant.

91 **minusculella** (Herrich-Schäffer, 1855) – *minusculus*, smallish: from the diminutive size of the moth.

92 **anomalella** (Goeze, 1783) – ἀνώμαλος (anōmalos), anomalous, inconsistent: Goeze studied the leaf-mining larva which has various adaptations, e.g. in the structure of the legs, in which it differs from those that feed externally.

93 **centifoliella** (Zeller, 1848) – *Rosa centifolia*, the form from which the cabbage rose was developed: the larva, however, prefers small-leaved forms.

94 **spinosissimae** (Waters, 1928) – *Rosa spinosissima*, now called *R. pimpinellifolia*, burnet rose: the larval foodplant.

95 **viscerella** (Stainton, 1853) – *viscera*, entrails: from the gut-like formation of the gallery mine in leaves of elm. The name was suggested to Stainton by J. W. Douglas.

96 **ulmiphaga** (Preissecker, 1942) – *Ulmus*, elm; φαγεῖν (phagein), to eat: from the larval foodplant. Placed on the British list through misidentification.

97 **malella** (Stainton, 1854) – *Malus*, apple: the larval foodplant.

98 **catharticella** (Stainton, 1853) – *Rhamnus catharticus*, buckthorn: the larval foodplant.

99 **hybnerella** (Hübner, 1796) – i.h.o. J. Hübner (1761–1826), the distinguished German entomologist and illustrator. Although it appears that Hübner bestowed the name in his own honour, it was probably proposed by another entomologist but first used in print by Hübner himself.

99a **mespilicola** (Frey, 1856) – *Mespilus*, the medlar genus: a larval foodplant, although in Britain it has been found only on *Sorbus torminalis* and *S. aria*, wild service and whitebeam.

100 **oxyacanthella** (Stainton, 1854) – *Crataegus oxyacantha*, now called *C. laevigata*, midland hawthorn: one of the larval foodplants.

= **aeneella** sensu auctt. – *aëneus*, brazen: from the gloss on the forewing. *S. aeneella* (Heinemann, 1862) is a synonym of 92 *S. anomalella*, but the name was misapplied by British authors to *S. oxyacanthella* when feeding on *Malus*.

101 **pyri** (Glitz, 1865) – *Pyrus communis*, pear: the larval foodplant.

102 **aceris** (Frey, 1857) – *Acer*, the maple genus, that of the larval foodplants.

103 **nylandriella** (Tengström, 1848) – i.h.o. W. Nylander (1822–99), a Finnish entomologist.

= **aucupariae** (Frey, 1857) – *Sorbus aucuparia*, rowan: the larval foodplant.

104 **magdalenae** (Klimesch, 1950) – i.h.o. Magdalen, late wife of Dr J. Klimesch, the distinguished contemporary Austrian entomologist.

= **nylandriella** sensu auctt. – applied to this species through misidentification; see 103.

105 **desperatella** (Frey, 1856) – *desperatus*, despaired of: from early difficulty in rearing the adult.

106 **torminalis** (Wood, 1890) – *Sorbus torminalis*, wild service: the larval foodplant.

107 **regiella** (Herrich-Schäffer, 1855) – *regius*, royal: from the brilliant purple and gold of the forewing.

108 **crataegella** (Klimesch, 1936) – *Crataegus*, hawthorn: the larval foodplant.

109 **prunetorum** (Stainton, 1855) – *prunetum*, gen. pl. *prunetorum* (late Lat.): a blackthorn thicket: from the larval foodplant.

110 **betulicola** (Stainton, 1856) – *Betula*, birch; *colo*, to frequent: from the larval foodplant.

f. **nanivora** (Petersen, 1930) – *Betula nana*, dwarf birch; *voro*, to devour: the larval foodplant.

111 **microtheriella** (Stainton, 1854) – μικρός (micros), small; θηρίον (thērion), a little creature: when the name was bestowed it was believed to be the smallest moth.

112 **luteella** (Stainton, 1857) – *luteus*, saffron yellow: from the yellowish fascia on the forewing.

113 **sakhalinella** Puplesis, 1984 – Yuzhno-Sakhalinsk, U.S.S.R.: the type locality.

= **distinguenda** sensu auctt. *nec* (Heinemann, 1862) – *distinguendus*, to be distinguished, separated: although Heinemann correctly distinguished this species from 112 *luteella*, he muddled his material and his type specimen is in fact the next species.

= **discidia** Schoorl & Wilkinson, 1986 – *discidium*, a separation: a replacement name with much the same meaning as *distinguendus* above.

114 **glutinosae** (Stainton, 1858) – *Alnus glutinosa*, alder: the larval foodplant.

115 **alnetella** (Stainton, 1856) – *Alnus glutinosa*, alder: the larval foodplant.

116 **lapponica** (Wocke, 1862) – *Lapponicus*, Lappish: the type locality is in northern Norway.

117 **confusella** (Wood, 1894) – *confusus*, confused: from the species having been confused with 116 *S. lapponica* until Wood noticed the different characters of the mines. 'From the purplish-tinged fuscous forewings' (Macleod): incorrect.

Enteucha Meyrick, 1915 – perhaps ἐν (en), in; τεύχω (teukho), to make, to produce by art: well-wrought.

118 **acetosae** (Stainton, 1854) – *Rumex acetosa*, common sorrel: a larval foodplant.

OPOSTEGIDAE (119)

Opostega Zeller, 1839 – ὤψ (ōps), face; στέγη (stegē), a roof: from the very large antennal eyecaps.

119 **salaciella** (Treitschke, 1833) – *salax*, lustful: possibly the type specimens found by Treitschke were *in copula*; cf. 122 *spatulella*. 'From the delight with which it flies round foodplant' (Macleod); the foodplant and life history are unknown.

120 **auritella** (Hübner, 1813) – *auritus*, eared: from the prominent antennal eyecaps which may fancifully be supposed to resemble ears.

121 **crepusculella** Zeller, 1839 – *crepusculum*, evening twilight: from the usual time of flight.

122 **spatulella** Herrich-Schäffer, 1855 – *spatula*, voluptuousness: to indicate close affinity with 119 *O. salaciella*, q.v. Macleod gives the same explanation as for *O. salaciella*: the foodplant of this species is likewise unknown.

TISCHERIOIDEA (123)

TISCHERIIDAE (123)

Tischeria Zeller, 1839 – i.h.o. Carl F. A. von Tischer (1777–1849), a German entomologist who suggested some of the names first used by Zeller and other writers. Tischer's 'von', given by Pickard *et al.* and Stainton (1855–73), is omitted by Gilbert (1977).

123 **ekebladella** (Bjerkander, 1795) – i.h.o. Count Claes Ekeblad (18th century), a member of the Swedish court who, in collaboration with Count Tessin, formed a collection used by Linnaeus as a source for *Systema Naturae*.

124 **dodonaea** Stainton, 1858 – *Dodonaeus*, of Dodona, the most ancient oracle in Greece, where responses were given from lofty oaks: oak (*Quercus* spp.) is the larval foodplant; cf. 1853 *dodoneata* and 2014 *dodonaea*.

125 **marginea** (Haworth, 1828) – *margineus*, of a margin, coined from *margo, marginis*, an edge: from the dark costa and terminal margin of the forewing.

125a **heinemanni** Wocke, 1871 – i.h.o. H. von Heinemann (1812–71), a German entomologist.

126 **gaunacella** (Duponchel, 1843) – *gaunacum*, a Persian or Babylonian fur prepared from weasel or mouse skins: from the coloration of the forewing.

127 **angusticollella** (Duponchel, 1843) – *angustus*, narrow; *collum*, the neck: from the deeply incised intersegmental divisions of the larva.

INCURVARIOIDEA (129)

INCURVARIIDAE (129)

INCURVARIINAE (129)

Phylloporia Heinemann, 1870 – φύλλον (phullon), a leaf; πορεία (poreia), a walking: from the long, narrow gallery made by the larva in a leaf of birch before it is expanded into the blotch from which a case is excised; the journey made by the larva dragging this case of leaf fragments behind it in search of a site for pupation is also a possible meaning.

128 **bistrigella** (Haworth, 1828) – *bis, bi-*, two; *striga*, a furrow or streak: from the two parallel pale fasciae on the forewing.

Incurvaria Haworth, 1828 – *incurvatus*, curved; from the curved maxillary palpi (*palpi anteriores falcatim inflexi*). 'From the slightly sickle-shaped forewings' (Macleod), a non-existent character invented without reference to Haworth's explicit diagnosis. Cf. 757 *Recurvaria*.

129 **pectinea** Haworth, 1828 – *pectineus*, adj. coined from *pecten*, a comb: from the unipectinate antenna of the male adult.

= **zinckenii** (Zeller, 1839) – i.h.o. Dr J. L. Zincken (early 19th century), a German entomologist and medical practitioner living in Brunswick.

130 **masculella** ([Denis & Schiffermüller], 1775) – *masculus*, masculine, male: perhaps from the unipectinate antenna, a very distinctive character found only in the male. The amendment *'musculella'* was proposed by Fabricius and adopted by most earlier authors; it could be derived from *musca*, a fly (Pickard *et al.*) or *musculus*, a small mouse (Macleod), but neither has any convincing application and it is obligatory to retain the name as first printed.

131 **oehlmanniella** (Hübner, 1796) – i.h.o. G. Oehlmann (d. *c.* 1815), a professional German insect dealer living in Leipzig who provided Hübner with the type specimen for figuring.

132 **praelatella** ([Denis & Schiffermüller], 1775) – *praelatus*, preferred (Pickard *et al.*), or, more probably, a prelate (Macleod), from the episcopal purple of the forewing.

PRODOXINAE

From the non-British genus *Prodoxus* Riley, 1880 – προ- (pro-), in front; d, phonetic insertion; ὀξύς (oxus), sharp: perhaps referring to the antennae which are finely ciliate and pointed in contrast to those of *Incurvaria* which are uni- or bipectinate and blunt.

Lampronia Stephens, 1829 – λαμπρός (lampros), bright: from the bright coloration of some of the species.

133 **capitella** (Clerck, 1759) – *caput, capitis*, the head: from the conspicuous ochreous yellow head of the adult.

134 **flavimitrella** (Hübner, 1817) – *flavus*, yellow; *mitra*, a turban: from the conspicuous yellow head of the adult.

135 **luzella** (Hübner, 1817) – i.h.o. J. F. Luz, a senior officer in the Austrian army who sent Hübner the type specimen for figuring.

136 **rubiella** (Bjerkander, 1781) – *Rubus idaeus*, raspberry: the larval foodplant.

137 **morosa** Zeller, 1852 – *morosus*, morose, gloomy: from the dingy coloration of the adult suggestive of such a temperament.

138 **fuscatella** (Tengström, 1848) – *fuscatus*, blackened: from the dark fuscous coloration of both fore- and hindwing.

= **tenuicornis** (Stainton, 1854) – *tenuis*, thin; *cornu*, a horn: from the slender antenna; cf. suggested interpretation of the subfamily name.

139 **pubicornis** (Haworth, 1828) – *pubes*, the soft hair which appears on the body at the age of puberty: *cornu*, a horn: from the shortly ciliate male antenna.

NEMATOPOGONINAE (140)

Nematopogon Zeller, 1839 – νῆμα, νήματος (nēma, nēmatos), a thread; πώγων (pōgōn), a beard: referring both to the long, thin antenna (νῆμα) and to the densely hairy labial palpus ('mit flaumhaarigen Palpen') (πώγων).

140 **swammerdamella** (Linnaeus, 1758) – i.h.o. J. J. Swammerdam (1637–80), a Dutch entomologist.

141 **schwarziellus** (Zeller, 1839) – i.h.o. C. Schwarz (d. 1810), a German entomologist.

= **panzerella** sensu auctt. – i.h.o. Dr G. W. F. Panzer (1755–1829), a German entomologist, author and illustrator, who lived at Hersbruch, near Nuremburg.

142 **pilella** ([Denis & Schiffermüller], 1775) – *pilus*, a hair: from the long, hairlike antenna. *'Pileus*, hairy, from top of head, scales near base of hindwings and hind legs' (Macleod): possible, at least in part, but less likely.

143 **metaxella** (Hübner, 1813) – i.h.o. L. & T. Metaxa (19th century), Italian naturalists.

143a **magna** (Zeller, 1878) – *magnus*, large: from its relative size.

ADELINAE (149)

Nemophora Illiger, 1798 – νῆμα (nēma), a thread; φορέω (phoreō), to carry: from the long, thread-like antennae. Illiger ascribes the name to Hoffmannsegg and both entomologists are usually cited as nomenclators; this is incorrect as Illiger was sole author of the work in which the name first appears. It was intended as a family name but as such it is a junior synonym of 149 *Adela*; later both were reduced to generic status.

144 **fasciella** (Fabricius, 1775) – *fascia*, a band: from the blackish violet fascia on the coppery forewing.

145 **minimella** ([Denis & Schiffermüller], 1775) – *minimus*, smallest: from its small size compared with related species.

146 **cupriacella** (Hübner, 1819) – *cupreus*, of copper: from the coppery forewing.

147 **metallica** (Poda, 1761) – *metallicus*, metallic: from the gloss on the forewing.

= **scabiosella** (Scopoli, 1763) – *Scabiosa columbaria*, small scabious: one of the foodplants.

148 **degeerella** (Linnaeus, 1758) – i.h.o. Baron Karl DeGeer (1720–78), an eminent Swedish naturalist.

Adela Latreille, 1796 – ἄδηλος (adēlos), unseen: from the larval habit of concealing itself in a portable case. Latreille gave this explanation himself, so Sodoffsky's (1837) derivation from a town in Africa is certainly wrong. Like 144 Nemophora, q.v., it was originally a family name and has been reduced to generic status but, as the senior name, it is used for the subfamily.

149 **cuprella** ([Denis & Schiffermüller], 1775) – *cupreus*, coppery: from the metallic colour of the forewing.

149a **violella** ([Denis & Schiffermüller], 1775) – from the 'brilliant dark violet-red adult', not from the foodplant since the life history was unknown.

150 **reaumurella** (Linnaeus, 1758) – i.h.o. R.A.F. de Réaumur (1683–1757), the French scientist and entomologist; cf. 185 *ferchaultella*.

151 **croesella** (Scopoli, 1763) – Croesus, King of Lydia, 560–546 B.C., famed for his wealth: from the richly coloured forewing, which is purple with a golden fascia; cf. 1035 *Croesia*.

152 **rufimitrella** (Scopoli, 1763) – *rufus*, red; *mitra*, a turban: from the red head of the imago.

153 **fibulella** ([Denis & Schiffermüller], 1775) – *fibula*, a clasp: probably because when the moth is at rest the pale dorsal spots, one on each forewing, merge and resemble a clasp holding the wings together.

HELIOZELIDAE (154)

Heliozela Herrich-Schäffer, 1853 – ἥλιος (hēlios), the sun; ζῆλος (zēlos), emulation, eagerness: from the adult's habit of flying in bright sunshine.

154 **sericiella** (Haworth, 1828) – *sericus*, silky: from the silky gloss on the forewing.

= 155 **stanneella** sensu auctt. – *stanneus*, made of tin: from the colour of the forewing. If this is a distinct species, it does not occur in Britain.

156 **resplendella** (Stainton, 1851) – *resplendeo*, to shine: from the gloss on the forewing.

157 **hammoniella** Sorhagen, 1885 – Hammonia, the ancient name of Hamburg: Sorhagen found the early stages on the turf moors near the city and reared the adult.

= **betulae** (Stainton, 1890) – *Betula*, the birch genus which contains the larval foodplants.

Antispila Hübner, 1825 – ἀντί (anti), opposite; σπῖλος (spīlos), a spot: from the two spots situated opposite to each other on the forewing.

158 **metallella** ([Denis & Schiffermüller], 1775) – *metallum*, metal: from the metallic gloss on the forewing.

= **pfeifferella** (Hübner, 1813) – i.h.o. J. B. Pfeiffer (early 19th century), a German naturalist who lived at Augsburg and sent this and other species to Hübner to be figured. 'After German entomologist C. Pfeiffer (19th century) (Macleod): not correct. That Pfeiffer was a professor at Mainz, and his dates were 1718–87, not 19th century.

159 **petryi** Martini, 1898 – i.h.o. Dr A. Petry (1858–1932), a German schoolmaster and microlepidopterist who lived at Nordhausen.

= **treitschkiella** sensu auctt. – i.h.o. F. Treitschke (1776–1842), the German entomologist who collaborated with Ochsenheimer and completed *Die Schmetterlinge von Europa* after the latter's death.

DITRYSIA

δι- (di-), two; τρῦπα, τρυσ- (trūpa, trus-), a hole: from the two genital orifices in the female genitalia; cf. Monotrysia, p. 43).

COSSOIDEA (162)

COSSIDAE (162)

ZEUZERINAE (161)

Phragmataecia Newman, 1850 – *Phragmites australis*, the common reed; οἰκέω (oikeō), to dwell: from the habits of the larva of the next species, which mines the stems of reed.

160 **castaneae** (Hübner, 1790) – *Castanea sativa*, the sweet chestnut: not the foodplant; Hübner must have been misinformed. The wings are not chestnut-coloured as supposed by Macleod; in any case, the use of the genitive precludes that interpretation.

Zeuzera Latreille, 1804 – a typographical error for '*Zenzera, zenzara*, the Italian for a gnat, from *zenzero*, ginger, on account of the pungency of its bite. Latreille first wrote Zenzères: he afterwards uses Zeuzères, which Agassiz [J. L. R. Agassiz (1807–73)] derives from ζεύγνυμι [zeugnumi], *to bind*' (Pickard *et al.*). This is probably a punning name in the Fabrician style. Latreille wished to draw attention to the powerful jaws of the larva which mines in the living wood of various trees, and based his name on that of an insect noted for its ability to bite. Whether Latreille himself or the printer was responsible for the error in spelling is unimportant.

161 **pyrina** (Linnaeus, 1761) – *Pyrus communis*, the pear: one of the species of tree in the living wood of which the larva mines.

COSSINAE (162)

Cossus Fabricius, 1793 – For the meaning of *cossus* see the specific name. Fabricius renamed the next species *ligniperda* and transferred the Linnaean specific name to his new genus, or rather family, for, though small, it had a wide application and included, for example, the last species. Fabricius may have thought that *cossus* as understood by Pliny (see below) referred to wood-boring larvae in general and not to a particular species. After the introduction of the Rule of Priority, it was obligatory to reinstate *cossus* Linnaeus, 1758 as the name of the species, and *Cossus* being the senior name for the genus, the strange combination *Cossus cossus* came into being. Fabricius used to form his supraspecific names from the Greek and there is a Greek word κόσσος (kossos), a box on the ear. Fabricius liked cryptic and punning names and it is possible that he had this Greek word in mind; however, the main reason for the name lies in the first part of my explanation.

162 **cossus** (Linnaeus, 1758) – *cossus*, a larva found under the bark of trees and eaten by the Romans (Pliny 17: 24, 37), but considered by some authorities to be that of the stag beetle (*Lucanus cervus*).

= **ligniperda** Fabricius, 1793 – *lignum*, wood; *perdo*, to destroy: from the larva which bores into living wood and renders it unfit for human use.

ZYGAENOIDEA (166)

ZYGAENIDAE (166)

PROCRIDINAE (163)

Adscita Retzius, 1783 – *adscitus*, enrolled, adopted: this is, in fact, a Linnaean name, but Retzius was the first to use it in a strictly generic sense. Linnaeus divided Sphinx, the hawk-moths, into four groups (see 370 *Sesia*), the first three of which he called '*Legitimae*' or true hawk-moths, and the fourth '*Adscitae*', a group of seven 'hangers-on' (15 in the 12th edition (1767)), which he attached to Sphinx in default of a better position; the group included 163 *statices* and 169 *filipendulae*. Retzius retained these two in *Adscita*, although he altered their specific names.

= **Procris** Fabricius, 1807 – Procris, daughter of Erechtheus and wife of Cephalus. Cephalus and Procris loved each other dearly, in spite of elaborate ploys by Aurora and Diana to estrange them. Diana gave Procris a magic spear that never missed its mark and she, in turn, gave it to her husband. As the result of a misunderstanding, he killed Procris with it whilst out hunting. The story is pervaded by the theme of hunting (see Ovid, *Metamorphoses*, book 7), and the English name 'forester' for the next species, in current use at the time of Fabricius' visits to this country, may have prompted his choice. The name is conserved in the title of the subfamily.

165 **globulariae** (Hübner, 1793) – from *Globularia*, a genus of plants of the order Selaginaceae (Pickard *et al.*); a name for the marsh marigold (*Caltha palustris*) (Spuler).

164 **geryon** (Hübner, 1813) – Geryon, a mythical three-headed monster slain by Hercules as one of his twelve Labours.

163 **statices** (Linnaeus, 1758) – *Statice armeria* (now *Armeria maritimum*), thrift: not the foodplant, nor stated to be so by Linnaeus, who quotes from Petiver and Ray to record that they had thought it to be a butterfly.

ZYGAENINAE (166)

Zygaena Fabricius, 1775 – ζύγαινα (zugaina), the hammer-headed shark. Linnaeus divided Sphinx into four sections, listed under 370 *Sesia*, q.v. Zygaena was Fabricius' name for the fourth section, a family rather than a genus, embracing a large number of diversified species (72 are listed by Fabricius (1793)). The first of these happened to be 167 *filipendulae*, and this accident may be the reason why, when the scope of Zygaena was reduced to that of a genus, the name became associated with the burnets. It has puzzled authors and emendations have been proposed. Macleod was in part right when he derived it from ζυγόν (zugon), a yoke, but was wrong when he continued 'perhaps from appearance of antennae with thick bent-over ends', because Fabricius did not erect the name to describe the burnets. He wanted a Greek word collateral with Sphinx and perhaps as enigmatic which hinted at the idea of *adscitus* or linkage and punningly chose Zygaena; it has no direct application to the genus as now constituted.

166 **exulans** (Hohenwarth, 1792) – *exulans*, pres. part. of *exulo*, to be an exile: from the remote type locality at 2,400m in the Austrian Alps.

subsp. **subochracea** White, 1872 – *sub-*, somewhat; *ochraceus*, pertaining to *ochra*, yellow earth: from the yellowish irroration in the female and the yellow patagium and legs.

167 **loti** ([Denis & Schiffermüller], 1775) – *Lotus corniculatus*, bird's-foot trefoil: the larval foodplant.

subsp. **scotica** (Rowland-Brown, 1919) – the subspecies occurring in Scotland.

168 **viciae** ([Denis & Schiffermüller], 1775) – *Vicia*, a genus of vetches used on the Continent, but in Britain meadow vetchling (*Lathyrus pratensis*) and common bird's-foot trefoil (*Lotus corniculatus*) are preferred.

subsp. **ytenensis** Briggs, 1888 – from Ytene, the ancient name of the district in which the New Forest, the type locality, is situated.

subsp. **argyllensis** Tremewan, 1967 – belonging to Argyll, the type locality.

169 **filipendulae** (Linnaeus, 1758) – *Spiraea filipendula*, now *Filipendula vulgaris*, dropwort: wrongly stated by Linnaeus to be the foodplant.

subsp. **stephensi** Dupont, 1900 – the British subspecies, named i.h.o. J. F. Stephens (1792–1852), the British entomological author.

170 **trifolii** (Esper, 1783) – *Trifolium*, the clover genus: it does not contain the foodplants of this species, though it does include those of certain congeners, e.g. 171 *Z. lonicerae*.

subsp. **decreta** Verity, 1926 – *decretus*, past part. of *decerno*, to separate: probably because the subspecies was 'separated' from subsp. *palustrella*, named by Verity in the same paper. The spots on the forewing are not necessarily more widely separated in this subspecies, though Verity may have supposed that they were.

subsp. **palustrella** Verity, 1926 – *paluster*, *palustris*, marshy, belonging to a marsh: probably to indicate affinity with the Continental subsp. *palustris* Oberthür, 1896, but an ill-chosen name, since the subspecies occurs on dry downland; subsp. *decreta*, on the other hand, is found in marshes.

171 **lonicerae** (Scheven, 1777) – *Lonicera*, the honeysuckle genus: it does not, however, include any larval foodplant.

subsp. **latomarginata** Tutt, 1899 – *latus*, broad; *marginatus*, margined: from the relatively broad black border of the hindwing.

subsp. **jocelynae** Tremewan, 1962 – i.h.o. Jocelyn, wife of R. F. Bretherton, the contemporary British entomologist who discovered the subspecies in the Isle of Skye.

subsp. **insularis** Tremewan, 1960 – *insularis*, belonging to, occurring in, an island: the subspecies occurring in the island of Ireland.

172 **purpuralis** (Brünnich, 1763) – *purpura*, purple: from the colour of the forewing.

subsp. **segontii** Tremewan, 1958 – Segontii (gen.), of Segontium, a Roman settlement near the present town of Caernarvon, in the region where this subspecies used to occur.

subsp. **caledonensis** Reiss, 1931 – Caledonia, Scotland, where the type locality is situated.

subsp. **sabulosa** Tremewan, 1960 – *sabulosus*, sandy: from the type locality situated on the sand-hills on the coast of Co. Clare, Ireland.

LIMACODIDAE (173)

Apoda Haworth, 1809 – ἀ-, alpha privative; πoῦς, ποδός (pous, podos), the foot: from the larva which is almost apodal. 'That greatest of all entomologists, Fabricius' had placed the next species in 14 Hepialus, but had suggested later (Fabricius, 1793) that it and its relatives should perhaps have a genus (i.e. family) of their own. Haworth obliged, both entomologists apparently unaware that Knoch had forestalled them (see 174 *Heterogenea*). 'From larva's habit of drawing its feet in when molested' (Macleod).

173 **limacodes** (Hufnagel, 1766) – λείμαξ (leimax), a garden, *Limax*, the slug genus; εἶδος, ὠδ- (eidos, ōd-), form: from the shape and gait of the almost apodal larva. The family takes its name from this species, the earliest to be named, and not from a genus as is the usual practice.

= **avellana** sensu auctt. – *avellana* Linnaeus, 1758, is a synonym of 981 *Archips rosana* Linnaeus, 1758, and was applied to this species through misidentification.

Heterogenea Knoch, 1783 – ἕτερος (heteros), different; γένος (genos, Lat. *genus*), stock, kind, a taxonomic group, in 1783 corresponding to our family or superfamily: from the anomalous structure of the larva and pupa, not conforming with any of the Linnaean groupings and therefore warranting a 'different family'. This name marks the first serious modification to be made to the Linnaean classification but it was undeservedly overlooked by Knoch's contemporaries; see p. 26.

174 **asella** ([Denis & Schiffermüller], 1775) – *Oniscus asellus*, a species of woodlouse: from the onisciform larva.

TINEOIDEA (239)

PSYCHIDAE (186)

TALEPORIINAE (181)

Diplodoma Zeller, 1852 – διπλοῦς (diplous), double; δῶμα (dōma), a house; from the double case constructed by the larva.

180 **herminata** (Geoffroy, 1785) – *herminatus*, Latinized from the French 'herminé', adorned with ermine, 'La Teigne à Bordure herminée', the tineid with the ermined border: not entirely appropriate, since the terminal area of the forewing is fuscous spotted with yellow, not white spotted with black.

Narycia Stephens, 1836 – Narycia, the Greek city reputed to be the birthplace of Ajax; also its colony in southern Italy: a 'geographical' name (p. 33).

175 **monilifera** (Geoffroy, 1785) – *monile*, a collar; *fero*, to bear: the black head and collar are diagnostic.

Dahlica Enderlein, 1912 – intended as a genus of Diptera: apparently Enderlein got himself into a muddle.

= **Solenobia** sensu auctt. – σωλήν (sōlēn), a pipe; βιόω (bioō), to live: from the tubular case tenanted by the larva. This name is correctly a synonym of *Taleporia* Hübner and its type species is 181 *T. tubulosa*.

176 **triquetrella** (Hübner, 1813) – *triquetrus*, triangular: from the shape of the larval case.

177 **inconspicuella** (Stainton 1849) – *inconspicuus*, not conspicuous: either because the larva lives concealed in its case, or because it is seldom observed.

178 see below 181

179 **lichenella** (Linnaeus, 1761) – λειχήν (leikhēn), lichen: the larvae were first found feeding on lichen on the walls of a church in Sweden.

Taleporia Hübner, 1825 – ταλαιπωρία (talaipōria), hard work: from the laborious task experienced by the larva in dragging along its bulky case.

180 see below 174

181 **tubulosa** (Retzius, 1783) – *tubulus*, a small pipe: from the elongate larval case.

Bankesia Tutt, 1899 – i.h.o. E. R. Bankes (1861–1929), a British entomologist.

178 **douglasii** (Stainton, 1854) – i.h.o. J. W. Douglas (1814–1905), a British entomologist.

= 182 **conspurcatella** sensu auctt. – *conspurcatus*, defiled: from the dirty, unattractive appearance of the adult. Not a British species.

183 see below 185

PSYCHINAE (186)

Luffia Tutt, 1899 – i.h.o. W. A. Luff (1851–1910), a Channel Islands collector.

184 **lapidella** (Goeze, 1783) – *lapis, lapidis*, a stone: from the situations in which the larva is found feeding on lichen.

185 **ferchaultella** (Stephens, 1850) – i.h.o. R. A. Ferchault de Réaumur (1683–1757), the distinguished French naturalist and scientist; cf. 150 *reaumurella*.

Bacotia Tutt, 1899 – i.h.o. A. W. Bacot (1866–1922), a British entomologist.

183 **sepium** (Speyer, 1846) – *sepes*, gen. pl. *sepium*, a fence: from the situation in which the larval cases may be found.

Proutia Tutt, 1899 – i.h.o. L. B. Prout (1864–1943), a British entomologist.

188 **betulina** (Zeller, 1839) – *Betula*, birch: not the foodplant, but the larva can be found feeding on lichens growing on the trunks.

= **eppingella** Tutt, 1900 – Epping Forest, Essex, the type locality of what was then thought to be a distinct species.

Psyche Schrank, 1801 – ψύχη (psukhē), the soul; also a butterfly or moth from the analogy between metamorphosis and resurrection. The soul is the spirit of man liberated from the impurities of the flesh. In the second half of the volume in which the name Psyche first appears (*Fauna boica* 2 (2): 96 (1802)) Schrank discusses parthenogenesis and that phenomenon, together with the meaning of 186 *casta* (see below), may have prompted this name for a family in which some of the members refrain from sexual activity. Psyche was at first suprageneric with the same coverage as the present Psychidae.

186 **casta** (Pallas, 1767) – *castus*, chaste: from the occurrence of parthenogenesis in the family, though not in this species. P. S. Pallas (1741–1811), who was Professor of Natural History in the Imperial Academy of Sciences at St Petersburg, seems to have been the first naturalist to record parthenogenesis; writing on this subject, Schrank (1802, see above) gives, though without full reference, the following quotation from Pallas, '*sine habito cum masculis commercio fecunda ova parit*' (it produces fertile eggs without having had sexual intercourse with males). The Aristotelian theory of abiogenesis (spontaneous generation), which dispenses with the necessity of both parents, still had its adherents well into the 18th century, so the absence of one of them would not have seemed unduly revolutionary to Pallas' contemporaries. 'Spotless, from absence of marking on wings' (Macleod); this would have been a reasonable interpretation had not Pallas been aware of parthenogenesis.

187 **crassiorella** Bruand, 1853 – *crassus*, comparative *crassior*, thick, thicker: from the larval case which is stouter than that of the previous species; the adult is also bigger.

Epichnopterix Hübner, 1825 – ἐπίχνοος (epikhnoos), a wool-like covering; πτέρυξ (pterux), a wing: from the hair-like scales on the forewing.

188 see above 186

189 **plumella** ([Denis & Schiffermüller], 1775) – *pluma*, the downy part of a feather: from the specialized wing-scaling.

Whittleia Tutt, 1899 – i.h.o. F. G. Whittle (1854–1921), a British entomologist who had extensive knowledge of the next species on the Essex salt-marshes.

190 **retiella** (Newman, 1847) – *rete*, a net: from the reticulate forewing.

OIKETICINAE

From the non-British genus *Oiketicus* Guilding, 1827 – οἰκητικός (oikētikos), accustomed to a fixed dwelling: from the larval case.

Acanthopsyche Heylaerts, 1881 – ἄκανθα (akantha), a prickle; ψύχη (psukhē), here a member of the Psychidae: from the very long tibial spur of the foreleg; see MBGBI 2: 147, fig. 74.

191 **atra** (Linnaeus, 1767) – *ater*, black: from the colour of the hair-scales covering the head and body.

Pachythelia Westwood, 1848 – παχύς (pakhus), thick; θῆλυς (thēlus), female: from the robust, wingless female.

192 **villosella** (Ochsenheimer, 1810) – *villosus*, hairy, shaggy: from the hair-like scaling.

Canephora Hübner, 1822 – κανηφόρος (kanēphoros), bearing a basket: from the case carried by the larva. Maidens with baskets on their heads in religious processions at Athens were called 'Κανηφόροι' (Kanēphoroi). Hübner first proposed this name in his *Tentamen* [1806].

= **Lepidopsyche** Newman 1860 – λεπίς, λεπίδος (lepis, lepidos), a scale; ψύχη (psukhē), here a moth of the family Psychidae: from the fact that the next species is less thinly scaled than most other psychids.

193 **hirsuta** (Poda, 1761) – *hirsutus*, shaggy: from the somewhat roughly-scaled wings.

= **unicolor** (Hufnagel, 1766) – *unicolor*, of one colour: from the unicolorous forewing. There is no valid evidence for occurrence in Britain.

Thyridopteryx Stephens, 1835 – θύριον (thurion), a window; πτέρυξ (pterux), a wing: from the hyaline wings.

194 **ephemeraeformis** (Haworth, 1803) – ἐφήμερος (ephēmeros), living only for a day, *Ephemera*, a genus of mayflies; *forma*, shape, appearance: from a fancied resemblance of the adult to a mayfly.

Sterrhopterix Hübner, 1825 – στερρός (sterrhos), hard, stiff; πτέρυξ (pterux), a wing: from the wing structure.

195 **fusca** (Haworth, 1809) – *fuscus*, dusky: from the wing coloration.

TINEIDAE (239)

SCARDIINAE

From the genus *Scardia* Treitschke, 1830 – perhaps from σκαίρω, σκαρδ- (skairo, skard-), to frisk, suggested by 196 *choragella*, q.v., or from Scardus, the name of a range of mountains in Albania (Macleod, under *Microscardia*).

Morophaga Herrich-Schäffer, 1854 – the genus was erected to accommodate a single species, *morella* Duponchel, 1834, the name apparently a Latinized form of the French word 'morille', a mushroom, a morel; φαγεῖν (phagein), to eat.

196 **choragella** ([Denis & Schiffermüller], 1775) – χορηγός, χοραγός (khorēgos, khoragos), leader of the chorus which, in Greek drama, danced as well as sang: perhaps from 'dancing' behaviour of the adult overlooked by modern British entomologists, or because its relatively large size renders it the leader of the dance or pack.

= **boleti** (Fabricius, 1777) – *Boletus*, a genus of fungi with a pore-like surface instead of gills: a larval pabulum.

EUPLOCAMINAE (197)

Euplocamus Latreille, 1809 – εὖ (eu), well, goodly; πλόκαμος (plokamos), a lock or curl of hair: presumably descriptive of an adult character.

197 **anthracinalis** (Scopoli, 1763) – *anthracinus*, coal-black: from the ground colour of the forewing. There is no reliable evidence for occurrence in Britain.

DRYADAULINAE (198)

Dryadaula Meyrick, 1893 – δρυάς (druas), a wood-nymph, a dryad; δαυλός (daulos), shaggy: from the thick scaling on the vertex of the head and possibly from an association with woodland.

198 **pactolia** Meyrick, 1902 – Pactolus, a river of Lydia with proverbially golden sands: from golden markings on the forewing.

TEICHOBIINAE (199)

Psychoides Bruand, 1853 – ψύχη (psukhē), the soul, a moth, especially of the Psychidae; εἶδος (eidos), form, appearance: from the resemblance of the next species to the Psychidae, the family in which Bruand first placed it.

= **Teichobia** Herrich-Schäffer, 1855 – τεῖχος (teikhos), a wall; βιόω (bioō), to live: from a situation in which the ferns on which the larva feeds may be found.

199 **verhuella** Bruand, 1853 – i.h.o. Q. M. R. Verhuell (mid-19th century), a Dutch entomologist; *verhuellella* Stainton, 1854 is an unjustified emendation.

200 **filicivora** (Meyrick, 1937) – *Dryopteris filix-mas*, male fern; *voro*, to devour: from a larval foodplant.

MEESSIINAE (202)

Tenaga Clemens, 1862 – probably a meaningless neologism; τέναγος (tenagos), a shoal or shallow, seems to have no application.

= **Lichenovora** Petersen, 1957 – λειχήν, λιχήν (leikhēn, likhēn), lichen; *voro*, to devour: from the larval pabulum.

201 **nigripunctella** (Haworth, 1828) – *niger*, black; *punctum*, a spot: from the blackish spots scattered over the forewing.

248 **pomiliella** Clemens, 1862 – probably a meaningless neologism; derivation from *poma*, a fruit, seems to have no application.

Eudarcia Clemens, 1860 – almost certainly a meaningless neologism. The imaginative may see a link with εὐδαρκής, εὐδερκής (eudarkēs, euderkēs), bright-eyed. The character which Clemens italicizes in his generic diagnosis is that the cell of the hindwing is open; there could conceivably be a link between cell and *ocellus*, an eye, and between bright-eyed and open-eyed.

= **Meessia** Hofmann, 1898 – i.h.o. A. Mees (d. 1915), a German entomologist.

202 **richardsoni** (Walsingham, 1900) – i.h.o. N. M. Richardson (1855–1925), the British entomologist who discovered the species at Portland, Dorset.

Infurcitinea Spuler, 1910 – *in*, no, un-, *furca*, a fork; *tinea*, a member of the Tineidae: from the wing venation.

203 **argentimaculella** (Stainton, 1849) – *argentum*, silver; *maculum*, a spot: from the markings on the forewing.

204 **albicomella** (Herrich-Schäffer, 1851) – *albus*, white; *coma*, a hair of the head: from the white head of the adult.

Ischnoscia Meyrick, 1895 – ἰσχνός (iskhnos), dry; σκιά (skia), a shadow: perhaps because the species favour dry, shady habitats.

205 **borreonella** (Millière, 1874) – Borréon in the Alpes-Maritime (France) is the type locality.

Stenoptinea Dietz, 1905 – στενός (stenos), narrow; p, phonetic insertion; *tinea*, a member of the Tineidae: from the narrow forewing.

= **Celestica** Meyrick, 1917 – κήλησις (kēlēsis), enchantment, probably influenced by *coelestis*, heavenly: from the attractive colour of the next species.

206 **cyaneimarmorella** (Millière, 1854) – κυάνεος (kuaneos), *cyaneus*, dark blue, sea-blue; μάρμαρος (marmaros), *marmor*, marble: from the sprinkling of silvery blue scales on the forewing.

= **angustipennis** (Herrich-Schäffer, 1854) – *angustus*, narrow; *pennae* (pl.), a wing: from the elongate forewing.

MYRMECOZELINAE (207)

Myrmecozela Zeller, 1852 – μύρμηξ, μύρμηκος (murmēx, murmēkos), an ant; ζῆλος (zēlos), zeal: from the habitat of the larva in a nest of ants and its implied fondness for them.

207 **ochraceella** (Tengström, 1848) – *ochraceus*, ochreous: from the colour of the forewing.

Ateliotum Zeller, 1839 – ἀτελείωτος (ateleiōtos), uncompleted: from the shape of the hindwing which is rounded, not pointed, so appearing truncate.

208 **insularis** (Rebel, 1896) – *insularis*, pertaining to an island: the type locality is in the Canary Islands.

= **horrealis** (Meyrick, 1937) – *horreum*, a storehouse: the type locality (under this name) was a London warehouse.

209 }
210 } see below 214

Haplotinea Diakonoff & Hinton, 1956 – ἁπλόος (haploös), single; *tinea*, clothes moth: from the larval chaetotaxy, 'distinguished . . . from all Tineina . . . by the unisetose instead of bisetose SV group of the meso- and metathorax'.

211 **ditella** (Pierce, Metcalfe & Diakonoff, 1938) – *dis, ditis* (*dives*), rich: from the gloss on the forewing.

212 **insectella** (Fabricius, 1794) – *'habitat in insectis ex Africa missis'*, it lives on insects sent from Africa: the larval pabulum. *'Insectus*, cut up, from numerous small yellowish spots or dashes on forewings' (Macleod).

Cephimallota Bruand, 1851 – κεφαλή (kephalē), the head; μαλλός (mallos), a lock of wool: from the rough-haired head of the adult.

213 **angusticostella** (Zeller, 1851) – *angustus*, narrow; *costa*, rib, the anterior margin of a wing: from the narrow ochreous edging of the forewing costa.

Cephitinea Zagulyaev, 1964 – perhaps κηφήν (kēphēn), a drone bee; *tinea*, a clothes moth: from a supposed resemblance.

214 **colongella** Zagulyaev, 1964 – *co-*, equal, *longus*, long: perhaps with elongate wings resembling some other species.

SETOMORPHINAE (209)

Setomorpha Zeller, 1852 – σής (sēs), a moth, a clothes moth; μορφή (morphē), shape: indicative of affinity with *Tinea*.

209 **rutella** Zeller, 1852 – *rutellum*, a small shovel: from the spatulate terminal segment of the labial palpus. *'Ruta*, rue, from foodplant in France' (Macleod): the larva feeds on stored products and the type localities are in Africa.

Lindera Blanchard, 1852 – i.h.o. J. Linder (1830–69), a coleopterist who lived at Strasburg.

210 **tessellatella** Blanchard, 1852 – *tessellatus*, chequered: from the forewing pattern.

NEMAPOGONINAE (215)

Nemapogon Schrank, 1802 – νῆμα (nēma), a thread; πώγων (pōgōn), a beard: from the conspicuous bristles on segment 2 of the labial palpus.

215 **granella** (Linnaeus, 1758) – *granum*, a grain, a seed: stated correctly by Linnaeus to frequent granaries.

216 **cloacella** (Haworth, 1828) – *cloaca*, a sewer – *'habitat in cloacis'*, it lives in drains: maybe, but it prefers fungus growing on wood.

216a **inconditella** (Lucas, 1956) – *inconditus*, irregular, confused: from the variable irroration and strigulation of the forewing.

= **heydeni** Petersen, 1957 – i.h.o. Senator C. H. G. von Heyden (1793–1866), a German entomologist.

217 **wolffiella** Karsholt & Nielson, 1976 – i.h.o. N. L. Wolff (1900–78), a Danish entomologist.

= **albipunctella** (Haworth, 1828) *nec* ([Denis & Schiffermüller], 1775) – *albus*, white; *punctum*, a spot: from the white spots on the forewing.

218 **variatella** (Clemens, 1859) – *variatus*, variegated: from the forewing markings.
= **personella** (Pierce & Metcalfe, 1934) – *persona*, a mask as worn by actors: the species had been confused with 216 *N. cloacella* which it could be said to mimic or impersonate.
219 **ruricolella** (Stainton, 1849) – *rus, ruris*, the country; *colo*, to inhabit: from the habitat.
220 **clematella** (Fabricius, 1781) – *Clematis vitalba*, traveller's-joy, *'Habitat in Angliae Clematide'*, it lives on *Clematis* in Britain: however, this is not the foodplant.
= **arcella** sensu auctt. – *arcus*, a bow: from the curved fascia on the forewing.
221 **picarella** (Clerck, 1759) – *pica*, a magpie: from the black-marked, white forewing.

Archinemapogon Zagulyaev, 1964 – ἀρχι- (arkhi-), chief; the genus 215 *Nemapogon*: the wingspan is greater than in most *Nemapogon* species. Derivation from ἀρχή (archē), beginning, origin, may have been intended, to indicate priority over *Nemapogon* in evolution or systematic position.

222 **yildizae** Koçak, 1981 – perhaps i.h.o. an entomologist or friend: Koçak gives no explanation.
= **laterella** (Thunberg, 1794) *nec* ([Denis & Schiffermüller], 1775) – *later*, a brick: from the coloration of the forewing.

Nemaxera Zagulyaev, 1964 – apparently a portmanteau word formed from 215 *Nemapogon* and 224 *Triaxomera* indicating affinity to the two genera.

223 **betulinella** (Fabricius, 1787) – *Piptoporus betulinus*: a bracket fungus which is one of the principal foodplants.
= **corticella** (Curtis, 1834) – *cortex*, bark: the larva sometimes feeds in rotten wood, but not strictly on bark.

Triaxomera Zagulyaev, 1959 – I have been unable to find a convincing derivation. The name is likely to be based on some structural character or characters, probably of the distinctive genitalia, and may be telescopically formed like 223 *Nemaxera* and 226 *Triaxomasia*, but not from them, being senior to both.

224 **parasitella** (Hübner, 1796) – *parasitus*, a parasite: from the fact that the fungus eaten by the larva is parasitic on trees.
225 **fulvimitrella** (Sodoffsky, 1830) – *fulvus*, tawny; *mitra*, a mitre: from the red head of the adult.

Triaxomasia Zagulyaev, 1964 – apparently a portmanteau word formed from 224 *Triaxomera* and *Neurothaumasia* Le Marchand, 1934, a genus not represented in Britain but which follows *Triaxomasia* in Leraut (1980), indicating affinity to the two genera.

226 **caprimulgella** (Stainton, 1851) – *Caprimulgus europaeus*, the nightjar or goatsucker: from a fancied resemblance in colour to that bird.

TINEINAE (239)

Monopis Hübner, 1825 – μονώψ, μονῶπος (monōps, monōpos), one-eyed: from the single hyaline spot in the disc of the forewing.

227 **laevigella** ([Denis & Schiffermüller], 1775) – *levigo, laevigo*, to pulverize: the authors describe the adult as dusted with gold.
= **rusticella** (Hübner, 1796) – *rusticus*, of the country: from the habitat.
228 **weaverella** (Scott, 1858) – i.h.o. R. Weaver (19th century), a British entomologist who lived near Birmingham. See 43 *Fomoria weaveri*.
229 **obviella** ([Denis & Schiffermüller], 1775) – *obvius*, lying in the way, obvious: from the yellow dorsal streak which makes the moth conspicuous.
= **ferruginella** (Hübner, 1813) – *ferrugineus*, rusty-coloured: from the colour of the head of the adult.
230 **crocicapitella** (Clemens, 1859) – *croceus*, yellow; *caput, capitis*, the head: from the colour of the head of the adult.
231 **imella** (Hübner, 1813) – *imus*, lowest: relevance obscure; perhaps simply the bottom specimen in the row Hübner had before him for figuring.
232 **monachella** (Hübner, 1796) – μοναχός (monakhos), solitary (adj.), a monk (n.): from the fuscous and white coloration of the forewing, suggesting a monk's habit; cf. 2033 *monacha*.
233 **fenestratella** (Heyden, 1863) – *fenestratus*, windowed: from the resemblance of the hyaline spot on the forewing to a window.

Trichophaga Ragonot, 1894 – θρίξ, τριχός (thrix, trikhos), hair; φαγεῖν (phagein), to eat: from the larval pabulum which includes hair in birds' nests, fur, etc.

234 **tapetzella** (Linnaeus, 1758) – *tapete*, a carpet, tapestry: from the larval pabulum, as correctly stated by Linnaeus.

235 **mormopis** Meyrick, 1935 – Μορμώ (Mormō), a bugbear; ὤψ (ōps), a face: from a fancied ugly face in the wing-pattern, from its unpleasant habits as a pest of furs, etc., or from both.

= **percna** Corbet & Tams, 1943 – πέρκνος (perknos), Lat. *percnus*, dusky and also the name of a species of eagle: either from the dark wing-base, or from the elongate forewing, suggestive of a bird of prey.

Tineola Herrich-Schäffer, 1853 – dim. of *Tinea*, a clothes moth.

236 **bisselliella** (Hummel, 1823) – *bisellium*, a double seat occupied by a single person, a seat of honour: from the larval pabulum which includes the upholstery of chairs. Hummel should have used only one 's'.

Niditinea Petersen, 1957 – *nidus*, a nest; *Tinea*, a clothes moth: from the larval habit of feeding in birds' nests.

237 **fuscella** (Linnaeus, 1758) – *fuscus*, dusky: from the colour of the forewing.

= **fuscipunctella** (Haworth, 1828) – *fuscus*, dusky; *punctum*, a spot: from the dark stigmata of the forewing.

238 **piercella** (Bentinck, 1935) – i.h.o. F. N. Pierce (1861–1943), the British entomologist who first recognized this to be a distinct species from his study of the genitalia.

Tinea Linnaeus, 1758 – *tinea*, a gnawing worm, applied to various larvae including those of Lepidoptera destructive to clothes. Linnaeus' family embraced all moths that rested with involute wings (Pl. III), groups such as the Crambinae and Galleriinae being included as well as the Tineoidea. The application of the name has been progressively eroded.

239 **columbariella** Wocke, 1877 – *columbarium*, a pigeon-house: a situation in which the larvae may be found.

240 **pellionella** Linnaeus, 1758 – *pellis*, a skin, a hide; *pellio*, *pellionis*, a furrier: from the larval habit of feeding in fur and other material of animal origin.

241 **lanella** Pierce & Metcalfe, 1934 – *lana*, wool: from a foodstuff of the larva.

242 **translucens** Meyrick, 1917 – *translucens*, allowing light to shine through: from the subcostal hyaline patch on the forewing.

= **metonella** Pierce & Metcalfe, 1934 – abbreviated from μετονομάζω (metonomazō), to call by a new name: the species occurring in Britain had been misidentified as *T. merdella* Zeller, a distinct species not found in this country, and a new name was therefore needed.

243 **dubiella** Stainton, 1859 – *dubius*, doubtful: from the difficulty in correct determination; Stainton writes 'the male is *excessively* like *pellionella*'.

= **turicensis** Müller-Rutz, 1920 – *Turicensis*, belonging to, occurring at, Turicum, the Latin name for Zürich.

244 **flavescentella** Haworth, 1828 – *flavescens*, inclining to be yellow: from the pale ochreous coloration of the forewing.

245 **pallescentella** Stainton, 1851 – *pallescens*, inclining to be pale: from the pale greyish brown coloration of the forewing.

246 **semifulvella** Haworth, 1828 – *semi-*, half; *fulvus*, tawny, reddish yellow: from the ferruginous coloration of the distal half of the forewing.

247 **trinotella** Thunberg, 1794 – *tri-*, three; *nota*, a mark: from the three dark stigmata on the forewing.

= **ganomella** Treitschke, 1833 – perhaps a syncopated form of '*ganomelella*', from γάνος (ganos), brightness; μέλας (melas), black: from the conspicuous black stigmata on the bright ochreous yellow forewing.

247a **fictrix** Meyrick, 1914 – *fictrix*, fem. of *fictor*, one who makes, feigns or counterfeits (cf. fiction); from resemblance to another species.

247b **murariella** (Staudinger, 1859) – *murus*, a wall: from the habitat.

248 see below 201

Ceratophaga Petersen, 1957 – κέρας, κέρατος (keras, keratos), a horn; φαγεῖν (phagein), to eat: from the larval feeding habits.

249 **orientalis** (Stainton, 1878) – *orientalis*, eastern: first described from adults reared from larvae feeding in buffalo horns imported from Singapore.

250 **haidarabadi** Zagulyaev, 1966 – Hyderabad, India is the type locality.

HIEROXESTINAE (278)

Oinophila Stephens, 1848 – οἶνος (oinos), wine; φιλέω (phileō), to love: from occurrence in wine-cellars and on the corks of wine-bottles.

277 **v-flava** (Haworth, 1828) – v-, shape of marking; *flavus*, yellow: from the acutely angled yellowish fascia on the forewing.

Opogona Zeller, 1853 – ὤψ (ōps), a face; γωνία (gōnia), an angle: from the appearance of the labial palpi which are directed outwards at an angle.

= **Hieroxestis** Meyrick, 1892 – ἱερός (hieros), holy; ξεστός (xestos), polished: relevance obscure; holystone was a sandstone used by seamen to polish decks, so it may be that some species are of that colour or appear polished.

278 **sacchari** (Bojer, 1856) – *Saccharum officinarum*, sugar-cane: the larval foodplant.

279 **antistacta** Meyrick, 1937 – *antiae*, the hair of the forehead; *stacta*, myrrh-oil: the species differs from others in the genus in lacking erect scale-tufts on the vertex of the head, as if they had been plastered down with hair-oil. 'Wakely's Bentwing; Gr. *anti*, over, *stactos*, dropping; from way in which apex of forewings turns upwards' (Macleod): an entertaining misinterpretation. Heslop (1947) had given the name 'bentwing' to all members of his Lyonetiinae (= Cemiostominae + Lyonetiinae + Bedelliinae), although the apex of the forewing is flexed upwards and the name strictly applicable only in the Cemiostominae, and he had wrongly placed this species in the subfamily. Macleod, who had looked at Heslop's name but not the moth or Meyrick's description, jumped to the wrong conclusion. He got σταχτός wrong, too, because it means 'dropping' in the sense 'dripping', not 'drooping'.

OCHSENHEIMERIIDAE (251)

Ochsenheimeria Hübner, 1825 – i.h.o. F. Ochsenheimer (1767–1822), the distinguished German actor, playwright and entomologist.

251 **mediopectinellus** (Haworth, 1828) – *medius*, middle; *pecten*, *pectinis*, a comb: from the thick scaling in the central part of the antenna.

252 **urella** Fischer von Röslerstamm, 1842 – οὖρος (ouros), *ūrus*, the aurochs, the extinct wild ox: from the distinctive antennae which resemble a bull's horns.

= **bisontella** Lienig & Zeller, 1846 – *bison*, the aurochs, the wild ox: see above.

253 **vacculella** Fischer von Röslerstamm, 1842 – *vaccula*, a heifer; from the horn-like antennae.

LYONETIIDAE (262)

CEMIOSTOMINAE (254)

Leucoptera Hübner, 1825 – λευκός (leukos), white; πτερόν (pteron), a wing: from the shining white ground colour of the forewing.

= **Cemiostoma** Zeller, 1848 – κημός (kēmos), a muzzle; στόμα (stoma), the mouth: from the large antennal eyecaps which cover much of the frons or face.

254 **laburnella** (Stainton, 1851) – *Laburnum anagyroides*: the larval foodplant.

255 **wailesella** (Stainton, 1858) – i.h.o. G. Wailes (1802–82), a British entomologist who lived at Castle Eden Dene, Co. Durham.

256 **spartifoliella** (Hübner, 1813) – *Spartium junceum*, Spanish broom: a larval foodplant; *folium*, a leaf: the larva, however, mines the bark of twigs, not leaves.

257 **orobi** (Stainton, 1869) – *Orobus tuberosus*, now called *Lathyrus montanus*, bitter vetch: the larval foodplant.

258 **lathyrifoliella** (Stainton, 1865) – *Lathyrus sylvestris*, narrow-leaved everlasting pea; *folium*, a leaf: the larva mines the leaves of this foodplant.

259 **lotella** (Stainton, 1858) – *Lotus*, the bird's-foot trefoil genus: the larval foodplant.

260 **malifoliella** (Costa, 1836) – *Malus*, the apple; *folium*, a leaf; the larva mines a leaf of this foodplant.
= **scitella** (Zeller, 1839) – *scitus*, beautiful, elegant: in praise of the adult.

Paraleucoptera Heinrich, 1918 – παρα- (para-), alongside; *Leucoptera*, the preceding genus, q.v.: a genus close to *Leucoptera*.

261 **sinuella** (Reutti, 1853) – *sinus*, a fold or purse: probably from the blotch mine made by the larva.

LYONETIINAE (262)

Lyonetia Hübner, 1825 – i.h.o. Pierre Lyonet (1706–89), the distinguished French naturalist; he later became a naturalized Dutchman.

262 **prunifoliella** (Hübner, 1796) – *Prunus*, the plum and cherry genus; *folium*, a leaf: the larvae mine leaves of trees belonging in this genus.

263 **clerkella** (Linnaeus, 1758) – i.h.o. K. A. Clerck (1710–65), a Swedish entomologist, artist and disciple of Linnaeus, who spelt his name correctly in 1767, when it was too late! The species was discovered by Clerck.

BEDELLIINAE (264)

Bedellia Stainton, 1849 – i.h.o. G. Bedell (1805–77), a British entomologist.

264 **somnulentella** (Zeller, 1847) – *somnulentus*, sleepy: perhaps a reference to the fact that the adult overwinters, though Zeller took his first specimen on the 3rd April and so could not be sure of this; possibly a metonymic reference to the dusky coloration.

BUCCULATRICIDAE (265)

Bucculatrix Zeller, 1839 – *buccula*, a cheek, a mouth, a beaver, i.e. that part of the helmet which protects the mouth and cheeks: from the large antennal eyecaps which cover much of the frons.

265 **cristatella** Zeller, 1839 – *cristatus*, crested: from the erect scales on the vertex of the head.

266 **nigricomella** Zeller, 1839 – *niger*, black; *coma*, the hair of the head: from the black vertical tuft.

267 **maritima** Stainton, 1851 – *maritimus*, maritime: from the habitat.

268 **capreella** Krogerus, 1952 – *caprea*, the wild she-goat, the ibex: the habitat of both is similar; in Britain the moth is confined to the Scottish Highlands.

269 **artemisiella** Herrich-Schäffer, 1855 – *Artemisia vulgaris*, mugwort, the larval foodplant; however, the single British specimen, possibly misidentified, was reared from yarrow (*Achillea millefolium*).

270 **frangutella** (Goeze, 1783) – *Frangula alnus*, alder buckthorn: the larval foodplant.
= **frangulella**, a spelling correction for an obvious misprint, but not permissible under I.C.Z.N. rules.

271 **albedinella** Zeller, 1839 – *albedo, albedinis*, whiteness: from the whitish ground colour of the forewing.

272 **cidarella** Zeller, 1839 – *cidaris*, a headdress, a tiara: from the ochreous vertical tuft, conspicuous against the blackish wings.

273 **thoracella** (Thunberg, 1794) – *thorax, thoracis*, the breast, the thorax: from the conspicuous yellow thorax which is concolorous with the forewing, as stated in the original description.

274 **ulmella** Zeller, 1848 – *Ulmus*, elm: in Britain the foodplant is oak (*Quercus* spp.), but Zeller assured Stainton that he had reared this species from elm.

275 **bechsteinella** (Bechstein & Scharfenberg, 1805) – i.h.o. J. M. Bechstein (1757–1810), a German entomologist and the senior nomenclator.
= **crataegi** Zeller, 1839 – *Crataegus*, hawthorn, the genus of the larval foodplants.

276 **demaryella** (Duponchel, 1840) – i.h.o. Dr M. Demary (early 19th century), the first secretary of the Entomological Society of France.

277, 278, 279 } see below 250

GRACILLARIIDAE (280)

GRACILLARIINAE (280)

Caloptilia Hübner, 1825 – καλός (kalos), beautiful; πτίλον (ptilon), a feather, a wing: from the attractive appearance of the moths.

= **Gracillaria** Haworth, 1828 – *grăcilis*, slender, meagre, plain, unadorned, generally in a pejorative sense: from the slender abdomen and/or the often unicolorous forewing. The adjective from *grātia*, grace, is *grātiosus*.

280 **cuculipennella** (Hübner, 1796) – *cuculus*, a cuckoo; *penna*, a feather, *pennae* (pl.), a wing: from a fancied resemblance in coloration.

281 **populetorum** (Zeller, 1839) – *populetum*, gen. pl. *populetorum*, a poplar thicket: birch, however, is the foodplant.

282 **elongella** (Linnaeus, 1761) – *elongus*, very long: from the long forewing, not from the elongate roll made by the larva in a leaf of alder, since the life history was unknown in 1761.

283 **betulicola** (Hering, 1928) – *Betula*, birch, the larval foodplant; *colo*, to inhabit.

284 **rufipennella** (Hübner, 1796) – *rufus*, red; *pennae* (pl.), a wing: from the coloration of the forewing.

285 **azaleella** (Brants, 1913) – *Azalea*, generic name formerly used for certain species of *Rhododendron*: from the larval foodplant.

286 **alchimiella** (Scopoli, 1763) – Latinized from the Arabic 'al-kimia', alchemy: from the golden markings on the forewing, *'Felices Alchemistae si tinctura hac solari sua saturate metalla possent!'* (blessed indeed would alchemists have been had they been able to doctor their metals with the full richness of this dye!). *'Alchemilla*, lady's-mantle, but not really its foodplant' (Macleod).

287 **robustella** Jäckh, 1972 – *robustus*, oaken: oak (*Quercus robur*) is the foodplant.

288 **stigmatella** (Fabricius, 1781) – *stigmatus*, branded: from the conspicuous costal blotch of the forewing.

289 **falconipennella** (Hübner, 1813) – *falco, falconis*, a falcon; *pennae* (pl.), a wing: from a fancied resemblance in coloration or from the elongate forewing.

290 **semifascia** (Haworth, 1828) – *semi-*, half; *fascia*, a band: from the obsolescent fascia representing the costal blotch on the forewing.

291 **hemidactylella** ([Denis & Schiffermüller], 1775) – ἡμι- (hēmi-), half; δάκτυλος (dactulos), a finger, applied to the 'plumes' or wing-divisions of the Pterophoridae: half like a plume moth, the main resemblance being the narrow hindwing and long cilia.

292 **leucapennella** (Stephens, 1835) – λευκός (leukos), white; *pennae* (pl.), a wing: descriptive of the ground colour of the typical form; in f. *aurantiella* Peyerimhoff, 1871, the ground colour is reddish: *aurum*, gold; *antiae*, the forelock: golden-haired.

293 **syringella** (Fabricius, 1794) – *Syringa vulgaris*, lilac: a favourite foodplant.

Aspilapteryx Spuler, 1910 – ἀ-, alpha privative; σπίλος (spilos), a spot; πτέρυξ (pterux), a wing: from the absence of a costal blotch on the forewing of the next species. However, the wing is speckled with black and the initial 'a' may accordingly be interpreted as the 'alpha intensive', emphasizing rather than negating the sense of the word that follows and giving the sense 'with heavily spotted wing'. According to the lexicon, the alpha intensive is rare and more favoured by grammarians than classical authors themselves.

294 **tringipennella** (Zeller, 1839) – *Tringa*, the sandpiper genus of birds; *pennae* (pl.), a wing: from a fancied resemblance in coloration.

Calybites Hübner, 1822 – καλυβίτης (kalubitēs), living in a hut: from the cone spun and inhabited by the larva on a leaf of its foodplant.

295 **hauderi** (Rebel, 1906) – i.h.o. F. Hauder (1860–1923), who discovered the species at Kirschdorf in Austria.

= **pyrenaeella** (Chrétien, 1908) – from the Pyrenees, where Chrétien independently discovered it two years later than Hauder.

296 **phasianipennella** (Hübner, 1813) – *phasianus*, a pheasant; *pennae* (pl.), a wing: from a fancied resemblance in coloration.

297 **auroguttella** (Stephens, 1835) – *aurum*, gold; *gutta*, a spot: from the golden spots on the forewing.

Micrurapteryx Spuler, 1910 – μικρός (mikros), small; οὐρά (oura), a tail; πτέρυξ (pterux), a wing: from a small dark tuft near the apex of the forewing of the next species.

298 **kollariella** (Zeller, 1839) – i.h.o. V. Kollar (1797–1860), a Viennese entomologist. The species was added to the British list on insufficient evidence.

Parectopa Clemens, 1860 – παρέκτοπος (parektopos), somewhat out of the way: possibly from the furcation of veins on both fore- and hindwing, which Clemens describes in detail. 'From larva's habit of deserting original mine and making another near it' (Macleod); Clemens was aware of this habit and the explanation is possible, but less likely. The new mine is in a different leaf.

299 **ononidis** (Zeller, 1839) – *Onōnis, Onōnidis*, the restharrow genus: a larval foodplant, though in Britain clover (*Trifolium* spp.) is more usual.

Parornix Spuler, 1910 – παρα- (para-), alongside; the genus *Ornix* Treitschke, 1833, from ὄρνις (ornis), a bird: *Ornix* originally included a wide range of 'feathery-winged' Microlepidoptera such as the Gracillariidae and 489 Coleophoridae, q.v., but was later restricted to this genus; however, because of its early association with *Coleophora*, Spuler introduced this replacement name.

300 **loganella** (Stainton, 1848) – i.h.o. R. F. Logan (1827–87), a Scottish entomologist.

301 **betulae** (Stainton, 1854) – *Betula*, the birch genus: the larva feeds on birch.

302 **fagivora** (Frey, 1861) – *Fagus*, the beech genus; *voro*, to devour: from the larval foodplant.

302a **carpinella** (Frey, 1863) – *Carpinus*, the hornbeam genus: the larva feeds on hornbeam.

303 **anglicella** (Stainton, 1850) – *Anglicus*, English: the type locality is in England.

304 **devoniella** (Stainton, 1850) – Dawlish, south Devon, is the type locality.

305 **scoticella** (Stainton, 1850) – *Scoticus*, Scottish: the type locality is in Scotland.

306 **alpicola** (Wocke, 1876) – *Alpes*, the Alps; *colo*, to inhabit: the type localities are Alpine.

307 **leucostola** Pelham-Clinton, 1964 – λευκός (leukos), white; στολή (stolē), a garment: from the almost immaculate white forewing.

308 **finitimella** (Zeller, 1850) – *finitimus*, very close, very closely related: from close resemblance to the next species.

309 **torquillella** (Zeller, 1850) – *torquis*, a collar: from the prothoracic plate of the larva, which consists of four transversely placed black spots. 'From black line round back of head' (Macleod); no such line exists, the name being descriptive of the larva, not the adult.

Callisto Stephens, 1834 – καλός, superlative κάλλιστος (kalos, kallistos), beautiful, most beautiful. Callisto was also the name of one of Diana's nymphs. Zeus had an affair with her and then to avoid the jealousy and wrath of his wife Hera, he metamorphosed her into a she-bear. Hera sought revenge by causing Artemis (Diana) to slay Callisto in the chase. Zeus, however, had the last word by placing Callisto in the stars as Arctos, the Bear, and so immortalized her.

310 **denticulella** (Thunberg, 1794) – *denticulus*, a little tooth: from the small subtriangular white spots on the costa and dorsum of the forewing.

310a **coffeella** (Zetterstedt, 1839) – from the dark brown, coffee-coloured forewing.

Acrocercops Wallengren, 1881 – ἄκρον (akron), a point; κέρας (keras), a horn; ὤψ (ōps), a face: from the projecting tuft on segment 2 of the labial palpus which is present in some members of the genus.

311 **imperialella** (Zeller, 1847) – *imperialis*, pertaining to an emperor: from the richly marked forewing.

= 312 **hofmanniella** sensu auctt. – i.h.o. O. Hofmann (1835–1900), a German entomologist. Placed on the British list through misidentification, the true *A. hofmanniella* (Schleich, 1867) not having been found in Britain.

313 **brongniardella** (Fabricius, 1798) – i.h.o. A. Brongniart (1770–1847), a French entomologist.

Leucospilapteryx Spuler, 1910 – λευκός (leukos), white; σπίλος (spilos), a spot; πτέρυξ (pterux), a wing: from the white fasciae on the forewing of the next species, which are generally interrupted so as to form spots.

314 **omissella** (Stainton, 1848) – *omissus*, omitted: the name was proposed to Stainton by Douglas when the two entomologists found to their surprise that the species was not figured or named in any Continental text-book.

LITHOCOLLETINAE (315)

Phyllonorycter Hübner, 1822 – φύλλον (phullon), a leaf; ὀρυκτήρ (oruktēr) (rare), a digging, ὀρύκτης (oruktēs), a digger: a leaf-mine, a leaf-miner.

= **Lithocolletis** Hübner, 1825 – λιθοκόλλητος (lithokollētos), mosaic work: from the attractive variegated pattern of the forewings.

315 **harrisella** (Linnaeus, 1761) – probably i.h.o. Moses Harris (1731–88). The first fascicle of *The Aurelian* was published in December, 1758 and may well have been seen and admired by Linnaeus. Moses Harris' uncle, another Moses Harris, was also a collector and an 'Aurelian', but as he never attained to international fame, it is unlikely that the species was named after him.

316 **roboris** (Zeller, 1839) – *Quercus robur*, gen. *roboris*, the oak: the larval foodplant.

317 **heegeriella** (Zeller, 1846) – i.h.o. E. Heeger (d. 1866), an Austrian entomologist.

318 **tenerella** (Joannis, 1915), a replacement name for *tenella* Zeller, 1846 (preoccupied) – *tener*, delicate, *tenellus*, rather delicate, delicate little . . . : from the attractive appearance of the moth.

319 **saportella** (Duponchel, 1840) – i.h.o. M. le Comte de Saporta who discovered many new species in the south of France, but perhaps not this one, since its type locality is Paris.

320 **quercifoliella** (Zeller, 1839) – *Quercus*, the oak genus; *folium*, a leaf: the larvae mine leaves of oak.

321 **messaniella** (Zeller, 1846) – Messina (formerly Messana) in Sicily is the type locality. Stainton (1857*) expressed shame that this common British species had been left for a German to discover when on holiday in the Mediterranean.

321a **platani** (Staudinger, 1870) – *Platanus*, the plane-tree genus: from the larval foodplant.

322 **muelleriella** (Zeller, 1839) – i.h.o. one of several German entomologists named Müller; Zeller does not specify which one he had in mind.

323 **oxyacanthae** (Frey, 1856) – *Crataegus oxyacantha*, now called *C. laevigata*, midland hawthorn: a larval foodplant.

324 **sorbi** (Frey, 1855) – *Sorbus*, the genus of rowan and whitebeam, both larval foodplants.

325 **mespilella** (Hübner, 1805) – *Mespilus germana*, medlar, a foodplant on the Continent; in Britain it feeds on *Pyrus* and *Sorbus* spp.

326 **blancardella** (Fabricius, 1781) – i.h.o. S. Blankaart (late 18th century), a Dutch entomologist.

327 **cydoniella** ([Denis & Schiffermüller], 1775) – *Cydonia oblonga*, quince: a larval foodplant, but in Britain *Malus* is more frequently utilized.

328 **junoniella** (Zeller, 1846) – from Juno, the wife of Jupiter: Zeller gives no reason for this choice of name and there is probably no entomological application. 'From resemblance of spot at apex of forewings to eye of peacock's tail, peacock being sacred to Juno' (Macleod).

329 **spinicolella** (Zeller, 1846) – *Prunus spinosa*, blackthorn, the larval foodplant; *colo*, to inhabit.

= **pomonella** (Zeller, 1846) – *pomus*, a fruit-tree, especially the apple: the name was misapplied to this species which does not feed on *Malus*.

330 **cerasicolella** (Herrich-Schäffer, 1855) – *Prunus cerasus*, the dwarf cherry, a larval foodplant; *colo*, to inhabit.

331 **lantanella** (Schrank, 1802) – *Viburnum lantana*, the wayfaring-tree: a larval foodplant.

332 **corylifoliella** (Hübner, 1796) – *Corylus avellana*, hazel; *folium*, a leaf: the larva never feeds on hazel but 342 *P. coryli* does; both make upperside mines and Hübner appears to have confused the two. Stainton (1857*) explained the mistake by regarding Hübner's figure as 'representing rather the whole group than any single species'.

332a **leucographella** (Zeller, 1850) – λευκός (leukos), white; γραφή (graphē), a marking or writing: from the sharply defined white pattern on an orange-brown ground colour. Zeller attributes the name to Kollar; he himself seldom derived specific names from Greek sources.

333 **viminiella** (Sircom, 1848) – *vimen*, an osier twig: osier is a foodplant.

334 **viminetorum** (Stainton, 1854) – *viminetum*, gen. pl. *viminetorum*, an osier bed: osier is the larval foodplant.

335 **salicicolella** (Sircom, 1848) – *Salix*, *Salicis*, the willow and sallow genus; *colo*, to inhabit: from the larval foodplant.

336 **dubitella** (Herrich-Schäffer, 1855) – *dubito*, to doubt: from uncertainty over whether this was a distinct species.

337 **hilarella** (Zetterstedt, 1839) – *hilaris*, merry, lively: the latter sense was probably intended; if the former, an example of the pathetic fallacy.

= **spinolella** (Duponchel, 1840) – i.h.o. Marchese Maximilian Spinola (1780–1857), an Italian entomologist.

338 **cavella** (Zeller, 1846) – *cavus*, hollow: in his description Zeller writes '*lineola apicis alba excavata*', the white apical strigula excavate or curved. There can be no reference to the very large, 'inflated' mine, since Zeller did not know the life history.

339 **ulicicolella** (Stainton, 1851) – *Ulex, Ulicis*, the gorse genus; *colo*, to inhabit: from the larval foodplant.

340 **scopariella** (Zeller, 1846) – *Cytisus* (=*Sarothamnus*) *scoparius*, broom: the larval foodplant.

340a **staintoniella** (Nicelli, 1853) – i.h.o. H. T. Stainton (1822–92), the distinguished British microlepidopterist: von Nicelli was returning a compliment, since 359 *P. nicellii* had recently been named in his honour by Stainton.

341 **maestingella** (Müller, 1764) – probably i.h.o. an entomologist named Maesting, since Müller spelt the name with a capital letter.

342 **coryli** (Nicelli, 1851) – *Corylus*, the hazel genus: from the larval foodplant.

343 **quinnata** (Geoffroy, 1785) – *quinque*, five: 'La Teigne blanche à cinq bandes brunes', the white tineid with five brown fasciae. Modern authors would prefer to describe it as a brown moth with four sets of white strigulae.

344 **strigulatella** (Zeller, 1846) – *strigula*, a small linear mark: from the presence of such marks on the forewing.

345 **rajella** (Linnaeus, 1758) – i.h.o. John Ray (1627–1705), the distinguished British botanist and entomologist. Latin has no letter 'y' and the 'j' is the consonantal 'i' as in *Plebejus*; it is better to pronounce the name 'ray-ella' than as if it were derived from an Indian ruler. Hübner (1796) spelt the name '*rayella*'.

= **alnifoliella** (Hübner, 1796) – *Alnus*, the alder genus; *folium*, a leaf: the larvae mine the leaves of alder.

346 **distentella** (Zeller, 1846) – *distentus*, swollen: from the moth's relatively large size and broad strigulae. Not from the distended mine, since the life history was not known to Zeller.

347 **anderidae** (Fletcher, 1885) – *Anderida*, the Latin name for the part of east Sussex in which the type locality, Abbot's Wood, is situated.

348 **quinqueguttella** (Stainton, 1851) – *quinque*, five; *gutta*, a spot: the forewing has five, as opposed to the more usual four, costal strigulae.

349 **nigrescentella** (Logan, 1851) – *nigrescens*, blackish: the forewings are darker than those of most other members of the genus.

350 **insignitella** (Zeller, 1846) – *insignitus*, distinguished, striking: from the attractive appearance of the moth. 'Marked, from having forewings more marked than *L. nigrescentella*' (Macleod); an anachronism, since *insignitella* was named five years before *nigrescentella*.

351 **lautella** (Zeller, 1846) – *lautus*, literally well-washed (past part. of *lavo*, to wash); hence, spruce, bright, splendid: the moth is one of the most beautiful in the genus.

352 **schreberella** (Fabricius, 1781) – i.h.o. Professor J. C. D. von Schreber (1739–1810), a German entomologist whose lectures Fabricius attended at Leipzig.

353 **ulmifoliella** (Hübner, 1817) – *Ulmus*, the elm genus: a blunder, since the larvae mine leaves of birch. Stainton (1857*) suspected a confusion of identity, 'Hübner's figure is not unexceptionable, the 2nd and 3rd pairs of spots being represented as fasciae, as well as the first pair'.

354 **emberizaepenella** (Bouché, 1834) – *Emberiza*, the bunting genus of birds; *pennae* (pl.), a wing (*emberizaepenella* was a typographical error): from a fancied resemblance in coloration. Bouché described the species in *Ornix*; see 489 *Coleophora*.

355 **scabiosella** (Douglas, 1853) – *Scabiosa columbaria*, small scabious, is the larval foodplant.

356 **tristrigella** (Haworth, 1828) – *tri-*, three; *striga*, a streak: the forewing has two fasciae and costal and dorsal strigulae placed distally which almost unite to form a third fascia.

357 **stettinensis** (Nicelli, 1852) – Stettin (now Szczecin) is the type locality.

358 **froelichiella** (Zeller, 1839) – i.h.o. F. A. G. Frölich (19th century), a German entomologist.

359 **nicellii** (Stainton, 1851) – i.h.o. Graf von Nicelli (19th century), a German entomologist. See also 340a *P. staintoniella*.

360 **kleemannella** (Fabricius, 1781) – i.h.o. C. F. K. Kleemann (1735–89), a German entomologist.

361 **trifasciella** (Haworth, 1828) – *tri-*, three; *fascia*, a band: the markings of the forewing include two complete fasciae and costal and dorsal strigulae which almost unite to form a third.

362 **acerifoliella** (Zeller, 1839) – *Acer*, the maple genus; *folium*, a leaf: the larvae mine leaves of *Acer campestris*, field maple.

= **sylvella** (Haworth, 1828) *nec* (Hübner, 1813) – *silva, sylva*, a wood: the moth occurs in, but is not restricted to, a woodland habitat.

363 **platanoidella** (Joannis, 1920) – *Acer platanoides*, Norway maple: the larval foodplant.

364 **geniculella** (Ragonot, 1874) – *geniculum*, a little knee: from the geniculate fascia on the forewing.

365 **comparella** (Duponchel. 1843) – *compar*, like another: expressing resemblance, possibly to the next species. Duponchel cites Fischer von Röslerstamm as the nomenclator and since it was not his own name he offers no explanation for it.

366 **sagitella** (Bjerkander, 1790) – *sagitta*, an arrow: from the arrow-headed markings on the forewing.

PHYLLOCNISTINAE (367)

Phyllocnistis Zeller, 1848 – φύλλον (phullon), a leaf; κνίζω (knizō), to scrape: from the epidermal mine which only 'scratches the surface' of the leaf.

367 **saligna** (Zeller, 1839) – *Salix*, the sallow and willow genus; *lignum*, wood: from the larval mine which is mainly in the bark of twigs, most often on purple willow (*Salix purpurea*).

368 **unipunctella** (Stephens, 1834) – *unus*, one; *punctum*, a spot: from the small black spot at the apex of the forewing.

369 **xenia** Hering, 1936 – ξένος (xenos), a guest; *xenium*, a gift to a guest: the type locality is in Spain and Hering, a German entomologist, was on a visit to the country when he discovered the species.

SESIOIDEA (370)

SESIIDAE (370)

SESIINAE (370)

Sesia Fabricius, 1775 – σής (sēs), a moth or its larva. Linnaeus (1758) had divided Sphinx into four sections or subfamilies, viz.:–

1) *Legitimae alis angulatis*, i.e. 'genuine' hawk-moths with angled wings (e.g. 1980 *Smerinthus ocellata*).

2) *Legitimae alis integris*, i.e. 'genuine' hawk-moths with 'entire' wings (e.g. 1972 *Agrius convolvuli*).

3) *Legitimae alis integris, ano barbato*, i.e. 'genuine' hawk-moths with 'entire' wings, but also an anal tuft (see below).

4) *Adscitae habitu & larva diversae*, i.e. 'adopted' species, differing in appearance and larval character (see 163 *Adscita* and 166 *Zygaena*).

Sesia was Fabricius' name for subfamily 3, which included the clearwings, bee hawk-moths and the humming-bird hawk-moth. Most of these had names that indicated resemblance to another kind of insect, as 1983 *fuciformis*, with the form of a drone; so *Sesia fuciformis* implied 'the moth resembling a drone'. Later Latreille (1804) and Fabricius (1807) disagreed over the application, Latreille restricting the name to the clearwings and Fabricius to the bee-hawks (p. 32): Latreille's interpretation had priority, and now the application has been still further limited to this one genus. '*Ses*, clothes-moth' (Macleod), without further comment; correct, but certainly not in this context.

In fourteen of the specific names which follow, the name of an insect or group of insects is followed by the termination '*-formis*' from *forma*, form, indicating a fancied resemblance. This explanation will not be repeated.

370 **apiformis** (Clerck, 1759) – *apis*, a bee; *forma*.

371 **bembeciformis** (Hübner, 1797) – *bembex*, a sand wasp; *forma*.

PARANTHRENINAE (372)

Paranthrene Hübner, 1819 – παρά (para), beside, like; ἀνθρήνη (anthrēnē), a hornet: from a fancied resemblance.

372 **tabaniformis** (Rottemburg, 1775) – *tabanus*, a gad-fly, a horse-fly; *forma*.

Synanthedon Hübner, 1819 – σύν (sun), with, close to; ἀνθηδών (anthēdōn), the flowery one, i.e. the bee: from resemblance.

373 **tipuliformis** (Clerck, 1759) – *tipula*, a crane-fly, a daddy-long-legs; *forma*.
= **salmachus** (Linnaeus, 1758) – meaning not traced. Linnaeus spells the name with a capital 'S', indicating that it is a noun; possibly a little-known mythological character or perhaps from a genus of plants (then the genitive singular of a 4th declension noun). The suppression of this name in favour of the junior *tipuliformis* is in accordance with a ruling of the I.C.Z.N. (Opinion 1288), Clerck's name being the one familiar to fruit-farmers and used in all horticultural text-books.

374 **vespiformis** (Linnaeus, 1761) – *vespa*, a wasp; *forma*.

375 **spheciformis** ([Denis & Schiffermüller], 1775) – σφήξ (sphēx), a sand-wasp; *forma*.

376 **scoliaeformis** (Borkhausen, 1789) – *Scolia*, a genus of dagger-wasps; *forma*.

377 **flaviventris** (Staudinger, 1883) – *flavus*, yellow; *venter*, *ventris*, the belly: from the yellow-banded abdomen of the adult.

378 **andrenaeformis** (Laspeyres, 1801) – *Andrena*, a genus of mining bees; *forma*.
= **anthraciniformis** (Esper, 1798) – *Anthrax*, a genus of Diptera, itself named from *anthrax*, *anthracis*, a burning coal, cinnabar; *forma*. The retention of the junior name is in accordance with Opinion 1287 of the I.C.Z.N.

379 **myopaeformis** (Borkhausen, 1789) – *Myopa*, a genus of Diptera, a gad-fly; *forma*.

380 **formicaeformis** (Esper, 1783) – *formica*, an ant; *forma*.

381 **culiciformis** (Linnaeus, 1758) – *culex*, a mosquito; *forma*.

Bembecia Hübner, 1819 – *bembex*, a sand-wasp: from a fancied resemblance.

382 **scopigera** (Scopoli, 1763) – *scopae*, twigs, a besom, a broom; *gero*, to carry: from the anal tuft, '*scopa terminali binis flavis lineolis in medio notata*' (the anal brush marked with a pair of small central yellow lines). '*Scopos*, mark, *gero*, to bear: from orange mark on forewings' (Macleod), an explanation at variance with Scopoli's diagnosis and a Greek-Latin hybrid.

383 **muscaeformis** (Esper, 1783) – *musca*, a fly; *forma*.

384 **chrysidiformis** (Esper, 1783) – *Chrysis*, a genus of ruby-tailed wasps; *forma*.

CHOREUTIDAE (389)

Most members of this family were described by the early systematists in Tortrix or Pyralis *usu* Fabricius, 1775 (see p. 29). Guenée gave them family rank in the Pyraloididae, a name derived from the Fabrician nomenclature, not from affinity with the Pyralidae.

Anthophila Haworth, 1811 – ἄνθος (anthos), a flower; φιλέω (phileō), to love: from the habits of the adults which are diurnal and are often to be seen moving jerkily on flower-heads, especially those of the Umbelliferae.

385 **fabriciana** (Linnaeus, 1767) – i.h.o. J. C. Fabricius (1745–1808), the Danish entomologist who was Linnaeus' most distinguished pupil.

Tebenna Billberg, 1820 – τήβεννα (tēbenna), a robe of state: from the metallic ornamentation on the forewings of some of the species originally included in the genus.

386 **bjerkandrella** (Thunberg, 1784) – i.h.o. C. Bjerkander (1735–95), a Swedish entomologist who specialized in weevils. It is likely that all British records of this species refer to the next.

386a **micalis** (Mann, 1857) – *mico*, to move quickly in a jerky manner: from the behaviour of the adults; cf. 389 *Choreutis*. The termination *-alis* indicates that Mann supposed the species belonged to the Pyralidae. Fabricius described the next two species in Pyralis at the time when he was using that name instead of, and with the same meaning as, Tortrix; Guenée had named the family the Pyraloididae. Quite understandably, Mann got muddled.

Prochoreutis Diakonoff & Heppner, 1980 – προ- (pro-), instead of, a prefix here indicating affinity; the genus 389 *Choreutis*, q.v.: a genus closely related to *Choreutis*.

387 **sehestediana** (Fabricius, 1777) – i.h.o. Graf O. R. Sehestedt (1757–1838), a Norwegian

entomologist and pupil and friend of Fabricius. Some authors (e.g. Haworth and Curtis) misspell the name *schestediana*.

= **punctosa** (Haworth, 1811) – *punctosus*, adj. coined from *punctum*, a small spot: from the silvery specks on the forewing.

388 **myllerana** (Fabricius, 1794) – i.h.o. the memory of O. F. Müller (1730–84), a Danish naturalist and friend of Fabricius who describes him as '*Endomastracorum scrutator oculatissimus*, a most observant student of insectivores, the subject of one of his earlier works. He is, however, better known to lepidopterists today for his *Fauna Insectorum Fridrichsdalina* (1764), a local list covering the neighbourhood of Copenhagen and containing the first descriptions of a number of new species. Although Fabricius described this and the previous species in Pyralis, it was at the time when he was using that name for the Tortricidae, as the termination *-ana* clearly indicates.

Choreutis Hübner, 1825 – χορεύτης (khoreutēs), a dancer, the chorus in Greek drama being dancers as well as singers: the moths fly by day and settle on flowers where they make quick, jerky movements. Macleod attributes the choice of name to the 'mazy flight of some species', but it is the behaviour of the moths after they have settled that is being described. The Choreutidae and Glyphipterigidae were formerly regarded as a single family and the English name 'fanner' given to the latter; Macleod explains 'fanner' correctly.

= **Eutromula** Frölich, 1828 – εὖ (eu), well, intensive prefix; τρομέω (tromeō), to tremble: like *Choreutis*, referring to the quivering movements of the adults when they have settled.

389 **pariana** (Clerck, 1759) – *par*, a pair: from the pair of fasciae on the forewing.

390 **diana** (Hübner, 1822) – *dius, divus*, divine: in approbation.

YPONOMEUTOIDEA (424)

GLYPHIPTERIGIDAE (391)

Glyphipterix Hübner, 1825 – γλυφίς (gluphis), a notch; πτέρυξ (pterux), a wing: from the indentation in the termen of the forewing of most of the species.

391 **simpliciella** (Stephens, 1834) – *simplex, simplicis*, simple: from the relatively uncomplicated fuscous and white pattern of the forewing.

= **fischeriella** (Zeller, 1839) – i.h.o. G. Fischer von Waldheim (1770–1853), a German entomologist.

392 **schoenicolella** Boyd, 1858 – *Schoenus nigricans*, black bog-rush, the larval foodplant: *colo*, to inhabit.

393 **equitella** (Scopoli, 1763) – *eques, equitis*, a knight: probably, like 397 *thrasonella*, taken from a character in a classical Latin play and without entomological significance. It follows *thrasonella* in *Entomologia carniolica* and, according to Scopoli, occurs with it.

= **minorella** Snellen, 1882 – *minor*, smaller: there is uncertainty, not yet fully resolved, over the number of species in this complex; Heinemann & Wocke bestowed the name *majorella* in 1877 on one population, and Snellen *minorella* on another five years later.

394 **forsterella** (Fabricius, 1781) – i.h.o. J. R. Forster (1729–98), a cosmopolitan entomologist of Polish birth but English ancestry. He lived much of his life in England and became a personal friend of Fabricius. He accompanied Captain Cook in 1772 as senior naturalist to his second expedition and was given an honorary degree by Oxford University. Later he returned to the Continent and became Professor of Natural History at Halle. See also 1002 *forsterana*.

395 **haworthana** (Stephens, 1834) – i.h.o. A. H. Haworth (1767–1833), the distinguished British entomological author who had recently died.

396 **fuscoviridella** (Haworth, 1828) – *fuscus*, dusky brown; *viridis*, green: from the forewing which is fuscous with an obscure greenish tinge.

397 **thrasonella** (Scopoli, 1763) – Thraso, the bragging soldier in Terence's play *The Eunuch* (161 B.C.): without entomological relevance. See also 393 *equitella*.

DOUGLASIIDAE (398)

Tinagma Zeller, 1839 – τίναγμα (tinagma), a shake, a jerk: from the erratic flight of the adults in sunshine.

398 **ocnerostomella** (Stainton, 1850) – from a general similarity to the genus 444 *Ocnerostoma* Zeller, but separable from it, Stainton writes, by the longer palpi. Stainton found it a hard species to fit into the systematic list and later erected the genus *Douglasia* Stainton, 1854 to accommodate it, i.h.o. his friend and collaborator J. W. Douglas (1814–1905), a name conserved in the family title. The derivation given under *Ocnerostoma* (q.v.) has no application here, since in this species the mouth parts are functional.

399 **balteolella** (Fischer von Röslerstamm, 1840) – *balteolus*, a small girdle: from the white fascia on each forewing of the female; when the wings are folded, the two merge and resemble a belt.

HELIODINIDAE (400)

Heliodines Stainton, 1854 – ἥλιος (hēlios), the sun; δινήεις (dinēëis), a whirling: from the flight of the adults in sunshine; cf. 398 *Tinagma*.

400 **roesella** (Linnaeus, 1758) – i.h.o. A. J. Rösel von Rosenhof (1705–59), a German entomologist frequently cited in *Systema Naturae*. From 1746 onwards he edited a monthly magazine devoted to the study of insects as a hobby (*Die monatliche herausgegenbenen Insekten-Belustigung*). A French work based on the journal and containing reproductions of Rösel's fine colour and black-and-white illustrations has recently been published (*Les insectes de A. J. Rösel von Rosenhof*).

YPONOMEUTIDAE (424)

ARGYRESTHIINAE (401)

Argyresthia Hübner, 1825 – ἀργυρός (arguros), silver; ἐσθής (esthēs), dress: from the metallic gloss on the forewings of most species.

401 **laevigatella** Herrich-Schäffer, 1855 – *lēvigatus*, *laevigatus*, smooth: from the smooth, glossy texture of the forewing.

402 **illuminatella** Zeller, 1839 – *illuminatus*, illuminated, embellished: from the gloss on the forewing. There is no reliable evidence for its occurrence in Britain.

403 **glabratella** (Zeller, 1847) – *glaber*, smooth: from the gloss on the forewing.

404 **praecocella** Zeller, 1839 – *praecox*, early: from the moth's emergence in May, before most other members of the genus.

405 **arceuthina** Zeller, 1839 – ἀρκευθίνος (arkeuthinos), belonging to ἄρκευθος (arkeuthos), juniper: the larval foodplant.

406 **abdominalis** Zeller, 1839 – abdomen, the belly: from the abdomen of the adult which is pale with a faint reddish tinge.

407 **dilectella** Zeller, 1847 – *dilectus*, dear: Zeller is calling the species 'a dear little moth'.

408 **aurulentella** Stainton, 1849 – *aurulentus*, of the colour of gold: from the indistinct golden streaks on the forewing.

409 **ivella** (Haworth, 1828) – from the markings on the forewing which are roughly in the shape 'IV'; Haworth used the upper and lower case, 'Iv', adding to the verisimilitude. Authors emended the name to *quadriella*.

409a **trifasciata** Staudinger, 1871 – *tri-*, three; *fascia*, a band: from the three fasciae on the forewing.

410 **brockeella** (Hübner, 1813) – i.h.o. J. K. Brock (early 19th century), a German entomologist.

411 **goedartella** (Linnaeus, 1758) – i.h.o. J. Gödart (1620–68), a Dutch entomologist.

412 **pygmaeella** ([Denis & Schiffermüller], 1775) – *pygmaeus*, a pygmy: from the small size of the moth, although this is not notable amongst its congeners.

413 **sorbiella** (Treitschke, 1833) – *Sorbus aucuparia*, rowan: the larval foodplant.

414 **curvella** (Linnaeus, 1761) – *curva*, a curve: from the curved fascia on the forewing. The name was formerly applied to 421 through misidentification.

= **arcella** (Fabricius, 1777) – *arcus*, a bow: from the same character as *curvella* above; formerly applied to 220 *Nemapogon clematella* through misidentification.

= **cornella** sensu auctt. – *cornu*, a horn: from the same feature as *curvella* above; a synonym of 450 *Scythropia crataegella* and here a misidentification.

415 **retinella** Zeller, 1839 – *rete*, a net: from the reticulate forewing.

416 **glaucinella** Zeller, 1839 – *glaucus*, bluish grey: from the ground colour of the forewing.

417 **spinosella** Stainton, 1849 – *Prunus spinosa*, blackthorn: the larval foodplant.
= **mendica** (Haworth, 1828) – an adaptation and misidentification of *mendicella* Hübner, 1796. *Mendicus*, beggarly, needy: 'perhaps from resemblance of forewings with white and black markings to dress of mendicant friar' (Macleod) Alternatively, the name might refer to the paucity or complete absence of the rich metallic gloss characteristic of the genus.

418 **conjugella** Zeller, 1839 – *conjugo*, to yoke together: perhaps from the pattern when the moth is at rest with the wings folded, the dark costal half of each forewing being linked by a dark bar extending across the white dorsal half. Pickard *et al.* and Macleod derive from *conjux*, a spouse, but fanciful names, apparently without entomological application, are not found in the other species named by Zeller in the same paper.

419 **semifusca** (Haworth, 1828) – *semi-*, half; *fuscus*, dusky: from the contrasting costal and dorsal halves of the forewing which are fuscous and white respectively; cf. 418.

420 **pruniella** (Clerck, 1759) – *Prunus cerasus*, the cherry-plum, is the foodplant.

421 **bonnetella** (Linnaeus, 1758) – i.h.o. C. Bonnet (1720–93), a Swiss entomologist who studied, in particular, plant lice; some of his publications were printed in Britain.
= **curvella** sensu auctt. – see 414.

422 **albistria** (Haworth, 1828) – *albus*, white; *stria*, a streak: from the white dorsal streak of the forewing.

423 **semitestacella** (Curtis, 1833) – *semi-*, half; *testaceus*, brick-coloured: from the forewing which has the costal half ferruginous brown and the dorsal half white.

YPONOMEUTINAE (424)

Yponomeuta Latreille, 1796 – ὑπονομεύω (huponomeuō), to make underground mines. The name has nothing to do with the habits of the larvae as supposed by Macleod; Latreille expressly states that they are not leaf-miners and feed in a communal web spun over the foodplant. His diagnosis is almost wholly descriptive of the labial palpus and he may be comparing it fancifully to a miner's pickaxe. Fifty years later his compatriot Guenée was to compare the palpus of another genus to a billhook (1273 *Dichrorampha*, q.v.). The amendment *Hyponomeuta* Sodoffsky, 1837 is orthographically correct but inadmissible; Latreille would not have pronounced the 'h' himself and so omitted it.

424 **evonymella** (Linnaeus, 1758) – *Euonymus*, the spindle genus: the foodplant of some related species but not of this one. The members of the genus are easily confused and Linnaeus gives the correct foodplant as well as *Euonymus*.

425 **padella** (Linnaeus, 1758) – *Prunus padus*, bird-cherry: the foodplant of the previous species but not this one. Linnaeus states that it also lives on fruit-trees, which applies to the next species.

426 **malinellus** Zeller, 1838 – *malinus*, pertaining to an apple-tree: the correct larval foodplant.

427 **cagnagella** (Hübner, 1813) – a typographical error corrected by Hübner in 1822 to *cagnatella* (still incorrect) and by Treitschke in 1832 to *cognatella*, which is undoubtedly what Hübner had intended. *Cognatus*, related: having close affinity with the other members of this vexed group.

428 **rorrella** (Hübner, 1796), a typographical error corrected to *rorella* by Hübner in 1822 – *ros, roris*, dew: from the sprinkling of black spots on the forewing.

429 **irrorella** (Hübner, 1796) – *irroro*, to sprinkle with dew: from the sprinkling of black spots on the forewing.

430 **plumbella** ([Denis & Schiffermüller], 1775) – *plumbum*, lead: from the somewhat leaden ground colour of the forewing.

431 **sedella** Treitschke, 1832 – *Sedum telephium*, orpine: the larval foodplant.
= **vigintipunctata** (Retzius, 1783), unavailable – *viginti*, twenty; *punctatus*, spotted: from the pattern of the forewing.

Euhyponomeuta Toll, 1941 – εὐ (eu), prefix here signifying affinity; the genus 424 *Yponomeuta* (= *Hyponomeuta*): a closely related genus.

432 **stannella** (Thunberg, 1794) – *stannum*, tin: from the colour of the forewing.

Kessleria Nowicki, 1864 – i.h.o. H. F. Kessler (1816–97), a German entomologist.

433 **fasciapennella** (Stainton, 1849) – *fascia*, a stripe; *pennae* (pl.), a wing: from the forewing fascia.

434 **saxifragae** (Stainton, 1868) – *Saxifraga*, the saxifrage genus, which contains the larval foodplants.

Zelleria Stainton, 1849 – i.h.o. P. C. Zeller (1808–83), the distinguished German microlepidopterist and friend of Stainton.

435 **hepariella** Stainton, 1849 – ἧπαρ (hēpar), the liver: from the liver-coloured forewing.

Pseudoswammerdamia Friese, 1960 – ψεῦδος (pseudos), a falsehood; the genus 437 *Swammerdamia*, q.v.: a related genus.

436 **combinella** (Hübner, 1786) – *combino*, to unite: perhaps from the convolute wings with the apices pressed together when the moth is at rest.

Swammerdamia Hübner, 1825 – i.h.o. J. J. Swammerdam (1637–80), a Dutch entomologist.

437 **caesiella** (Hübner, 1796) – *caesius*, bluish grey: from the colour of the forewing.

437a **passerella** (Zetterstedt, 1839) – *passer*, a sparrow: first described in *Ornix*, see 489 *Coleophora*, and therefore bearing a bird's name; cf. 566 *sternipennella* and 572 *laripennella* of the same author and date.

= **nanivora** Stainton, 1871 – *Betula nana*, dwarf birch; *voro*, to devour: from the larval foodplant.

438 **pyrella** (Villers, 1789) – *Pyrus communis*, the pear: a larval foodplant.

439 **compunctella** (Herrich-Schäffer, 1855) – *compunctus*, pricked, speckled: from the markings on the forewing.

Paraswammerdamia Friese, 1960 – παρά (para), beside, a prefix indicating affinity; the genus 437 *Swammerdamia*, q.v.: a related genus.

440 **albicapitella** (Scharfenberg, 1805) – *albus*, white; *caput*, *capitis*, the head: from the white head of the adult.

= **spiniella** (Hübner, 1809) – *Prunus spinosa*, blackthorn: the larval foodplant.

441 **lutarea** (Haworth, 1828) – *lutareus*, muddy: from the colour of the forewing.

Cedestis Zeller, 1839 – κηδεστής (kēdestēs), a relation by marriage: to express affinity with 401 *Argyresthia* with which it was paired by Zeller.

442 **gysseleniella** Zeller, 1839 – i.h.o. J. V. G. Gysselin (19th century), an Austrian collector.

443 **subfasciella** (Stephens, 1834) – *sub-*, moderating prefix; *fascia*, a stripe: from the fascia on the forewing which is, in fact, quite well developed.

Ocnerostoma Zeller, 1847 – ὀκνηρός (oknēros), unready, useless; στόμα (stoma), the mouth: the haustellum is degenerate.

444 **piniariella** Zeller, 1847 – *Pinus sylvestris*, Scots pine: the larval foodplant.

445 **friesei** Svensson, 1966 – i.h.o. G. Friese, a contemporary German entomologist.

Roeslerstammia Zeller, 1839 – i.h.o. J. E. Fischer von Röslerstamm (1783–1866), a German entomologist.

446 **pronubella** ([Denis & Schiffermüller], 1775) – *pronuba*, a bridesmaid: the hindwing is light yellow, offering similarity to 2107 *Noctua pronuba*, the large yellow underwing; cf. also 985 *pronubana* and see 2452.

447 **erxlebella** (Fabricius, 1787) – i.h.o. J. C. P. Erxleben (1744–77), a German naturalist.

Atemelia Herrich-Schäffer, 1853 – ἀτημέλεια (atēmeleia), carelessness: perhaps whimsical self-rebuke for his failure to realize sooner that 448 and three other non-British species required a separate genus.

448 **torquatella** (Zeller, 1846) – *torquatus*, adorned with a collar: from the neck-tufts which form a collar of yellowish scales.

Prays Hübner, 1825 – πραΰς (praüs), gentle, tame: relevance obscure, but the antithesis to the next genus; perhaps given for some subjective reason. 'Soft, from smooth head' (Macleod); unlikely, because πραΰς does not mean soft to the touch.

449 **fraxinella** (Bjerkander, 1784) – *Fraxinus excelsior*, ash: the larval foodplant.

= **curtisella** (Donovan, 1793) – i.h.o. W. Curtis (1746–99), author of *Flora Londiniensis*, etc. Although best known as a botanist, he was also author of *A short history of the brown-tail moth* (1782). John Curtis (1791–1862), author of *British entomology*, was aged two when the name was bestowed.

Scythropia Hübner, 1825 – σκυθρωπός (skuthrōpos), of sad or angry countenance: in sense antithetical to the last genus and named as a pair to it.

450 **crataegella** (Linnaeus, 1767) – *Crataegus*, the hawthorn genus: the larval foodplant.
= **cornella** (Fabricius, 1775) – see 414 *Argyresthia curvella*.

PLUTELLINAE (464)

Ypsolopha Latreille, 1796, malformed and emended to *Hypsilophus* by Stainton – ὑψίλοφος (hupsilophos), high-crested: from the falcate forewing of 452 *Y. nemorella* and 453 *Y. dentella*, which gives the impression of an upwards-thrusting anal crest when the wings are folded, a character described in Latreille's brief diagnosis. The name gave rise to puzzlement; Pickard *et al.* left it unexplained and Macleod wrote 'from hairy head' though it is, in fact, smooth-scaled. Latreille had not yet observed the tuft on segment 2 of the labial palpus in, for example, 461 *Y. ustella*; when he did so, it became the motif for another name, *Cerostoma* Latreille, 1802 (κέρας (keras), a horn; στόμα (stoma), the mouth), now reduced to synonymy. Ypsolopha was conceived with broad family status, 345 *Phyllonorycter rajella* being one of the species cited as a typical member (Latreille, 1804); see p. 26.

451 **mucronella** (Scopoli, 1763) – *mucro, mucronis*, a sharp point: from the apex of the forewing which is produced into an acute point.

452 **nemorella** (Linnaeus, 1758) – *nemus, nemoris*, a grove: from the habitat.

453 **dentella** (Fabricius, 1775) – *dens, dentis*, a tooth: from the falcate apex of the forewing which projects upwards like a tooth when the wings are folded; see MBGBI 7(2), Pl. A, fig. 8. The generic name is a Greek rendering of the idea which prompted this one.

454 **asperella** (Linnaeus, 1761) – *asper*, rough: from the raised scale-tufts on the forewing.

455 **scabrella** (Linnaeus, 1761) – *scaber*, rough: from the raised scale-tufts on the forewing.

456 **horridella** (Treitschke, 1835) – *horridus*, shaggy: from the raised scale-tufts on the forewing.

457 **lucella** (Fabricius, 1775) – *lucus*, a grove: from the habitat, '*habitat in arboretis*', it lives in woodland; cf. 452 *nemorella*. Macleod's explanation '*lux*, light, from bright appearance' is not correct.

458 **alpella** ([Denis & Schiffermüller], 1775) – *Alpes*, the Alps: the type locality is the Vienna district and later Fabricius stated that it occurred on the Austrian mountains, but mountains are not its typical habitat, as supposed by Macleod.

459 **sylvella** (Linnaeus, 1767) – *silva, sylva*, a wood: from the habitat (cf. 452 and 457).

460 **parenthesella** (Linnaeus, 1761) – παρένθεσις (parenthesis), an insertion; from the whitish costal blotch which interrupts the reddish brown ground colour.

461 **ustella** (Clerck, 1759) – *uro*, to burn, past part. *ustus*, burnt: in some of the many colour forms the forewings could fancifully be regarded as having a scorched appearance.

462 **sequella** (Clerck, 1759) – *sequens*, following: no explanatory text accompanies Clerck's beautiful figure of this species; perhaps, being gravelled for a name, he just called it 'the next'.

463 **vittella** (Linnaeus, 1758) – *vitta*, a band: from the black streak extending along the dorsum. A streak in this position is normally referred to as a 'vitta'; cf. the terms used in describing forms of 1054 *Acleris cristana*.

Plutella Schrank, 1802 – Pickard *et al.* (p. 82) first derived from πλοῦτος (ploutos), wealth, but later (Addenda and Corrigenda, p. 114) from πλυτός (plutos), washed, 'from the smudged appearance of the insects, the markings running into one another'; Macleod chose their second interpretation. However, the name should then have been '*Plytella*' (see p. 10). Much more probably from Πλούτων (Ploutōn), Pluto, god of the Nether World. Of Schrank's twenty-one generic names in current use, nine are taken from Greek gods or mythical figures, two of them, 1626 *Maniola* and 2166 *Hadena*, directly connected with the Underworld. The name was originally suprageneric; it has no entomological application.

464 **xylostella** (Linnaeus, 1758) – *Lonicera xylosteum*, fly honeysuckle: given wrongly as the foodplant by Linnaeus, though he adds that it looks like a smaller version of a *Brassica*-feeding species figured by Rösel (see 400).

465 **porrectella** (Linnaeus, 1758) – *porrectus*, outstretched: from the antennae which are extended forwards when the moth is at rest.

465a **haasi** Staudinger, 1883 – i.h.o. E. Haase (1857–94), a German entomologist. J. A. Haas (late 18th century) is less likely.

Rhigognostis Staudinger, 1857 – ῥῖγος (rhīgos), cold; γνώστης (gnōstēs), one who knows: the moths know cold because they overwinter in the adult stage.

466 **senilella** (Zetterstedt, 1839) – *senilis*, aged: from the hoary aspect of the vertex of the head, where white and fuscous scales are mingled.

467 **annulatella** (Curtis, 1832) – *anulatus, annulatus*, ringed: the pale dorsal streak on the forewing has its upper margin undulate with the result that when the wings are folded there appear to be two rings.

468 **incarnatella** (Steudel, 1873) – *in-*, intensive prefix; *caro, carnis*, flesh: from the flesh-pink suffusion and markings on the forewing.

Eidophasia Stephens, 1842 – εἶδος (eidos), beautiful form; φάσις (phasis), appearance: from the striking appearance of the next species.

469 **messingiella** (Fischer von Röslerstamm, 1840) – i.h.o. Herr Messing (*c.* 1800–*c.* 1870), the court and state music director at Neustrelitz in northern Germany. He was also an amateur entomologist and discovered this species.

ORTHOTAELIINAE (470)

Orthotaelia Stephens, 1834 – ὀρθός (orthos), straight; τέλος (telos), end: from the straight, squared termen of the forewing; see MBGBI 7(2), Pl.A, fig. 9.

470 **sparganella** (Thunberg, 1794) – *Sparganium*, the bur-reed genus, provides some of the foodplants.

ACROLEPIINAE (476)

Digitivalva Gaedike, 1970 – *digitus*, a finger; *valva*, the valve or 'clasper' of the male genitalia: from a digital process on the valva.

471 **perlepidella** (Stainton, 1849) – *perlepidus*, very pretty: descriptive of the adult.

472 **pulicariae** (Klimesch, 1956) – *Pulicaria dysenterica*, common fleabane: the larval foodplant.

Acrolepiopsis Gaedike, 1970 – the genus 476 *Acrolepia* (q.v.); ὄψις (opsis), face, appearance: from close affinity between the genera.

473 **assectella** (Zeller, 1839) – *assector*, to follow, to accompany: Zeller gives no explanation; in his paper it follows *Digitivalva granitella* (Treitschke), with which 472 was formerly confused.

474 **betulella** (Curtis, 1838) – *Betula*, the birch genus: birch is not the foodplant and in any case the life history was unknown when Curtis bestowed the name.

475 **marcidella** (Curtis, 1850) – *marcidus*, withered: the adult overwinters and the coloration may have suggested a withered leaf to Curtis.

Acrolepia Curtis, 1838 – ἄκρον (acron), the top; λεπίς (lepis), a scale: from the 'smooth, scaly head' of the adult (Curtis); 'from rough top of head' (Macleod, who had not consulted Curtis or looked at the moth).

476 **autumnitella** Curtis, 1838 – *auctumnus*, autumn: the second generation of the adult occurs from October to April.

= **pygmeana** (Haworth, 1811) *nec* (Hübner, 1799) – *pygmaeus*, a dwarf: Haworth regarded it as a tortricid (termination '*-ana*') and as such it would be very small. The name cannot be used as it is a junior homonym.

EPERMENIIDAE (481)

Phaulernis Meyrick, 1895 – φαῦλος (phaulos), paltry, poor; ἔρνος (ernos), a shoot: from the forewing of the next species, originally the only one in the genus, which has a single, small scale-tooth on the dorsum of the forewing; in 481 *Epermenia* there are two or four.

477 **dentella** (Zeller, 1839) – *dens, dentis*, a tooth; from the scale-tooth on the dorsum of the forewing.

478 **fulviguttella** (Zeller, 1839) – *fulvus*, yellow; *gutta*, a spot: from the yellow spots on the forewing.

Cataplectica Walsingham, 1894 – καταπλέχω (kataplekhō), to entwine: from the larval habit of feeding amongst spun seeds.

479 **farreni** Walsingham, 1894 – i.h.o. W. Farren (1836–87), the Cambridge entomological dealer who discovered the species.

480 **profugella** (Stainton, 1856) – *profugus*, a fugitive: Stainton ascribes the name to Zeller, who had taken a specimen at Hermsdorf in Germany; he gives no explanation. Macleod's explanation is apocryphal.

Epermenia Hübner, 1825 – perhaps ἐπί (epi), upon; 'Αρμενία (Armenia), whence *mus Armenius, armenius, Mustela erminea* Linnaeus, 1758, the ermine: from the scale-tufts on the dorsal margin of the forewing which are compared to the ermine trimming on the robes of a judge; cf. 180 *herminata* and 2489 *Herminia*. '*Epi*, upon, *hermeneia*, interpretation' [of dreams, etc.] (Macleod); this seems pointless.

481 **illigerella** (Hübner, 1813) – i.h.o. J. C. W. Illiger (1775–1825), an Austrian entomologist; see p. 29.

482 **insecurella** (Stainton, 1849) – *insecurus*, uncertain: from uncertainty whether it was distinct from the previous species.

483 **chaerophyllella** (Goeze, 1783) – *Chaerophyllum temulentum*, rough chervil: the larva feeds on Umbelliferae, but is not recorded on this species in Britain.

484 **aequidentellus** (Hofmann, 1867) – *aequus*, equal; *dens, dentis*, a tooth: from the four scale-tufts of equal size on the dorsum of the forewing.

SCHRECKENSTEINIIDAE (485)

Schreckensteinia Hübner, 1825 – i.h.o. R. von Schreckenstein (d. 1808), a German entomologist who used to provide Hübner with Microlepidoptera for figuring.

485 **festaliella** (Hübner, 1819) – *festum*, a feast or holiday, *festalis* (late Lat.), pertaining to a feast: although the name is without obvious entomological significance, Hübner may have had some personal reason for associating the moth with a festive occasion.

GELECHIOIDEA (800)

COLEOPHORIDAE (489)

Augasma Herrich-Schäffer, 1853 – αὔγασμα (augasma), lustre: from the gloss on the forewing.

486 **aeratella** (Zeller, 1839) – *aeratus*, covered with copper: from the coppery reflections on the forewing.

Metriotes Herrich-Schäffer, 1853 – μετριότης (metriotēs), moderation, = Lat. *modestia*: from Duponchel's name, now reduced to synonymy, for the next species.

487 **lutarea** (Haworth, 1828) – *lutarius*, pertaining to mud: from the ochreous grey, supposedly earthy, coloration of the forewing.

= **modestella** (Duponchel, 1839) – *modestus*, moderate, chaste: from the sober-coloured, unadorned, unicolorous forewing.

Goniodoma Zeller, 1849 – γωνία (gōnia), an angle; δῶμα (dōma), a house: from the roughly rectangular larval case of the type species, *G. auroguttella* Zeller, 1849. The British species, which was formerly confused with *G. auroguttella*, makes its case in a more or less tubular seedhead and Stainton therefore described it in *Coleophora*, not *Goniodoma*, 'its habitation not showing any angles'.

488 **limoniella** (Stainton, 1884) – *Limonium vulgare*, common sea-lavender: the larval foodplant.

Coleophora Hübner, 1822 – κολεός (koleos), a sheath; φορέω (phoreō), to carry: from the portable cases made by the larvae. An early synonym was *Ornix* Treitschke, 1833 (see 300 *Parornix*), bestowed because a number of the species had already been named from birds, e.g. 494 *coracipennella*, and this generic name encouraged the continuation of the practice; fourteen of the species on the British list are named from birds and others have the termination *-pennella* or *-pennis*, from *pennae* (pl.), a wing. The incorporation of bird names into those given to Microlepidoptera with feathery wings became a convention like the use of the names of mythological personages for species or of places for genera. No similarity was necessary between the bird and the insect.

489 **albella** (Thunberg, 1788) – *albus*, white: from the colour of the forewing.

= **leucapennella** (Hübner, 1796) – λευκός (leukos), white; *pennae* (pl.), a wing: from the white ground colour of the forewing.

490 **lutipennella** (Zeller, 1838) – *luteus*, yellow; *pennae* (pl.), a wing: either directly from the colour of the forewing which is concolorous with that of 492 *flavipennella*, q.v.; or from *Parus luteus* Gmelin, 1774, a synonym of the scientific name of the yellow wagtail. '*Lutum*, mud' (Macleod): incorrect.

491 **gryphipennella** (Hübner, 1796) – *gryps*, a griffin; *pennae* (pl.), a wing: since the griffin is a fabulous creature with the body of a lion and the wings of an eagle, any resemblance is pure fantasy; see genus above. '*Gyps* (not *gryps*), griffon-vulture, *penna*, feather; from forewings' (Macleod): his emendation is unnecessary, since *gryps* and γρύψ (grups) are good classical Latin and Greek words.

492 **flavipennella** (Duponchel, 1843) – *flavus*, yellow; *pennae* (pl.), a wing: from the ground colour of the forewing. However, the name may well be derived from *Motacilla flava* Linnaeus, 1758, the blue-headed wagtail; *M. flava flavissima* (Blyth, 1834) is the scientific name of the yellow wagtail, the familiar summer visitor to Britain.

493 **serratella** (Linnaeus, 1761) – *serratus*, serrated: the larval case is excised from the margin of a leaf of its foodplant (birch, elm, etc.), and consequently bears the serrations of that leaf.

494 **coracipennella** (Hübner, 1796) – *corax*, a raven, *Corvus corax* Linnaeus, 1758, the raven; *pennae* (pl.), a wing: from the dark fuscous forewing.

494a **prunifoliae** Doets, 1944 – *Prunus spinosa*, blackthorn; *folium*, a leaf: from the larval foodplant, on the leaves of which the larva feeds. The termination *-ae* is malformed; Doets must have supposed that *folia*, nom. neut. pl. of a second declension noun, was nom. fem. sing. of a first declension noun like *mensa*, a table, and wrote *foliae* either for the gen. sing., correctly *folii*, or for the nom. pl., correctly *folia*. Toll (39) and Barasch (498), qq.v., made the same mistake. The gen. sing. is correctly formed in 516 *trifolii* Curtis and the nom. pl. in 501 *siccifolia* Stainton and 1641 *ilicifolia* Linnaeus and 1642 *quercifolia* Linnaeus, these authors having been better Latinists.

495 **spinella** (Schrank, 1802) – *Prunus spinosa*, blackthorn: a foodplant, though hawthorn (*Crataegus* spp.) and apple (*Malus* spp.) are preferred.

= **cerasivorella** Packard, 1870 – *Prunus cerasus*, dwarf-cherry; *voro*, to devour: dwarf-cherry is a rare alternative foodplant in Britain but Packard was an American and it may be more widely used in his country.

496 **milvipennis** Zeller, 1839 – *milvus*, a kite, *Falco milvus* Linnaeus, 1758, the red kite; *pennae* (pl.), a wing, *pennis*, abl. pl., with the wing: Zeller's use of *-pennis* rather than the more familiar *-pennella* shows that his name refers to the elongate shape rather than the colour of the wing. He described the larva as feeding on elm, not birch, so the name may not have been intended for the species that now bears it.

496a **adjectella** Herrich-Schäffer, 1861 – *adjectus*, additional: from its having been a new addition to the Coleophoridae in 1861, when the name was bestowed.

497 **badiipennella** (Duponchel, 1843) – *badius*, brown, bay-coloured; *pennae* (pl.), a wing: from the colour of the forewing. Duponchel's names for the *Coleophora* with the termination *-pennella* generally have reference to a species of bird, but I have not traced the scientific name of a bird incorporating *badius* that is not junior to the name of this moth.

498 **alnifoliae** Barasch, 1934 – *Alnus*, the alder genus; *folium*, a leaf: from the larval foodplant. See 39 and 494a for the malformed termination *-ae*.

499 **limosipennella** (Duponchel, 1843) – *Scolopax limosa* Linnaeus, 1758, the black-tailed godwit; *pennae* (pl.), a wing: from a fancied resemblance. '*Limosus*, muddy; *penna*, feather; from colour of forewings' (Macleod): an interpretation less in keeping with Duponchel's practice in naming the Coleophoridae.

500 **hydrolapathella** Hering, 1924 – *Rumex hydrolapathum*, the great water-dock: the larval foodplant.

501 **siccifolia** Stainton, 1856 – *siccus*, dry; *folia* (pl.), leaves: from the larval case which resembles

withered leaves. Haworth (1802), in his list of terms for habitats, gives *siccifoliis* (ablative pl.) which he explains as 'withered leaves on oaks, etc., in spring' and it is likely that Stainton had this in mind. Haworth and Stainton were better Latinists than Toll (39), Doets (494a) and Barasch (498).

502 **trigeminella** Fuchs, 1881 – *trigeminus*, one of a set of triplets; from the close resemblance between this species and 496 *C. milvipennis* and 497 *C. badiipennella*.

503 **fuscocuprella** Herrich-Schäffer, 1854 – *fuscus*, dusky; *cupreus*, coppery: from the colour of the forewing.

504 **viminetella** Zeller, 1849 – *viminetum*, an osier bed: the larva feeds on *Salix* spp., sallows and willows.

505 **idaeella** Hofmann, 1869 – *Vaccinium vitis-idaea*, cowberry: the larval foodplant.

506 **vitisella** Gregson, 1856 – *Vaccinium vitis-idaea*, cowberry: the larval foodplant.

507 **glitzella** Hofmann, 1869 – i.h.o. C. T. Glitz (1818–89), a German entomologist.

508 **arctostaphyli** Meder, 1933 – *Arctostaphylos uva-ursi*, bearberry: the larval foodplant.

509 **violacea** (Ström, 1783) – *violaceus*, violet-coloured: from the forewing, though this is bronzy fuscous, virtually without any violet tinge.

= **hornigi** Toll, 1952 – i.h.o. J. von Hornig (1819–86), a Russian entomologist.

510 **juncicolella** Stainton, 1851 – *Juncus*, the rush genus; *colo*, to inhabit: from the type locality. Stainton discovered the species by sweeping in a rushy area at Kilmun, Argyll, Scotland; cf. 598 *Elachista kilmunella*. Later, when the foodplant was known, he described the name as 'a very inappropriate one' (Stainton, 1859*).

511 **orbitella** Zeller, 1849 – *orbus*, bereaved: from the sombre ground colour of the forewing, bereft of pattern.

512 **binderella** (Kollar, 1832) – i.h.o. C. F. Binder von Kriegelstein (late 18th – early 19th century), an Austrian entomologist.

513 **potentillae** Elisha, 1885 – *Potentilla*, the cinquefoil genus, includes some of the foodplants.

514 **ahenella** Heinemann, 1876 – *aëneus, aheneus*, of bronze; from the colour of the forewing.

515 **albitarsella** Zeller, 1849 – *albus*, white; *tarsus*, the five-jointed terminal section of an insect's leg, adapted from ταρσός (tarsos), the flat of the foot: the tarsi are white except towards the base and are conspicuous, especially in living specimens.

516 **trifolii** (Curtis, 1832) – *Trifolium*, the clover genus: the next three closely-related species feed on *Trifolium* spp., but this species on melilot (*Melilotus* spp.).

517 **frischella** (Linnaeus, 1758) – i.h.o. J. L. Frisch (1660–1743), a German entomologist.

= **alcyonipennella** sensu auctt. – *alcyon*, halcyon, the kingfisher; *pennae* (pl.), a wing: from the slightly metallic forewing.

518 **mayrella** (Hübner, 1813) – i.h.o. U. Mayer, a timber inspector living in Vienna, who sent Hübner the type specimen.

= **spissicornis** (Haworth, 1828) – *spissus*, thick; *cornu*, a horn; from the basal half of the antenna which is thickened with dense scales.

519 **deauratella** Lienig & Zeller, 1846 – *deauratus*, gilt: from the coppery gloss on the forewing.

520 **fuscicornis** Zeller, 1847 – *fuscus*, dusky; *cornu*, a horn: from the antenna which is wholly dark, not ringed or tipped with white as in some related species.

521 **conyzae** Zeller, 1868 – *Inula conyza*, ploughman's spikenard: the larval foodplant.

522 **lineolea** (Haworth, 1828) – *lineola*, a small line: from the streaks along the veins of the forewing.

523 **hemerobiella** (Scopoli, 1763) – ἡμερόβιος (hēmerobios), living for a day, name source for *Hemerobius*, a genus of lacewings: from a fancied resemblance. Scopoli gives an accurate description of the life history and certainly did not suppose that this species lived only for one day. It has a two-year life-cycle.

524 **lithargyrinella** Zeller, 1849 – λιθάργυρος (litharguros), lead monoxide, litharge: from the ground colour of the forewing.

525 **solitariella** Zeller, 1849 – *solitarius*, solitary; it was described by Zeller from a single specimen; the name is an unfortunate one, since the larvae are gregarious. Zeller had reared his moth from a case fixed to a blade of grass for pupation and did not know the complete life history.

526 **laricella** (Hübner, 1817) – *Larix decidua*, European larch: the larval foodplant.

527 **wockeella** Zeller, 1849 – i.h.o. M. F. Wocke (1820–1906), a German entomologist.

528 **chalcogrammella** Zeller, 1839 – χαλκός (khalkos), brass; γράμμα (gramma), a mark: the ground colour is deep yellow and the markings silver, edged fuscous, so the name should be interpreted as meaning 'lettering on brass'. Macleod is wrong in supposing that the markings themselves are brazen.

529 **tricolor** Walsingham, 1899 – *tri-*, three; *color*, colour: from the forewing which is yellow with fuscous-edged silver streaks.

530 **lixella** Zeller, 1849 – *Lixus*, a genus of weevils: perhaps from a supposed resemblance between the beak or rostrum of the beetle and the porrected antennae of the moth; Zeller gives no explanation.

531 **ochrea** (Haworth, 1828) – ὤχρα (okhra), yellow earth, ochre: from the ground colour of the forewing.

532 **albidella** ([Denis & Schiffermüller], 1775) – *albidus*, white: from the ground colour of the forewing.

533 **anatipennella** (Hübner, 1796) – *anas*, a duck, *Anas* Linnaeus, 1758, the duck genus; *pennae* (pl.), a wing: Hübner must have had a domestic duck in mind, since the forewing of the moth is white.

534 **currucipennella** Zeller, 1839 – *curruca*, a small bird that has been identified with a blackcap and a wagtail, *Motacilla Curruca* Linnaeus, 1758, the lesser whitethroat: there need be no resemblance (see 489 *Coleophora*).

535 **ardeaepennella** Scott, 1861 – *ardea*, a heron, *Ardea* Linnaeus, 1758, the genus of birds including the herons and egrets; *pennae* (pl.), a wing: the colour resemblance is to an egret, since the forewing of the moth is white.

536 **ibipennella** Zeller, 1849 – *ibis*, the ibis, *Ardea Ibis* Linnaeus, 1758, the cattle egret; *pennae* (pl.), a wing: from the moth's forewing which is white, like the predominant colour of the bird.

537 **palliatella** (Zincken, 1813) – *palliatus*, cloaked; from the conspicuous flaps adorning the larval case.

538 **vibicella** (Hübner, 1813) – *vibex*, the mark of a lash, a weal: from the longitudinal streaks on the forewing.

539 **conspicuella** Zeller, 1849 – *conspicuus*, striking: from the bold yellow, brown and white pattern of the forewing.

540 **vibicigerella** Zeller, 1839 – *vibex*, the mark of a lash, a weal; *gero*, to carry, to wear: from the longitudinal streaks on the forewing.

541 **pyrrhulipennella** Zeller, 1839 – πυρρoύλος (purrhoulos), a red-coloured bird, perhaps the bullfinch, *Loxia Pyrrhula* Linnaeus, 1758, the bullfinch: *pennae* (pl.), a wing: if there is any resemblance in colour between the bird and the moth (which need not be the case: see 489 *Coleophora*), it is in the bird's back, certainly not in its reddish pink breast.

542 **serpylletorum** Hering, 1889 – *Thymus serpyllum*, Breckland thyme: a larval foodplant. The termination *-etum*, gen. pl. *-etorum*, signifies a thicket or stand of the foodplant; cf. 109 *prunetorum*.

543 **vulnerariae** Zeller, 1839 – *Anthyllis vulneraria*, kidney vetch: the larval foodplant.

544 **albicosta** (Haworth, 1828) – *albus*, white; *costa*, a rib, the anterior margin of an insect's wing: from the white costal streak.

545 **saturatella** Stainton, 1850 – *saturatus*, full, rich in colour: from the ground colour which is dark brown, deeper than that of most related species.

546 **genistae** Stainton, 1857 – *Genista anglica*, petty whin: the larval foodplant.

547 **discordella** Zeller, 1849 – *discors*, disagreeing, discordant: from the distinction between this species and 548 *C. niveicostella* which precedes it and with which it is compared in Zeller's paper. 'Discordant, i.e. different from *C. saturatella*, in spite of resemblance' (Macleod): an anachronism, since *C. saturatella* was not named until a year after Zeller had published his paper, and there is no close resemblance.

548 **niveicostella** Zeller, 1849 – *niveus*, snowy; *costa*, a rib, the anterior margin of an insect's wing: from the conspicuous white costal streak.

549 **onosmella** (Brahm, 1791) – *Onosma echiodes*: a foodplant not occurring in Britain, where *Echium vulgare*, viper's bugloss, is utilized.

550 **silenella** Herrich-Schäffer, 1855 – *Silene*, the campion genus, to which the larval foodplants belong.

551 **otitae** Zeller, 1839 – *Silene otites*, Spanish catchfly: the recorded foodplant in Britain is *S. nutans*, Nottingham catchfly.

552 **lassella** Staudinger, 1859 – *lassus*, weary, faint: Staudinger states that it is similar to 587 *C. caespititiella*, but the ground colour is a much fainter loamy yellow.

553 **striatipennella** (Nylander, [1848]) – *striatus*, grooved, streaked; *pennae* (pl.), a wing: from the ochreous streaks along the veins of the forewing which become fuscous towards the apex. The termination *-pennella* suggests that Nylander may have had *Motacilla striata* Pallas, 1764, the spotted flycatcher, in mind.

554 **inulae** Wocke, 1876 – *Inula*, the present genus of *I. conyza*, ploughman's spikenard, and the former genus of *Pulicaria dysenterica*, common fleabane, both of which are foodplants.

555 **follicularis** (Vallot, 1802) – *folliculus*, a small bag: from the larval case.

= **troglodytella** (Duponchel, 1843) – τρωγλοδύτης (trōglodutēs), one who creeps into holes, a cave-dweller, *Motacilla Troglodytes* Linnaeus, 1758 (now *Troglodytes troglodytes troglodytes*), the wren: the most likely explanation is that the name is based on an analogy between the domed nest of the bird and the larval case, but Duponchel gives no indication that he knew the life history. However, he ascribes the authorship of the name to Fischer von Röslerstamm who may have done so.

556 **trochilella** (Duponchel, 1843) – τροχίλος (trokhilos), a small bird, possibly a wagtail or wren, *Motacilla Trochilus* Linnaeus, 1758, the willow-warbler: this name has the same history as 555 *troglodytella* above; Duponchel ascribes the name to Fischer von Röslerstamm and does not himself describe the life history, but here too there may be an analogy between the larval case and the domed nest, on the assumption that Fischer von Röslerstamm had reared the moth.

556a **linosyridella** Fuchs, 1881 – *Linosyris vulgaris*, goldilocks: the reported foodplant on the Continent, but in Britain it is found on sea-aster (*Aster tripolium*).

557 **gardesanella** Toll, 1954 – from the type locality which is San Vigilio, on the shores of Lake Garda, Italy.

= **machinella** Bradley, 1971 – i.h.o. W. Machin (1822–1894), British entomologist who discovered the species; the name he then gave it, *maritimella*, was preoccupied by 585 *maritimella* Newman.

558 **ramosella** Zeller, 1849 – *ramosus*, branching, branched: from the forewing pattern of pale, branched veins on a darker ground colour.

559 **peribenanderi** (Toll, 1952) – περί (peri), round about, close to; *benanderi* Kanerva, 1941, a synonym of 565 *saxicolella* (Duponchel, 1843), q.v.: drawing attention to close similarity and at the same time punning on Benander's Christian name 'Per'.

560 **paripennella** Zeller, 1839 – *parus*, a tit; *pennae* (pl.), a wing: any resemblance is remote.

561 **therinella** Tengström, 1848 – θερινός (therinos, adj.), summer; from the time of appearance of the adult.

562 **asteris** Mühlig, 1864 – *Aster tripolium*, sea-aster: the larval foodplant.

563 **argentula** (Stephens, 1834) – *argentum*, silver: from the forewing, which has silvery white streaks along the veins contrasting with the ochreous ground colour.

564 **virgaureae** Stainton, 1857 – *Solidago virgaurea*, goldenrod: the larval foodplant.

565 **saxicolella** (Duponchel, 1843) – *saxum*, a rock; *colo*, to inhabit: the species is widely distributed and not necessarily associated with rocky situations. This was another name received from Fischer von Röslerstamm (cf. 555 and 556).

= **benanderi** Kanerva, 1941 – i.h.o. P. Benander (1885–1976), a Swedish entomologist (cf. 559).

566 **sternipennella** (Zetterstedt, 1839) – *sterna*, a tern; *pennae* (pl.), a wing: the only possible resemblance between the moth and bird is in their elongate wings, but it was the generic name *Ornix* (see generic introduction) that influenced Zetterstedt's choice.

= **flavaginella** Lienig & Zeller, 1846 – *flavus*, yellow; the termination *-ago*, see 2271 *citrago*: from the yellowish markings on the forewing.

= **moeniacella** Stainton, 1887 – see under 574 *C. deviella*, the species to which this name has been applied by British authors.

567 **adspersella** Benander, 1939 – *adspersus*, sprinkled: from the fuscous scales scattered over the forewing.

568 **versurella** Zeller, 1849 – *versura*, a turning: Zeller gives the antennal annulations as an important diagnostic character and the name may refer to these rings or their alternation between white and fuscous. 'Perhaps from larva's habit of rotating its case so as to keep it at constant angle to surface of leaf to which it is attached' (Macleod): a triple blunder! The larva feeds on seeds and not leaves, a phyllophagous coleophorid larva spins its case to the leaf and cannot rotate it and Zeller did not know the life history.

569 **squamosella** Stainton, 1856 – *squamosus*, scaly: from the white scales scattered over the forewing.

570 **pappiferella** Hofmann, 1869 – *pappus*, the hairy seed of a plant; *fero*, to carry: the larva feeds in a seedhead of mountain everlasting (*Antennaria dioica*) and decorates its case with the pappus hairs.

571 **granulatella** Zeller, 1849 – *granum*, grain, a seed: from the pabulum. There is no reliable evidence for the occurrence of this species in Britain.

572 **vestianella** (Linnaeus, 1758) – *vestis*, a garment; *'habitat in vestimentis, quae destruit'*, it lives in clothes, which it destroys: although Linnaeus bestowed this name on a clothes moth, his specimens got muddled and the type turns out to be this coleophorid; the rules of the I.C.Z.N. require the type to take precedence over the description.

= **laripennella** (Zetterstedt, 1839) – *larus*, a sea-gull: see 566 *sternipennella*.

= **annulatella** Nylander, 1848 – *annulatus*, ringed, formed from *anulus* (*annulus*), a small ring: from the antennal annulations.

573 **atriplicis** Meyrick, 1928 – *Atriplex*, the orache genus, which includes the larval foodplants.

574 **deviella** Zeller, 1849 – *devius*, that which lies off the high-road (*'via'*), out of the way: possibly descriptive of the type locality which could have been a remote salt-marsh, but Zeller gives no explanation.

= **suaedivora** Meyrick, 1928 – *Suaeda maritima*, annual sea-blite; *voro*, to devour: from the larval foodplant.

= **moeniacella** sensu auctt. – Stainton first named a species in this group *muehligella* in honour of his friend G. C. Mühlig, but on learning that the name was preoccupied he rechristened it *moeniacella* from Francofurtum Moeniacum, Frankfurt-on-Main, which was where Mühlig lived (Stainton, 1887).

574a **aestuariella** Bradley, 1984 – *aestuarium*, a tide-place, an estuary: from the habitat. This derivation *has* to be right, since I myself proposed the name to Dr Bradley!

575 **salinella** Stainton, 1859 – *salinus*, pertaining to salt: from the habitat on salt-marshes.

576 **artemisiella** Scott, 1861 – *Artemisia maritima*, sea-wormwood, the larval foodplant.

577 **artemisicolella** Bruand, 1855 – *Artemisia vulgaris*, mugwort, the larval foodplant; *colo*, to inhabit.

578 **murinipennella** (Duponchel, 1844) – *murinus*, pertaining to the colour of mice; *pennae* (pl.), a wing: the forewing is ochreous grey.

579 **antennariella** Herrich-Schäffer, 1861 – *antenna*, a sail-yard, a yard-arm, thence the antenna of an insect: from the distinctive antenna which is whitish, ringed dark fuscous, and has the scape rough-scaled.

580 **sylvaticella** Wood, 1892 – *silvaticus* (*sylvaticus*), pertaining to a wood: a pun on the habitat and the foodplant – 'flies in woods (hence the name), in May. Larva on *Luzula sylvatica*'. I am tempted to suggest that Wood was also indulging in word-play on his own name – a woodland species feeding on wood-rush and discovered by Wood.

581 **taeniipennella** Herrich-Schäffer, 1854 – *taenia*, a band or fillet, a streak in paper (Pliny), etc.; pennae (pl.), a wing: from the forewing which has the veins marked with fine white lines.

582 **glaucicolella** Wood, 1892 – *Juncus glaucus*, now called *J. inflexus*, hard rush: an important larval foodplant.

583 **cratipennella** Clemens, 1864 – *crates*, a hurdle, a lattice; *pennae* (pl.), a wing: the wing is, in fact, streaked rather than reticulate.

= **tamesis** Waters, 1929 – *Tamesis*, the R. Thames: from the type locality (under this name) at

Binsey, near Oxford, where Waters found the species in a water-meadow beside the R. Thames.

584 **alticolella** Zeller, 1849 – *altus*, high; *colo*, to inhabit: Zeller gives the type locality as Schneeberge ('snow mountain'), but the species occurs at all elevations. 'Perhaps from larval case, which is usually attached to seed-head high up on rush' (Macleod).

585 **maritimella** Newman, 1873 – *Juncus maritimus*, sea-rush: the larval foodplant.

586 **adjunctella** Hodgkinson, 1882 – *adjunctus*, joined to: *Juncus*, rush: probably a pun, since the larva feeds with its case attached to a seed-head of *Juncus gerardi*, saltmarsh rush.

587 **caespititiella** Zeller, 1839 – *caespes, caespitis*, turf: referring loosely to the habitat; Zeller found it in woodland glades and it occurs in Britain in similar situations, feeding on various rushes (*Juncus* spp.).

588 **salicorniae** Wocke, 1876 – *Salicornia*, the glasswort genus, that of the larval foodplants.

589 **clypeiferella** Hofmann, 1871 – *clipeus, clypeus*, a round shield; *fero*, to bear: from the ring of chitinous spines on abdominal segment 1 of the adult, used to cut its way out of its cocoon and also, perhaps, to help it on its way out of the soil prior to expanding its wings. Macleod supposed that the abdomen was crested.

ELACHISTIDAE (593)

Perittia Stainton, 1854 – περισσός, περιττός (perissos, perittos), prodigious, strange: from Stainton's puzzlement over the proper place in the systematic list of the next species; it was a misfit and he first described it as a scythrid. The explanation offered by Macleod '*peritteia* [περιττεία], abundance, from prevalence' is certainly wrong because Stainton regarded the moth as a rarity and περιττεία is a hapax legomenon from the New Testament (2 Corinthians 8: 2), unlikely to be used as a name source.

590 **obscurepunctella** (Stainton, 1848) – *obscurus*, obscure; *punctum*, a spot: from the obsolescent discal stigma.

Mendesia Joannis, 1902 – i.h.o. M. C. Mendes (1874–1944), a French entomologist 'who discovered it' (Macleod). Genera cannot be 'discovered' and Mendes was born 80 years after the next species was named.

591 **farinella** (Thunberg, 1794) – *farina*, flour: from the white forewing.

Stephensia Stainton, 1858 – i.h.o. J. F. Stephens (1792–1852), the British entomological author.

592 **brunnichella** (Linnaeus, 1767) – i.h.o. M. T. Brünnich (18th century), a Danish entomologist; cf. 1155.

Elachista Treitschke, 1833 – ελάχιστος (elachistos), very small: though the species are small, many of the Nepticulidae, Lyonetiidae, etc., are smaller.

593 **regificella** Sircom, 1849 – *regificus*, royal, magnificent: from the golden-metallic markings on the forewing.

594 **gleichenella** (Fabricius, 1781) – i.h.o. W. F. von Gleichen (1717–83), a Viennese naturalist.

595 **biatomella** (Stainton, 1848) – *bi-*, two; *atomus*, something indivisible, too small to divide, an atom: from the two black spots on the forewing.

596 **poae** Stainton, 1855 – *Poa aquatica*, now *Glyceria maxima*, reed sweet-grass, reed-grass: the larval foodplant.

597 **atricomella** Stainton, 1849 – *ater*, black; *coma*, the hair of the head: from the dark brown (not black) scales on the vertex. The name draws attention to affinity with 600 *E. luticomella* which has a very similar life history.

598 **kilmunella** Stainton, 1849 – from Kilmun, Argyll, Scotland, the type locality.

598a **eskoi** Kyrki & Karvonen, 1985 – i.h.o. Professor Esko Suomalainen, the Honorary President of the Finnish Lepidopteral Society and the collector of one of the paratypes.

599 **alpinella** Stainton, 1854 – *alpinus*, pertaining to high mountains, the Alps: Stainton at first thought that this species was a variety of 598 *E. kilmunella*, q.v., which he had taken in a mountainous part of Scotland.

600 **luticomella** Zeller, 1839 – *luteus*, yellow; *coma*, the hair of the head: from the yellow scales on the vertex.

601 **albifrontella** (Hübner, 1817) – *albus*, white; *frons*, *frontis*, the forehead: from the white head of the adult.

602 **apicipunctella** Stainton, 1849 – *apex*, the tip of the wing; *punctum*, a spot: from a small silvery white subapical spot usually present on the forewing.

603 **subnigrella** Douglas, 1853 – *sub-*, almost, somewhat; *niger*, black: from the blackish forewing.

604 **orstadii** Palm, 1943 – i.h.o. E. T. Orstadius (1861–1939), a Swedish entomologist.

605 **pomerana** Frey, 1870 – Pomerania, a district now divided between East Germany and Poland: the type locality.

606 **humilis** Zeller, 1850 – *humilis*, literally on the ground, hence lowly, humble: either sense, or both, may have been intended.

607 **canapennella** (Hübner, 1813) – *canus*, hoary, grey; *pennae* (pl.), a wing: from the coloration of the male forewing.

= **pulchella** (Haworth, 1828) – *pulcher*, beautiful: from the markings of the female which are much more attractive than those of the male.

= **obscurella** Stainton, 1849 – *obscurus*, indistinct, obscure: from the obsolescent markings on the male forewing.

608 **rufocinerea** (Haworth, 1828) – *rufus*, reddish; *cinereus*, ashen: from the colour of the forewing.

609 **cerusella** (Hübner, 1796) – *cerussa*, white lead, ceruse: from the colour of the forewing.

610 **argentella** (Clerck, 1759) – *argentum*, silver: from the colour of the forewing.

611 **triatomea** (Haworth, 1828) – *tri-*, three; *atomus*, a speck (see 595 *biatomella*): from the forewing which has, however, only two distinct dark spots.

612 **collitella** (Duponchel, 1843) – *collitus*, smeared: from the suffused fasciae on the forewing.

613 **subocellea** (Stephens, 1834) – *sub-*, somewhat; *ocellus*, a small eye: from the scattered brown scales on the forewing which are sometimes grouped so as to form a pattern vaguely suggesting an ocellus.

614 **triseriatella** Stainton, 1854 – *tri-*, three; *series*, a row: from the scattered black scales on the forewing which are sometimes arranged in three longitudinal rows.

615 **dispunctella** (Duponchel, 1843) – *dis-*, apart, in dispersion; *punctum*, a spot: from the black scales scattered over the white forewing.

616 **bedellella** (Sircom, 1848) – i.h.o. G. Bedell (1805–77), a British entomologist.

616a **littoricola** Le Marchand, 1938 – *litus*, *litoris* (*littus*), the shore; *colo* to inhabit: the type localities are round the estuary of the R. Garonne on the French Atlantic coast.

617 **megerlella** (Hübner, 1810) – i.h.o. J. C. Megerle von Mühlfeld (1765–1832), a German entomologist.

618 **cingillella** (Herrich-Schäffer, 1855) – *cingillum*, a small girdle: from the straight median fascia on each forewing; when the wings are folded the two merge to resemble a girdle.

619 **unifasciella** (Haworth, 1828) – *unus*, one; *fascia*, a band: from the single yellowish white fascia on each forewing.

620 **gangabella** Zeller, 1850 – *gangaba*, a porter, a Persian word: Pickard *et al.* and Macleod give this derivation, the latter adding 'seems pointless'. The name was suggested to Zeller by Fischer von Röslerstamm, so it was not incumbent on him to explain it. Neither entomologist knew the life history, so if there is a meaning, it applies to the adult.

621 **subalbidella** Schläger, 1847 – *sub-*, somewhat; *albidus*, white: from the ochreous white forewing.

622 **revinctella** Zeller, 1850 – *revinctus*, bound: as in 618 *cingillella* and 619 *unifasciella*, the median fascia gives the appearance of a belt when the wings are folded.

= **adscitella** Stainton, 1851 – *adscitus*, enrolled, admitted: accepted as a species distinct from 617 *E. megerlella* and accordingly added to the systematic list. Stainton may have remembered the 'Adscitae' of Linnaeus (163, genus).

623 **bisulcella** (Duponchel, 1843) – *bi-*, two; *sulcus*, a furrow, track or trail: the 'furrows' are the fasciae, one on each forewing.

Biselachista Traugott-Olsen & Nielsen, 1977 – *bis-*, twice; *Elachista*, the genus, q.v.: a second genus within the Elachistidae.

624 **trapeziella** (Stainton, 1849) – τραπέζιον (trapezion), a small table, an irregular four-sided figure, a trapezium: from the figure that would be formed if the spots on the forewing were

connected to each other.

625 **cinereopunctella** (Haworth, 1828) – *cinereus*, ashen; *punctum*, a spot; from the greyish white spots on the forewing.

626 **serricornis** (Stainton, 1854) – *serra*, a saw; *cornu*, a horn, the antenna: from the serrate antenna of the adult.

627 **scirpi** (Stainton, 1887) – *Scirpus maritimus*, sea club-rush: the larval foodplant.

628 **eleochariella** (Stainton, 1851) – *Eleocharis*, the spike-rush genus, that of the larval foodplants.

629 **utonella** (Frey, 1856) – from the Latin name of Mt. Uetliberg, near Zürich, the type locality.

630 **albidella** (Nylander, 1848) – *albidus*, whitish: from the ground colour.

Cosmiotes Clemens, 1860 – κοσμιότης (kosmiotēs), propriety, decorum: this name, together with that of the type species, *C. illectella* Clemens, 1860 (*illectus*, alluring) shows that Clemens was much attracted by the genus.

631 **freyerella** (Hübner, 1825) – i.h.o. C. F. Freyer (1794–1885), a German entomologist.

632 **consortella** (Stainton, 1851) – *consors*, a comrade: from similarity to *C. freyerella* (= *nigrella* Hübner *nec* Fabricius), 'very like *Nigrella*, but the fascia placed more obliquely'.

633 **stabilella** (Stainton, 1858) – *stabilis*, fixed, stable: from lack of variation.

OECOPHORIDAE (651)

OECOPHORINAE (651)

Schiffermuelleria Hübner, 1825 – i.h.o. I. Schiffermüller (1727–1809), the distinguished Austrian entomological author who collaborated with Denis (see 638, genus).

634 **grandis** (Desvignes, 1842) – *grandis*, large, grand, sublime: from the spectacular markings, not from the size which is not exceptional.

635 **subaquilea** (Stainton, 1849) – *subaquilus*, brownish: from the colour of the forewing.

636 **similella** (Hübner, 1796) – *similis*, like: from resemblance to 649 *Esperia sulphurella*, not what one would expect but it is what Hübner says.

637 **tinctella** (Hübner, 1796) – *tinctus*, dyed: from the golden ochreous colour of the forewing, perhaps suggesting saffron dye.

Denisia Hübner, 1825 – i.h.o. M. Denis (1729–1800), the Austrian entomologist who collaborated with Schiffermüller (see 634, genus).

= **Chambersia** Riley, 1891 – i.h.o. V. T. Chambers (1831–83), an American entomologist.

638 **augustella** (Hübner, 1896) – *augustus*, majestic: from the pattern of the forewing which is distinctive and attractive, but hardly majestic; Macleod accordingly suggested 'from month of moth's occurrence', but this is even less likely since the moth, at any rate in Britain, appears in June.

638a **albimaculea** (Haworth, 1828) – *albus*, white; *macula*, a spot: from the pattern of the forewing.

Bisigna Toll, 1956 – *bi-*, two; *signum*, a marking: the most prominent feature in the pattern of the next species is the double silver fascia on the forewing.

639 **procerella** ([Denis & Schiffermüller], 1775) – *procer*, *proceris*, a chief, a noble: from the spectacular orange, silver-metallic and black forewing.

Batia Stephens, 1834 – βάτος (batos), a thorn-bush: perhaps because 641 *B. lambdella* 'was discovered in a furze bush' (Donovan, 1793). 'Larva occurs on hawthorn' (Macleod): factually incorrect for any species in the genus. Alternatively from Batia, daughter of Teucer, founder of the royal house of Troy. Stephens, who gives no explanation, may have been punning on both senses.

640 **lunaris** (Haworth, 1828) – *lunaris*, crescent-shaped: from the shape of the pretornal blackish mark on the forewing.

641 **lambdella** (Donovan, 1793) – λάμβδα (lambda), the Greek letter 'λ': from the shape of the pretornal blackish mark on the forewing; cf. 2032 *Arctornis l-nigrum*, but there the capital lambda, 'Λ', is intended.

641a **internella** Jäckh, 1972 – *inter*, between; n, phonetic insertion; *-ella*, the family termination: from the size of the adult which is intermediate between 640 *B. lunaris* and 641 *B. lambdella*.

642 **unitella** (Hübner, 1796) – *unitas*, unity: from the unicolorous forewing.

Dafa Hodges, 1974 – a meaningless neologism (R. W. Hodges, pers. comm.).

643 **formosella** ([Denis & Schiffermüller], 1775) – *formosus*, beautiful: from the attractive orange and ferruginous forewing.

Borkhausenia Hübner, 1825 – i.h.o. M. B. Borkhausen (1732–1807), a German entomologist.

644 **fuscescens** (Haworth, 1828) – *fuscescens*, tending to be dusky: from the fuscous forewing.

645 **minutella** (Linnaeus, 1758) – *minutus*, small: from its small size in relation to the species of 'Phalaena (Tinea)' already named by Linnaeus in 1758.

Telechrysis Toll, 1956 – τέλος (telos), the end; χρύσος (khrusos), gold: this should apply to the apex of the forewing, but Toll's type species has white-tipped cilia and so too has the next species.

646 **tripuncta** (Haworth, 1828) – *tri*, three; *punctum*, a spot: from the three white spots on the forewing.

Hofmannophila Spuler, 1910 – i.h.o. O. Hofmann (1835–1900), a German entomologist; φίλος (philos), loved, loving: it seems like a sick joke to suggest that Hofmann had any feeling other than aversion for 647 *H. pseudospretella!*

647 **pseudospretella** (Stainton, 1849) – ψεῦδος, ψευδο- (pseudos, pseudo-), falsehood, false; *Tinea spretella* [Denis & Schiffermüller] (? = 237 *Niditinea fuscella* (Linnaeus)), from *spretus*, despised: from its resemblance in habits and, to some extent, in appearance, to an unpopular clothes moth.

Endrosis Hübner, 1825 – ἔνδροσος (endrosos), bedewed: probably with reference to the wing pattern of certain species, but not the next which was not included in the genus by Hübner.

648 **sarcitrella** (Linnaeus, 1758) – σάρξ, σαρκός (sarx, sarkos), flesh: Linnaeus describes it only as a clothes moth, but it could eat the fur, wool or feathers while still attached to the corpse; or, by a stretch of the imagination, the whitish ochreous ground colour could be seen as flesh-coloured. Macleod derives from σαρκιτης (sarkitēs), an unknown precious stone, 'from mark near base of hindwings'; there is a barely visible hyaline patch in this position but it is not mentioned by Linnaeus.

Esperia Hübner, 1825 – i.h.o. E. J. C. Esper (1742–1810), a German entomologist.

649 **sulphurella** (Fabricius, 1775) – *sulfureus*, *sulphureus*, of sulphur: from the yellow hindwing.

650 **oliviella** (Fabricius, 1794) – i.h.o. G. A. Olivier (1756–1814), a French entomologist.

Oecophora Latreille, 1796 – οἶκος (oikos), a house; φορέω (phoreō), to carry: from the portable larval case borne by species in other genera, embraced by this name in its original usage for the whole of the Oecophoridae.

651 **bractella** (Linnaeus, 1758) – *bractea*, gold-leaf: from the extensive golden yellow pattern on the thorax and forewings.

Alabonia Hübner, 1825 – *ala*, a wing; *bonus*, good: from the spectacular coloration and pattern of the next species. Hübner's formation of a generic name from the Latin is so unusual as to throw doubt on my interpretation.

652 **geoffrella** (Linnaeus, 1767) – i.h.o. E. L. Geoffroy (1727–1810), a prominent French entomologist.

Aplota Stephens, 1834 – ἁπλότης (haplotēs), simplicity: from the relatively simple forewing pattern of the next species, described by Stephens as being 'of a plain, dingy hue, nearly destitute of any markings'. Stephens has dropped his aitch; cf. 668 *Enicostoma*, etc.

653 **palpella** (Haworth, 1828) – *palpus*, a stroking, whence the palpus of an insect regarded as an organ of touch: from the greatly elongate labial palpus.

Pleurota Hübner, 1825 – πλευρά (pleura), a rib, the side of a geometrical figure or the anterior margin of an insect's wing: from the conspicuously marked costa of the next species.

654 **bicostella** (Clerck, 1759) – *bi-*, twice; *costa*, the anterior margin of a wing: from the white costal and brown subcostal streak, giving an impression of two costas.

655 **aristella** (Linnaeus, 1767) – ἄριστος (aristos), best: there is no reliable evidence for the occurrence of this striking species in Britain.

Parocystola Turner, 1896 – παρά (para), beside, akin to; *Ocystola* Meyrick, 1885, an Australasian genus.

656 **acroxantha** (Meyrick, 1885) – ἄκρον (akron), the tip, extremity: ξανθός, (xanthos), yellow: from the yellow terminal cilia of the forewing.

Hypercallia Stephens, 1829 – ὑπέρ (huper), beyond measure; κάλλος (kallos), beauty: from the striking beauty of the next species.

657 **citrinalis** (Scopoli, 1763) – *citrus*, a citrus tree, such as the orange or lemon: from the forewing which is lemon-yellow with orange-crimson markings.

Carcina Hübner, 1825 – καρκίνος (karkinos), a crab, an ulcer: resemblance in either sense is obscure; Hübner included 657 *citrinalis* in the genus and refers to the bright colours of the moths in his diagnosis, so perhaps their purplish and yellow patterns suggested an ulcerous sore. The explanation of Macleod 'from flat body' may be dismissed because 'flat-body' belongs to the Depressariinae, not the Oecophorinae.

658 **quercana** (Fabricius, 1775) – *Quercus*, the oak genus: this polyphagous species includes oak amongst its foodplants.

Amphisbatis Zeller, 1870 – ἀμφισβητέω, ἀμφισβατέω (amphisbēteō, amphisbateō), to disagree: echoing the sense of 659 *incongruella*.

659 **incongruella** (Stainton, 1849) – *incongruus*, incongruous: described by Stainton in the genus *Butalis* Treitschke (now 911 *Scythris* Hübner), where it was a misfit; later he transferred it to *Oecophora* Latreille, where he still found it incongruous.

Pseudatemelia Rebel, 1910 – ψεῦδος (pseudos), falsehood, ψευδο- (pseudo-), false; 448 *Atemelia* Herrich-Schäffer, a genus of the Yponomeutidae: however, the genera have little in common.

660 **josephinae** (Toll, 1956) – named by Graf S. von Toll (1893–1961) i.h.o. his wife.

661 **flavifrontella** ([Denis & Schiffermüller], 1775) – *flavus*, yellow; *frons*, *frontis*, forehead: from the yellowish head of the adult.

662 **subochreella** (Doubleday, 1859) – *sub-*, somewhat; *ochra*, yellow earth, ochre: from the coloration of the forewing.

CHIMABACHINAE (663)

Diurnea Haworth, 1811 – *diurnus*, pertaining to the day: from the diurnal flight of 664 *D. phryganella*.

= **Chimabache** Hübner, 1825 – χεῖμα (kheima), winter; βάκχη (bakkhē), a bacchante, a reveller: from 664 *D. phryganella* which flies actively by day in late autumn. Zeller (1839) emended the spelling to *Chimabacche*.

663 **fagella** ([Denis & Schiffermüller], 1775) – *Fagus sylvatica*, the beech: one of the foodplants of this polyphagous species.

664 **phryganella** (Hübner, 1796) – *Phryganea*, a genus of the Phryganeidae, a family of caddis-flies (Trichoptera): from a supposed resemblance.

Dasystoma Curtis, 1833 – δασύς (dasus), thick, hairy; στόμα (stoma), the mouth: from the densely scaled labial palpus.

= **Cheimophila** sensu auctt. – χεῖμα (kheima), winter; φιλέω (phileō), to love: from the appearance in October–November of 664 *Diurnea phryganella*. *Cheimophila* Hübner, 1825 is a junior synonym of *Diurnea* Haworth, 1811.

665 **salicella** (Hübner, 1796) – *Salix*, the sallow and willow genus: this polyphagous species sometimes feeds on sallow.

DEPRESSARIINAE (669)

Semioscopis Hübner, 1825 – σημειοσκόπος (sēmeioskopos), a diviner: in his generic diagnosis Hübner mentions letter-like markings on the forewings of some species and he may have regarded these fancifully as runes.

666 **avellanella** (Hübner, 1793) – *Corylus avellana*, hazel: this is a possible foodplant, though not recorded as such in Britain.

667 **steinkellneriana** ([Denis & Schiffermüller], 1775) – i.h.o. Professor Steinkellner (18th century), a Jesuit priest and Viennese entomologist.

Enicostoma Stephens, 1829 – ἑνικός (henikos), singular, unique; στόμα (stoma), the mouth: from the labial palpus which is very long, segment 2 being thickened with dense scales and the tip of segment 3 being exposed. Stephens should have written '*Henicostoma*'; cf. 653 *Aplota*.

668 **lobella** ([Denis & Schiffermüller], 1775) – λοβός (lobos), a lobe: from the ovate forewing. '*Loba*, straw, from black tufts on forewings' (Macleod); unlikely, since straw is not black.

Depressaria Haworth, 1811 – *depressus*, pressed down, flat: a name given for two reasons, (1) from the flattened abdomen, Haworth being the originator of the vernacular name 'flat-body', and (2) from the way in which the wings are held flat in the resting position, this being the source of the names 683 *depressana* and 688 *applana*, qq.v. Haworth wrote '*corpus valde depressum, alae incumbentes, etiam depressae*' (the body strongly flattened, the wings resting upon the body, also held flat).

669 **discipunctella** Herrich-Schäffer, 1854 – *discus*, a quoit, the central or discal area of an insect's wing; *punctum*, a spot: from the elongate dark spots in the disc of the forewing.

670 **daucella** ([Denis & Schiffermüller], 1775) – *Daucus carota*, wild carrot: the foodplant of several related species but not this one.

671 **ultimella** Stainton, 1849 – *ultimus*, last, furthest: from the dark terminal area of the forewing, described by Stainton as 'entirely fuscous'.

672 **pastinacella** (Duponchel, 1838) – *Pastinaca sativa*, wild parsnip, one of the foodplants.
= **heracliana** sensu auctt. – see 688.

673 **pimpinellae** Zeller, 1839 – *Pimpinella*, the burnet saxifrage genus, which contains the larval foodplants.

674 **badiella** (Hübner, 1796) – *badius*, brown, bay-coloured: from the coloration of the forewing.
= 675 **brunneella** Ragonot, 1874 – *brunus* (late Lat.), brown: from the coloration of the forewing.

676 **pulcherrimella** Stainton, 1849 – *pulcherrimus*, very beautiful: in approbation.

677 **douglasella** Stainton, 1849 – i.h.o. J. W. Douglas (1814–1905), a British entomologist.

678 **weirella** Stainton, 1849 – i.h.o. J. J. Weir (1822–94), a British entomologist.

679 **emeritella** Stainton, 1849 – *emeritus*, a veteran, one who has served well: a praiseworthy species, distinguished for its yellow head.

680 **aegopodiella** (Hübner, 1826) – *Aegopodium podagraria*, ground elder: this, however, is not listed as a foodplant.
= **albipunctella** sensu auctt. – *albus*, white; *punctum*, a spot: from the conspicuously white second discal stigma.

681 **olerella** Zeller, 1854 – *olus*, *oleris*, a name applied to various garden herbs and vegetables: the actual foodplant, yarrow (*Achillea millefolium*), was unknown to Zeller in 1854.

682 **chaerophylli** Zeller, 1839 – *Chaerophyllum temulentum*, rough chervil: the larval foodplant.

683 **depressana** (Fabricius, 1775) (= *depressella* (Fabricius, 1798)) – *depressus*, flattened: a name whose history epitomizes the movement away from the Linnaean doctrine that the classification of moths should be based on the way in which their wings are folded in repose. According to Linnaeus, Noctuae rest with '*alis depressis*', with wings pressed down upon the body, Tineae with '*alis convolutis fere in cylindrum*', with involute wings, and Tortrices with '*alis obtusissimis ut fere retusis, planiusculis*', with wings so very blunt as to be almost retuse (see 2311) and flattish (Pl. III). The depressariines do not wrap their wings round the abdomen, so they were excluded from Tinea; they clearly were not Noctuae although they rest '*alis depressis*' and so by default they were placed in Tortrix (or Pyralis *usu* Fabricius, see p. 29) with the termination *-ana*. Fabricius probably had misgivings as early as 1775 and gave a name descriptive of the wing-folding to justify his taxonomic placing. By 1798 he and Latreille had broadened the basis of classification and he reallocated the moths to Tinea with the appropriate family termination, a change prohibited by future I.C.Z.N. rules. Later, when the terminology descriptive of wing-folding had been forgotten, entomologists wrongly supposed that the name referred to the flat body, also characteristic of the group.

684 **silesiaca** Heinemann, 1870 – *Silesiacus*, belonging to Silesia, a region partly in Czechoslovakia and partly in Poland, where the moth was first found.

Levipalpus Hannemann, 1953 – *lēvis*, smooth; *palpus*, the labial palpus: from the labial palpus which is more smoothly scaled than in other Depressariinae.

685 **hepatariella** (Lienig & Zeller, 1846) – *hepatarius*, pertaining to the liver: from the liver-coloured forewing.

Exaeretia Stainton, 1849 – ἐξαίρετος (exairetos), singular, remarkable: from the unusually shaped forewing of 687 *E. allisella*, at first the only species in the genus.

686 **ciniflonella** (Lienig & Zeller, 1846) – *cinis*, ash; *flo*, to blow; hence *ciniflo*, *ciniflonis*, a hair-curler, a heated curling-iron operated by the *cinerarius*, a slave who placed it in ashes which he stimulated by blowing. The explanation given by Macleod, 'from larva's occurrence in curled up leaves of birch', is impossible, since the life history was not known until many years after the species was named. Zeller ascribes the authorship to Madam Lienig who perhaps, being stumped for a name, mischievously called it after the hair-curler she was using at the time; 1181 *grandaevana* appears to be another whimsical name from the same authors. A more prosaic explanation is that the forewing has grey irroration as if ash had been blown over it and that the name is a play on words; however, there is no hint of this in the original description.

687 **allisella** Stainton, 1849 – i.h.o. T. H. Allis (1817–70), a Yorkshire entomologist who discovered the species.

Agonopterix Hübner, 1825 – ἀγώνιος (agōnios), without angle: πτέρυξ (pterux), a wing: from the rounded apex and termen of the forewing.

688 **heracliana** (Linnaeus, 1758) – *Heracleum sphondylium*, hogweed, a larval foodplant; *-ana*, the Tortrix termination: the resting posture with the wings held flat (*alis depressis*) excluded it from Tinea. Linnaeus' description belongs to 672 *Depressaria pastinacella*, a species which bore the name for 200 years; then it was discovered that the type specimen in Linnaeus' collection (held by the Linnean Society of London) was this species. Since the type specimen takes precedence over the description, the name was transferred.

= **applana** (Fabricius, 1777) – *ad*, towards, upon, nearly; *planus*, flat: a name given for the same reason as 683 *depressana*, drawing attention to the flatly held wings in the resting posture which required it to be placed in Tortrix (*-ana*) (Pyralis *usu* Fabricius). Macleod wrongly supposed that both the names referred to the 'flat body', the source of the English vernacular name of later origin and, in part, of the generic name 669 *Depressaria*, q.v. This is really the oldest name for this species, since *heracliana* was never intended for it.

689 **ciliella** (Stainton, 1849) – *cilium*, an eye-lash, *cilia* (pl.), the fringe on the wings of Lepidoptera: from the hindwing cilia, which are reddish-tinged.

690 **cnicella** (Treitschke, 1832) – κνῆκος (knēkos), Lat. *cnicus*, safflower (*Carthamus tinctoria*), an oriental member of the thistle family yielding red dye: the foodplant, at any rate in Britain, is sea-holly (*Eryngium maritimum*), so the reference is probably to the reddish ground colour. However, *Cnicus* is an old generic name for other thistle species, so the name may refer to a foodplant after all.

691 **purpurea** (Haworth, 1811) – *purpureus*, purple-coloured: from the crimson-purple forewing.

692 **subpropinquella** (Stainton, 1849) – *sub-*, near; 696 *A. propinquella*, q.v.

693 **putridella** ([Denis & Schiffermüller], 1775) – *putridus*, decaying, the colour of dead wood: the authors describe the moth as 'holzbraunlichter', the colour of light brown wood; cf. 2098 *putris* and 2206 *putrescens*. Macleod's 'from brown veins on forewings' misses the point: it is the ground colour that is described.

694 **nanatella** (Stainton, 1849) – *nanus*, a dwarf: from its relatively small size.

695 **alstromeriana** (Clerck, 1759) – i.h.o. C. Alströmer (d. 1792), a Swedish naturalist and, like Clerck, a pupil of Linnaeus. For the termination *-ana* see 683 *depressana*.

696 **propinquella** (Treitschke, 1835) – *propinquus*, near: from resemblance to another species, perhaps 702 *A. assimilella*, already named by Treitschke.

697 **arenella** ([Denis & Schiffermüller], 1775) – *arena*, sand: from the sandy colour of the forewing.

697a **kuznetzovi** Lvovsky, 1983 – i.h.o. either N. Y. Kuznetzov (1873–1948) or V. I. Kuznetzov (contemporary), both Russian entomologists.

698 **kaekeritziana** (Linnaeus, 1767) – Linnaeus included this species in Tortrix, the family he had chosen to receive names given to commemorate his students (see p. 20). The type locality

being Uppsala where Linnaeus taught, it is all the more likely that Kaekeritz was a deserving student and perhaps also the discoverer of the species.

= **liturella** ([Denis & Schiffermüller], 1775) – *litura*, a smear, a blot: from the rather ill-defined, diffused discal spot on the forewing.

699 **bipunctosa** (Curtis, 1850) – *bi-*, two; *punctum*, a dot: from the two subbasal dots, one on each forewing.

700 **pallorella** (Zeller, 1839) – *pallor*, paleness; from the pale ochreous forewing.

701 **ocellana** (Fabricius, 1775) – *ocellus*, a small eye: from the red-ringed, white second discal stigma on the forewing.

702 **assimilella** (Treitschke, 1832) – *assimilis*, similar: from similarity to another species, perhaps 697 *A. arenella* or 703 *A. atomella*.

703 **atomella** ([Denis & Schiffermüller], 1775) – *atomus*, an atom, a speck: referring either to the minute first discal stigma or to the sparse irroration of blackish scales on the forewing.

= **pulverella** (Hübner, 1825) – *pulvis, pulveris*, dust: from the irroration on the forewing.

704 **scopariella** (Heinemann, 1870) – *Cytisus (=Sarothamnus) scoparius*, broom: the larval foodplant.

705 **ulicetella** (Stainton, 1849) – *Ulex*, the gorse genus that includes the larval foodplant.

= 707 **prostratella** (Constant, 1884) – *Genista prostrata*, a species of greenweed not found in Britain: the larval foodplant. '*Prostratus*, spread out, from marking of forewings' (Macleod).

706 **nervosa** (Haworth, 1811) – *nervosus*, sinewy; *nervus*, a sinew or, as here, a vein or nervure on an insect's wing: from the blackish dots or scaling demarcating the veins on the forewing.

= **costosa** (Haworth, 1811) – *costa*, a rib: meaning similar to that of *nervosa*.

707 see 705

708 **carduella** (Hübner, 1817) – *Carduus*, a genus of thistles: from the larval foodplant.

709 **liturosa** (Haworth, 1811) – *litura*, a smear, a blot: from the supposedly smudged markings on the forewing.

710 **conterminella** (Zeller, 1839) – *conterminus*, bordering on, close to: probably, like 702 *assimilella*, 696 *propinquella* and 692 *subpropinquella*, indicating close affinity with another species, e.g. 709 *A. liturosa*, as suggested by Macleod. The suggestion of Pickard *et al*. that *conterminella* is derived from *terminus*, the end, and refers to the larval spinning at the tip of shoots of sallow (*Salix*) is improbable: Zeller received the name from Fischer von Röslerstamm and there is no evidence that he knew the life history.

711 **curvipunctosa** (Haworth, 1811) – *curvus*, curved; *punctum*, a spot: from the curved mark formed by the junction of the first discal spot and another placed costad and proximad of it.

712 **astrantiae** (Heinemann, 1870) – *Astrantia major*, astrantia: the principal foodplant.

713 **angelicella** (Hübner, 1813) – *Angelica sylvestris*, wild angelica: the principal foodplant.

714 **yeatiana** (Fabricius, 1781) – i.h.o. T. P. Yeates, a British entomologist who took his own life in 1782 'more esteemed as an entomologist than as a man' (Haworth, 1812).

715 **capreolella** (Zeller, 1839) – *capreola*, variously interpreted as a wild goat, chamois or roe-deer: from a supposed similarity in the colour of the forewing, which is pale fuscous.

716 **rotundella** (Douglas, 1846) – *rotundus*, round: from the rounded termen of the forewing. 'From larva's occurrence in rolled leaves of carrot' (Macleod); Douglas did not know the life history in 1846.

ETHMIINAE (717)

Ethmia Hübner, 1819 – ἠθμός (ēthmos), a sieve: 722 *pyrausta* is the only species included in the genus by Hübner; its wings are blackish grey with black stigmata, possibly suggestive of holes in a colander.

717 **terminella** Fletcher, 1938 – *terminus*, here the termen of the wing: from the terminal series of black dots on the forewing.

718 **dodecea** (Haworth, 1828) – *dodecim*, twelve: from the black dots on the forewing.

= **decemguttella** (Hübner, 1810) *nec* (Fabricius, 1794) – *decem*, ten; *gutta*, a spot. Neither Haworth nor Hübner counted correctly for there are, in fact, eleven black dots on each forewing.

719 **funerella** (Fabricius, 1787) – *funus, funeris*, a funeral: from the funereal implications of the black and white wing pattern.

720 **bipunctella** (Fabricius, 1775) – *bi-*, two; *punctum*, spot: there are in fact three black spots or projections from the blackish costal half into the white dorsal area of the forewing. Stainton (1873*) is therefore probably correct when he states that this is 'a name which no doubt arises from the two anterior spots on the thorax, the two posterior spots being frequently lost sight of when the insect is pinned'.

721 **pusiella** (Linnaeus, 1758) – probably *pusio*, a little boy: Linnaeus may have had a phrase from Appius (*fl.* 150 A.D.) in mind *'pusio bellissimus'*, a very comely lad, because this is a handsome species. The derivation given by Pickard *et al.* from *pusillus*, small, puny, is most improbable. Cf. 1955 *pusaria*.

722 **pyrausta** (Pallas, 1771) – πυραύστης (puraustēs), a moth that gets singed in the candle: perhaps a reference to the black markings on the forewing. Cf. 1361 *Pyrausta*, named thirty-one years later.

STATHMOPODINAE (877)

Stathmopoda Herrich-Schäffer, 1853 – σταθμός (stathmos), a balance; πούς, ποδός (pous, podos), the foot: from the posture of the adult, which rests with the hindleg projecting horizontally between the fore- and midlegs.

877 **pedella** (Linnaeus, 1761) – *pes, pedis*, the foot: for the same reason as the generic name.

GELECHIIDAE (800)

ARISTOTELIINAE (751)

Metzneria Zeller, 1839 – i.h.o. Herr Metzner (d. 1861), a German entomologist who lived at Frankfurt-on-Oder. See also 726, 1196.

723 **littorella** (Douglas, 1850) – *litus, litoris* (*littus, littoris*), the sea-shore: from the type locality on the south-east coast of the Isle of Wight.

724 **lappella** (Linnaeus, 1758) – *Arctium lappa*, greater burdock: the larval foodplant.

725 **aestivella** (Zeller, 1839) – *aestivus*, pertaining to the summer: from the time of the emergence of the adult.

726 **metzneriella** (Stainton, 1851) – see the generic name above.

727 **neuropterella** (Zeller, 1839) – νεύρον (neuron), a vein; πτέρον (pteron), a wing: from the brown scaling along the veins of the forewing.

727a **aprilella** (Herrich-Schäffer, 1854) – *aprilis*, the month of April: perhaps to express appreciation of the moth's fresh, spring-like beauty and at the same time to indicate affinity with 725 *M. aestivella*, q.v., but not to be taken literally, since the adult emerges in June and July. Cf. 2247 *aprilina*.

728 see below 734

Isophrictis Meyrick, 1917 – ἴσος (isos), equal; φρίσσω (phrissō), to bristle: from the uniform length of the scales on segment 2 of the labial palpus.

729 **striatella** ([Denis & Schiffermüller], 1775) – *stria*, a furrow, a streak: from the longitudinal whitish streaks in the disc and on the fold of the forewing.

Apodia Heinemann, 1870 – ἀ, alpha privative; πούς, ποδός (pous, podos), the foot: from the structure of the larva, which is almost apodal.

730 **bifractella** (Duponchel, 1843) – *bi-*, two; *fractus*, broken: from the orange postmedian fascia which is broken into a subapical and a tornal spot in most specimens.

Eulamprotes Bradley, 1971 – εὖ (eu), well, but here a prefix denoting little more than affinity; the genus *Lamprotes* Heinemann, 1870, from λαμπρότης (lamprotēs), brilliancy, reduced to synonymy because it was preoccupied by *Lamprotes* Reichenbach, 1811: from the shining forewing of some of the species in the genus.

731 **atrella** ([Denis & Schiffermüller], 1775) – *ater*, black: from the dark-coloured forewing.

731a **phaeella** Heckford & Langmaid, 1988 – φαίος (phaios), the hue of twilight, dusky: from the blackish forewing, the name being analogous to 731 *atrella*.

732 **unicolorella** (Duponchel, 1843) – *unus*, one; *color*, colour: from the dark, unicolorous forewing.

733 **wilkella** (Linnaeus, 1758) – i.h.o. B. Wilkes (d. 1749), one of the earliest British entomological

artists; see p. 15.

Argolamprotes Benander, 1945 – ἀργός (argos), shining; the genus *Lamprotes* Heinemann (see 731): from the shining forewing of the next species.

734 **micella** ([Denis & Schiffermüller], 1775) – *mico*, to shine, sparkle: from the glossy forewing.

Paltodora Meyrick, 1894 – perhaps from παλτόν (palton), a light spear; δόρυ (doru), the shaft of a spear, the spear itself. Meyrick erected the genus as monotypic to accommodate the next species. In his diagnosis he emphasises the lanceolate shape of the wings, 'forewings elongate, narrow, pointed', 'hindwings elongate-trapezoidal, apex pointed, produced'. 'Name of nymph' (Macleod): possible, but I cannot trace such a nymph. If Macleod is right, the name is probably a pun in the Fabrician style.

728 **cytisella** (Curtis, 1837) – *Cytisus* (= *Sarothamnus*) *scoparius*, broom: however, bracken (*Pteridium aquilinum*) is the real foodplant.

Monochroa Heinemann, 1870 – μόνος (monos), single; χρώς (khrōs), skin, a surface, its colour: from the unicolorous forewing of some of the member species.

735 **tenebrella** (Hübner, 1817) – *tenebrae*, darkness: from the dark purplish fuscous forewing.

736 **lucidella** (Stephens, 1834) – *lucidus*, bright: from whitish ochreous markings on the forewing, although these are not noticeably bright.

737 **palustrella** (Douglas, 1850) – *paluster*, *palustris*, marshy: from the principal habitat.

738 **tetragonella** (Stainton, 1885) – τέτρα- (tetra-), four; γωνία (gōnia), angle: from the positioning of four dark fuscous spots on the forewing.

739 **conspersella** (Herrich-Schäffer, 1854) – *conspersus*, besprinkled: from the scattered greyish scales on the dark fuscous forewing.

740 **hornigi** (Staudinger, 1883) – i.h.o. J. von Hornig (1819–86), a Russian entomologist; cf. 509.

740a **niphognatha** (Gozmány, 1953) – νιφάς (niphas), snow; γναθός (gnathos), a jaw: from the white labial palpus. The frons is also white.

741 **suffusella** (Douglas, 1850) – *suffusus*, overspread: from the brownish irroration on the ochreous forewing.

742 **lutulentella** (Zeller, 1839) – *lutulentus*, muddy: from the brownish fuscous forewing.

743 **elongella** (Heinemann, 1870) – *e-*, intensive prefix; *longus*, long: from the elongate stigmata on the forewing . . .

744 **arundinetella** (Stainton, 1858) – *arundinetum*, a bed of reeds: from the habitat, not the foodplant which Stainton knew to be lesser pond-sedge (*Carex acutiformis*).

744a **moyses** Uffen, [1991] – from the Old Testament national leader Moses, Moyses being an alternative spelling used in the Vulgate. During their sojourn in Egypt, the Israelites multiplied to such an extent that they became a threat to their Egyptian masters. Accordingly, Pharaoh ordered all male Hebrew infants to be thrown into the River Nile at birth. When Moses was born, his mother put him in an ark made of bulrushes which she placed among the flags growing by the river. There, as intended, he was found by Pharaoh's daughter who took pity on him and brought him up as her own son, 'and she called his name Moses: and she said, Because I drew him out of the water' (Exodus 2:10). Likewise this species was found as a larva mining vegetation growing in borrow-dykes and the leaves bearing the mines were 'drawn out of the water'. The only other names in the British list taken from biblical sources are 1197 *campoliliana* and, apparently, 1583 subsp. *delila*, qq.v.*

745 **divisella** (Douglas, 1850) – *divisus*, divided: from the forewing pattern, which has the costal area suffused ochreous whitish, contrasting with the brown ground colour of the rest of the wing.

Chrysoesthia Hübner, 1825 – χρύσος (khrusos), gold; ἐσθής (esthēs), clothing: from the yellow markings on the forewing of 747 *C. sexguttella*.

746 **drurella** (Fabricius, 1775) – i.h.o. Dru Drury (1725–1803), a British entomologist.
= **hermannella** sensu auctt. – i.h.o. J. Hermann (1738–1800), a German naturalist.

* The publication of the paper giving the first description of this species has been delayed and may therefore follow that of this book. Here there is no description, the foodplant is not named and the habitat is not peculiar to this species. The name will therefore rank in this context as *nomen nudum* and its authorship will not be usurped.

747 **sexguttella** (Thunberg, 1794) – *sex*, six; *gutta*, a drop, a spot: from the yellow spots on the forewing which vary in number: usually there are three on each wing.

Ptocheuusa Heinemann, 1870 – πτωχεύω (ptōkheuō), to be a beggar: echoing the sense of 748 *paupella* and its non-British relative *P. inopella* (Zeller, 1839).

748 **paupella** (Zeller, 1847) – *pauper*, poor, indigent: perhaps the pale, wan colour of the forewing suggested to Zeller an undernourished beggar.

Sitotroga Heinemann, 1870 – σῖτος (sitos), corn; τρώγω (trōgō), to chew: from the larval pabulum of the next species.

749 **cerealella** (Olivier, 1789) – Ceres, goddess of tillage and corn, *cerealis* (late Lat.), pertaining to corn: from the larval pabulum.

Psamathocrita Meyrick, 1925 – ψάμαθος (psamathos), sand of the sea; κριτός (kritos), picked out, chosen: Meyrick selected the next species, which is sandy-coloured, as the type species of the genus.

750 **osseella** (Stainton, 1861) – *osseus*, like bone: from the colour of the forewing.

750a **argentella** Pierce & Metcalfe, 1942 – *argentum*, silver: from the colour of the forewing.

Aristotelia Hübner, 1825 – i.h.o. the Greek philosopher, Aristotle (384–322 B.C.), whose works include *Historia Animalium* which gives the first scientific study of insects.

751 **subdecurtella** (Stainton, 1859) – *sub-*, somewhat like; *A. decurtella* (Hübner, 1813), a species not occurring in Britain: *decurtare*, to cut off; from the dorsal streak on the forewing which ends in a bar at the tornus.

752 **ericinella** (Zeller, 1839) – *Erica*, the heath genus: apparently, however, the larva feeds only on *Calluna*, the heather genus.

753 **brizella** (Treitschke, 1833) – *Briza*, quaking grass: there is no connection, this being a salt-marsh species with its larva feeding on thrift (*Armeria maritima*) and sea-lavender (*Limonium vulgare*).

Xystophora Wocke, 1876 – ξυστόν (xuston), something scraped or polished, like the shaft of a spear, but here denoting the abraded particles; φορέω, (phoreō), to carry: from the dense irroration of pale scales on the forewing.

754 **pulveratella** (Herrich-Schäffer, 1854) – *pulveratus*, dusted: from the irroration on the forewing.

GELECHIINAE (800)

Stenolechia Meyrick, 1894 – στενός (stenos), narrow; -lechia, aphaeretic form of the generic name 800 *Gelechia*, q.v.: a genus of the Gelechiidae with narrow, elongate forewings.

755 **gemmella** (Linnaeus, 1758) – *gemma*, a bud, a gem: probably in the latter sense and with reference to the moth's elegance. Macleod derives it from the former meaning, adding 'from larva's occurrence in buds of oak'; Linnaeus incorrectly describes it as a leaf-miner.

Parachronistis Meyrick, 1925 – παρά (para), contrary to (as in paradox); χρόνος (khronos), time: a replacement name for *Poecilia* Heinemann, 1870, which was preoccupied, i.e. mistimed. Macleod gives the right explanation but misspells *Poecilia*.

756 **albiceps** (Zeller, 1839) – *albus*, white; *caput*, the head: descriptive of the adult.

Recurvaria Haworth, 1828 – *recurvus*, bent back: from the shape of the labial palpus. Originally a genus of forty-seven species from the Oecophoridae and Gelechiidae, all with long, sickle-shaped palps ('*falcatim recurvi*'). Haworth placed this genus next to 129 *Incurvaria*, q.v. 'From wavy outer margin of forewings' (Macleod, who neglected to consult the original diagnosis and gave a non-existent character as he did in the case of *Incurvaria*).

757 **nanella** ([Denis & Schiffermüller], 1775) – *nanus*, a dwarf: from its small size for a gelechiid.

758 **leucatella** (Clerck, 1759) – λευκός (leukos), white: from the prominent white fascia on the forewing.

Coleotechnites Chambers, 1880 – κολεός (koleos), a sheath; τεχνίτης (technitēs) a craftsman: from the larval spinning; a portable case like that of the Coleophoridae (also from κολεός) is not made.

= **Pulicalvaria** Freeman, 1963 – *pulex*, a flea; *calvaria*, a skull: 'uncus subconical, resembling a flea's head (hence the name *Pulicalvaria*)'.

759 **piceaella** (Kearfott, 1903) – *Picea*, the spruce genus, which includes the larval foodplants.

Exoteleia Wallengren, 1881 – ἐξ-, ἐξο- (ex-, exo-), outside; the genus *Teleia* Heinemann, now 765 *Teleiodes*, q.v.: the next species was taken out of that genus.

760 **dodecella** (Linnaeus, 1758) – δώδεκα (dōdeka), twelve: in his description Linnaeus states that each forewing has three pairs of fuscous spots.

Athrips Billberg, 1820 – ἀθρόος (athroos), in heaps, in masses; ἴψ (ips), a worm, a larva: from the gregarious larvae of some of the species.

= **Rhynchopacha** Staudinger, 1871 – ῥύγχος (rhugkhos), a snout; παχύς (pakhus), thick: from the second segment of the labial palpus, which is thickened with scales.

761 **tetrapunctella** (Thunberg, 1794) – τέτρα- (tetra-), four; *punctum*, a spot: from the markings on the forewing which include three, sometimes four, fuscous spots.

761a **rancidella** (Herrich-Schäffer, 1854) – *rancidus*, stinking, rank: application obscure and Herrich-Schäffer offers no explanation; perhaps the moth was reared from foodplant that went rotten. The Duke of Newcastle told me that this happened to the flowers in particular when he reared the species on cotoneaster.

762 **mouffetella** (Linnaeus, 1758) – i.h.o. Thomas Mouffet (1553–1604), physician and one of the earliest British entomologists, having been editor and part author of *Theatrum Insectorum*, published posthumously in 1634. Linnaeus gives repeated references to Mouffet in *Systema Naturae*.

Xenolechia Meyrick, 1895 – ξένος (xenos), a guest, a foreigner; the genus 800 *(Ge)lechia*: a genus having links with *Gelechia* but distinct from it; cf. 755 *Stenolechia* Meyrick, 1894.

763 **aethiops** (Humphreys & Westwood, 1845) – *Aethiops*, an Ethiopian: from the dark-coloured forewing.

Pseudotelphusa Janse, 1958 – ψευδο- (pseudo-), false; *Telphusa* Chambers, an American genus formerly used for the species now in 765 *Teleiodes*. Telphusa (Thelphusa) was the name of an Arcadian nymph.

764 **scalella** (Scopoli, 1763) – *scala*, a ladder: from the transverse fasciae; *'fasciolae illae scalas repraesentant'*, those little fasciae look like the rungs of a ladder.

Teleiodes Sattler, 1960 – *Teleia* Heinemann, 1870 (τέλειος (teleios), perfect, unblemished: from the attractive appearance of the member species), a name proposed for this group but preoccupied by *Teleia* Hübner, 1825, a synonym of 1038 *Acleris* Hübner, 1825; εἶδος, ὠδ- (eidos, ōd-), form: a genus that resembles *Teleia* Heinemann.

765 **vulgella** (Hübner, 1813) – *vulgus*, a crowd, the common people: from the abundance of the species.

766 **scriptella** (Hübner, 1796) – *scriptum*, something written: from the dark markings on the forewing regarded as written characters.

767 **decorella** (Haworth, 1812) – *decorus*, seemly, handsome: from the moth's attractive appearance.

768 **notatella** (Hübner, 1813) – *notatus*, marked, distinguished: from the distinctive pattern of the forewing.

769 **wagae** (Nowicki, 1860) – i.h.o. A. Waga (1799–1890), a French entomologist.

770 **proximella** (Hübner, 1796) – *proximus*, very close, very similar: from resemblance to another species, but which one is uncertain since this name is senior to those of most likely candidates.

771 **alburnella** (Zeller, 1839) – *alburnus*, a kind of white fish: from the whitish forewing. 'Fish, perhaps punning reference to German entomologist Fischer von Röslerstamm, who discovered it' (Macleod); a misrepresentation of fact, since Zeller ascribes the name to Carl von Tischer and makes no reference to Fischer von Röslerstamm.

772 **fugitivella** (Zeller, 1839) – *fugitivus*, fugitive: Zeller describes three forms of this species so the name may refer to the 'fugitive' markings which are liable to absent themselves.

773 **paripunctella** (Thunberg, 1794) – *par*, a pair; *punctum*, a spot: from the black spots on the forewing which are arranged in pairs; Zeller (1839) named this species *triparella* since there are three pairs of these spots.

774 **luculella** (Hübner, 1813) – *luculus*, a little grove: probably from the habitat or type locality. Macleod suggests an allusion to the moth's habit of resting on tree-trunks.

775 **sequax** (Haworth, 1828) – *sequax*, following: this species 'follows', i.e. resembles, 834 *Caryocolum tricolorella* (Haworth, 1812) which Haworth redescribed as *Recurvaria contigua* in 1828, placing *sequax* in the same genus.

Teleiopsis Sattler, 1960 – the genus *Teleia*, see 765 *Teleiodes*; ὄφις (opsis), the look, appearance: a genus similar to *Teleia*.

776 **diffinis** (Haworth, 1828) – *diffinis*, unrelated, different: a word coined to indicate difference from 779 *affinis* Haworth, 1828. Cf. *similis*, *assimilis* and *dissimilis* Linnaeus (p. 18) and 2316 *affinis* and 2317 *diffinis* Linnaeus.

Bryotropha Heinemann, 1870 – βρύον (bruon), moss; τροφή (trophē), food: from the foodplant of many of the species in the genus.

777 **basaltinella** (Zeller, 1839) – *basaltes*, basalt: from the blackish grey ground colour.

778 **umbrosella** (Zeller, 1839) – *umbrosus*, shady: from the dark ground colour of the forewing.

779 **affinis** (Haworth, 1828) – *affinis*, akin: from similarity to 776 *Teleiopsis diffinis*, which was given a correlated name by Haworth in the same work.

780 **similis** (Stainton, 1854) – *similis*, similar: i.e. similar to 779 *B. affinis*.

781 **mundella** (Douglas, 1850) – *mundus*, neat, elegant: perhaps over-complimentary.

782 **senectella** (Zeller, 1839) – *senectus*, old age: probably from greyish irroration suggesting a grizzled head of hair.

783 **boreella** (Douglas, 1851) – βορέας (boreas), the north wind: the species has a northern distribution.

784 **galbanella** (Zeller, 1839) – *galbanus*, yellowish: from the colour of the fascia and the irroration on the forewing.

785 **figulella** (Staudinger, 1859) – *figulus*, a potter: from the reddish, terracotta ground colour of the forewing. Status in Britain uncertain.

786 **desertella** (Douglas, 1850) – *desertus*, waste, desert: from the habitat on sandhills.

787 **terrella** ([Denis & Schiffermüller], 1775) – *terra*, earth, the ground: from its occurrence in open grassland, where it keeps close to the ground. 'From colour of forewings' (Macleod): possible, but less likely.

788 **politella** (Stainton, 1851) – *politus*, polished; from the gloss on the forewing.

789 **domestica** (Haworth, 1828) – *domesticus*, pertaining to the house: the larva feeds on moss, frequently that growing on the walls of houses; it does not live indoors.

Chionodes Hübner, 1825 – χιονώδης, χιονειδής (khionōdes, khioneides), snowy: from the white scales surrounding the stigmata on the forewing.

790 **fumatella** (Douglas, 1850) – *fumatus*, smoked: from the dusky, blackish brown ground colour.

791 **distinctella** (Zeller, 1839) – *distinctus*, separate, distinct: the markings are more strongly expressed than those of 787 *Bryotropha terrella* with which Zeller compares it in his description. 'From distinct marking of type specimen, though usually indistinct' (Macleod): not what Zeller wrote.

Mirificarma Gozmány, 1955 – *mirificus*, wonderful; *arma* (pl.), tools, weapons: from the remarkable aedeagus, which Gozmány describes as 'very long, in a long tube of [*sic*] vinculum, with an accompanying and very long filamental prong'.

792 **mulinella** (Zeller, 1839) – *mulinus*, pertaining to a mule: from a supposed colour resemblance.

793 **lentiginosella** (Zeller, 1839) – *lentiginosus*, freckled: from the ochreous spots and edging to the discal stigmata on the forewing.

Lita Treitschke, 1833 – λιτός (litos), plain, unadorned: from the relatively simple pattern of some of the species.

794 **sexpunctella** (Fabricius, 1794) – *sex*, six; *punctum*, a spot: from the five pale fasciae which, by inversion, suggest six dark bands on the forewing.

= **virgella** (Thunberg, 1794) – *virga*, a rod, a coloured stripe on a garment: like Fabricius, Thunberg saw the ground colour as pale and the dark areas as forming fasciae or stripes across the forewing.

795 **solutella** (Zeller, 1839) – *solutus*, unbound, loose, dissolved: Zeller described the forewing as having a sprinkling of grey scales and a very indistinct terminal fascia which may have seemed 'washed-out' to him.

Aroga Busck, 1914 – probably a meaningless neologism.

796 **velocella** (Zeller, 1839) – *velox*, swift: Zeller may have observed the behaviour of the adult which rests on the ground and makes short, rapid sallies downwind.

Neofaculta Gozmány, 1955 – νέος (neos), new; *Faculta* Busck, 1939, a Nearctic genus of the Gelechiidae: from affinity.

797 **ericetella** (Geyer, 1832) – *Erica*, the heath genus: from the larval foodplant.

Neofriseria Sattler, 1960 – νέος (neos), new; *Friseria* Busck, 1939, a Nearctic genus of the Gelechiidae: from affinity.

798 **peliella** (Treitschke, 1835) – πηλός (pēlos), mud: from the fuscous, somewhat muddy ground colour of the forewing.

799 **singula** (Staudinger, 1876) – *singula*, individual, separate: from the recognition that it was distinct from the previous species which it resembles closely.

Gelechia Hübner, 1825 – γῆ (gē), the earth; λέχος (lekhos), a bed; so γηλεχής (gēlekhēs), sleeping on the ground: from the fact that most of the species in the original, more extensive genus feed on low-growing plants as larvae and fly close to the ground as adults.

800 **rhombella** ([Denis & Schiffermüller], 1775) – *rhombus*, a rhomb, an equilateral parallelogram other than a square: from the figure made if the black spots on the forewing were joined.

801 **scotinella** Herrich-Schäffer, 1854 – σκοτεινός (skoteinos), dark-coloured: from the coarse, dark fuscous irroration on the forewing.

801a **senticetella** Staudinger, 1859 – *sentis*, a brier, a thorn-bush, *senticetum*, a thorn-brake: from the habitat.

802 **sabinella** Zeller, 1839 – *Juniperus sabina*, savin: the larval foodplant.

802a **sororculella** (Hübner, 1817) – *sororcula*, a little sister: probably from affinity with another species.

803 **muscosella** Zeller, 1839 – *muscosus*, mossy: the larva, however, feeds in the catkins of Salicaceae and is not one of the moss-feeding gelechiids.

804 **cuneatella** Douglas, 1852 – *cuneatus*, wedge-shaped: from the form of the dark markings on the forewing.

805 **hippophaella** (Schrank, 1802) – *Hippophae rhamnoides*, sea-buckthorn: the larval foodplant.

806 **nigra** (Haworth, 1828) – *niger*, black: from the dark fuscous forewing.

807 **turpella** ([Denis & Schiffermüller], 1775) – *turpis*, base: from the plain, dingy coloration of the forewing.

Platyedra Meyrick, 1895 – πλατύς (platus), wide; ἕδρα (hedra), a seat, foundation: what is notably broad is the hindwing and Meyrick may have regarded it as the foundation as it is *under* the forewing. 'From broad base of antennae' (Macleod): not a prominent feature nor mentioned by Meyrick.

808 **subcinerea** (Haworth, 1828) – *sub-*, somewhat; *cinereus*, ash-coloured: from the greyish ochreous ground colour.

Pexicopia Common, 1958 – *pexus*, woolly; *copia*, abundance: from the rough-scaled labial palpus, from the long-haired pecten on the antenna, or from both.

809 **malvella** (Hübner, 1805) – from the Malvaceae, the mallow family, to which the larval foodplant, marsh mallow (*Althaea officinalis*), belongs.

Scrobipalpa Janse, 1951 – *scrobis*, a ditch, a trench; *palpus*, the palp: from the second segment of the labial palpus which is deeply furrowed beneath.

810 **suaedella** (Richardson, 1893) – *Suaeda*, the sea-blite genus: the larva feeds on its members.

811 **samadensis** (Pfaffenzeller, 1870) – from Samaden in south-eastern Switzerland: the type locality.

subsp. **plantaginella** (Stainton, 1883) – *Plantago coronopus*, buck's-horn plantain and *P. maritima*, sea-plantain: the larval foodplants.

812 **instabilella** (Douglas, 1846) – *instabilis*, changeable: from variability in the colour and pattern of the forewing.

813 **salinella** (Zeller, 1847) – *salinae* (pl.), a saltern, a salt-works: from the habitat on saltings.

814 **ocellatella** (Boyd, 1858) – *ocellatum*, something small with eye-like marks: from the dark-centred ochreous stigmata on the forewing.

814a **pauperella** (Heinemann, 1870) – *pauper*, poor, indigent: from the poverty of distinct markings on the forewing.
= **klimeschi** sensu auctt. – i.h.o. Dr J. Klimesch, the contemporary Austrian microlepidopterist.

815 **nitentella** (Fuchs, 1902) – *nitens*, shining: the suitability of the name is not obvious, since the moth is no more glossy than its congeners.

816 **obsoletella** (Fischer von Röslerstamm, 1841) – *obsoletus*, worn-out, obsolete: in closely related species there are two dark spots near the base of the forewing, but they are obsolete in this species; Fischer von Röslerstamm gives this as the distinctive character. 'From indistinct bar across forewings, which is sometimes obsolete' (Macleod): an inaccurate statement.

817 **clintoni** Povolny, 1968 – i.h.o. E. C. Pelham-Clinton, later Duke of Newcastle (1920–88), who discovered the species.

818 **atriplicella** (Fischer von Röslerstamm, 1841) – *Atriplex*, the orache genus: the larva feeds on *Atriplex* spp.

819 **costella** (Humphreys & Westwood, 1845) – *costa*, a rib, the anterior margin of an insect's wing: from the dark fuscous costal blotch on the forewing.

820 **artemisiella** (Treitschke, 1833) – *Artemisia*, the wormwood genus: the larva really feeds on thyme (*Thymus* spp.). Treitschke, citing von Tischer as his authority, gave the foodplant as field-wormwood (*Artemisia campestris*).

821 **murinella** (Herrich-Schäffer, 1854) – *murinus*, mouse-like, mousy: from its diminutive size, Herrich-Schäffer describing it as being easily the smallest member of the genus.

822 **acuminatella** (Sircom, 1850) – *acuminatus*, pointed: from the sharply pointed forewing, especially noticeable in the female.

822a **stangei** (Hering, 1889) – i.h.o. G. Stange, a schoolmaster who lived in Mechlenburg and discovered the species.

Scrobipalpula Povolný, 1964 – diminutive of 810 *Scrobipalpa*, q.v.

823 **psilella** (Herrich-Schäffer, 1854) – ψιλός (psilos), stripped bare, bald: Herrich-Schäffer describes it as being less heavily spotted than 824 *Gnorimoschema streliciella* and this may be the 'bareness' he had in mind.

823a **tussilaginis** (Frey, 1867) – *Tussilago farfara*, colt's-foot: the larval foodplant.

Gnorimoschema Busck, 1900 – γνώριμος (gnōrimos), well-known; σχῆμα (skhēma), form, appearance: a new name, perhaps, for a familiar species.

824 **streliciella** (Herrich-Schäffer, 1854) – Neustrelitz, Germany: the type locality, where it may have been discovered by Messing (see 469) or Demarné (see 1135).

Phthorimaea Meyrick, 1902 – φθοριμαῖος (phthorimaios), destructive: from the feeding habits of 825 *P. operculella*, the type species, which is a pest of potato.

825 **operculella** (Zeller, 1873) – *operculum*, a cover: from the very large valvae, covering the rest of the male genitalia; an early example of the diagnostic use of the genitalia.

Caryocolum Gregor & Povolný, 1954 – κάρυον (karuon), a nut, a pod; *colo*, to inhabit: the Caryophyllaceae, the family which includes campions, pinks, chickweed, etc., is also named from the same root and includes the foodplants of this genus.

826 **vicinella** (Douglas, 1851) – *vicinus*, neighbouring, near: from close relationship with other species, 830 *C. fraternella* and 834 *C. tricolorella* being cited by Douglas.

827 **alsinella** (Zeller, 1868) – Alsineae, a former subfamily name for the chickweeds.
subsp. **semidecandrella** (Threlfall, 1887) – *Cerastium semidecandrum*, little mouse-ear: the larval foodplant.

828 **viscariella** (Stainton, 1855) – *Lychnis viscaria*, sticky catchfly: a larval foodplant.

829 **marmoreum** (Haworth, 1828) – *marmoreus*, resembling marble: from the variegated forewing.

830 **fraternella** (Douglas, 1851) – *fraternus*, brotherly, closely related: from close relationship with other species, 798, 826, 831 and 834 being those mentioned by Douglas.

831 **proximum** (Haworth, 1828) – *proximus*, very close: from similarity to 834 *C. tricolorella* and 775 *Teleiodes sequax*, both of them species named by Haworth.

832 **blandella** (Douglas, 1852) – *blandus*, pleasant, charming: from the moth's neat appearance.

833 **junctella** (Douglas, 1851) – *junctus*, linked, joined: from the confluence of the two anterior

stigmata with an oblique black bar extending from the costa.

834 **tricolorella** (Haworth, 1812) – *tri-*, three; *color*, colour: from the forewing which is ferruginous with black and white markings.

835 **blandulella** (Tutt, 1887) – *blandulus*, pleasant little . . . : indicating affinity with 832 *C. blandella*.

836 **kroesmanniella** (Herrich-Schäffer, 1854) – i.h.o. D. W. Krössmann (mid-19th century). He was an army schoolmaster at Hanover.

837 **huebneri** (Haworth, 1827) – i.h.o. J. Hübner (1761–1826), the German entomologist and entomological artist who died in the previous year.

= **knaggsiella** (Stainton, 1866) – i.h.o. Dr H. G. Knaggs (1832–1908), a British entomologist.

ANACAMPSINAE (852)

Nothris Hübner, 1825 – νωθρός (nōthros), sluggish: possibly from the behaviour of the moths, but Hübner gave no indication of this.

838 **verbascella** (Hübner, 1796) – *Verbascum pulverulentum*, hoary mullein: the larval foodplant.

839 **congressariella** (Bruand, 1858) – because Bruand first announced and exhibited it at a congress held in June, 1857.

Reuttia Hofmann, 1898 – i.h.o. C. Reutti (1830–94), a German entomologist.

840 **subocellea** (Stephens, 1834) – *sub-*, somewhat; *ocellus*, a little eye: from the black apical dot on the forewing.

Sophronia Hübner, 1825 – σώφρων (sōphrōn), sober, staid: from the soft, shaded coloration of the type species, *Tinea illustrella* Hübner (*in-*, not, un-; *lustro*, to illumine), which is not found in Britain.

841 **semicostella** (Hübner, 1813) – *semi-*, half; *costa*, a rib, the anterior margin of an insect's wing: from the white costal streak which extends to just beyond the middle of the forewing.

842 **humerella** ([Denis & Schiffermüller], 1775) – *humerus*, the shoulder: from the white costal streak extending along the basal half of the forewing ('the shoulder').

Aproaerema Durrant, 1897 – ἀ, alpha privative; προαιρέω (proaireō), to prefer, to choose before: Durrant translates his name 'not the thing chosen before', this being his second choice of name, the first having been preoccupied.

843 **anthyllidella** (Hübner, 1813) – *Anthyllis vulneraria*, kidney vetch: the larval foodplant.

Syncopacma Meyrick, 1925 – συγκοπή (sugkopē), cutting short; ἀκμή (akmē), a point: from the very short, filiform maxillary palpi.

844 **larseniella** (Gozmány, 1957) – i.h.o. C. C. R. Larsen (1846–1920), a Danish entomologist.

845 **sangiella** (Stainton, 1863) – i.h.o. J. Sang (1828–87), a north-country British entomologist.

846 **vinella** (Bankes, 1898) – i.h.o. A. C. Vine (1844–1917), a British entomologist who collected the larvae, from which the type material was bred, at Ditchling, Sussex, and sent them to Bankes.

847 **taeniolella** (Zeller, 1839) – *taeniola*, a little band: from the white fasciae, one on each forewing, which merge to form an apparent girdle when the wings are folded.

848 **albipalpella** (Herrich-Schäffer, 1854) – *albus*, white; *palpus*, the palp: from the colour of the labial palpus.

848a **suecicella** (Wolff, 1958) – *Suecica*, Swedish: from the country in which the species was first recognized.

849 **cinctella** (Clerck, 1759) – *cinctus*, a girdle: from the character which gave 847 (q.v.) its name. Fabricius describes the adult as rotating on the same spot before settling down to rest and Scopoli named it *vorticella* for the same reason; it is unlikely, however, that Clerck, an artist depicting a preserved specimen, was aware of this habit.

850 **polychromella** (Rebel, 1902) – πολύς (polus), many; χρῶμα (khrōma), colour: descriptive of the adult.

Acanthophila Heinemann, 1870 – ἄκανθα (akantha), a thorn, a prickly plant: φίλος (philos), dear, in composition loving: the genus is monotypic and the single member feeds on lichen, so this generic name is either based on misinformation regarding the life history or refers only to the habitat.

851 **alacella** (Zeller, 1839) – *alacer*, lively: from the behaviour of the adult.

Anacampsis Curtis, 1827 – ἀνάκαμψις (anakampsis), a bending back, flexure: from the indentation of the subterminal fascia on the forewing of 853 *A. populella*.

852 **temerella** (Lienig & Zeller, 1846) – *temerē*, by chance: possibly the larvae from which Madam Lienig bred the type material were obtained through serendipity – she liked cryptic names, cf. 686 *ciniflonella*. Pickard *et al.* derive the name from *temerarius*, occurring by chance, implying the explanation given above. Macleod has '*temero*, to stain, from marking of forewings'; this explanation is also possible, but the reference would be to the blackish ground colour rather than to the markings which are indistinct. Cf. 1958 *temerata*.

853 **populella** (Clerck, 1759) – *Populus*, the poplar genus, which includes some of the larval foodplants.

854 **blattariella** (Hübner, 1796) – *Verbascum blattaria*, moth mullein: the foodplant, however, is birch. '*Blatta*' is a generalized Latin word for various insects including the beetle, cockchafer, moth and cockroach (*Blatta* Linnaeus, 1758, is the cockroach genus) and Hübner need not have been thinking of the plant.

Acompsia Hübner, 1825 – ἄκομψος (akompsos), unadorned: from the next species which has a unicolorous grey-brown forewing.

855 **cinerella** (Clerck, 1759) – *cinereus*, ashy: from the coloration of the forewing.

CHELARIINAE (858)

Anarsia Zeller, 1839 – ἀνάρσιος (anarsios), incongruous: from difference, especially in the labial palpus, from all other gelechiids.

856 **spartiella** (Schrank, 1802) – *Spartium junceum*, Spanish broom: the larva feeds also on *Ulex*, *Cytisus* (= *Sarothamnus*) and *Genista*.

857 **lineatella** Zeller, 1839 – *lineatus*, having straight lines: from the longitudinal streaks on the forewing.

Hypatima Hübner, 1825 – ὕπατος (hupatos), highest, best; τιμάω (timao), to honour: the next species, which is the type species, is neat but hardly worthy of such high praise.

= **Chelaria** Haworth, 1828 – χηλή (khēlē), a cloven hoof, a crab's claw, a forked instrument: from the structure of the labial palpus, which has a projecting tuft on its ventral surface at the apex of segment 2.

858 **rhomboidella** (Linnaeus, 1758) – *rhomboides* (ῥομβοειδής), a four-sided figure whose opposite sides and angles are equal, a rhomboid (cf. 800 *rhombella*): '*alis fuscis: macula rhombea nigra*' (wings fuscous: a black rhomboid spot), with reference to the costal blotch of the forewing.

Psoricoptera Stainton, 1854 – ψωρικός (psōrikos), mangy; πτέρον (pteron), a wing: from the tufts of raised scales on the forewing, giving this appearance. ψώρα (psōra), properly the itch or scab, was once used in the sense 'moth' by Nicander, a medical writer of the 2nd century B.C., according to a gloss by a scholiast.

859 **gibbosella** (Zeller, 1839) – *gibbosus*, humped, gibbous: from the scale-tufts on the forewing, from the tuft on segment 3 of the labial palpus, or from both.

Mesophleps Hübner, 1825 – μέσος (mesos), middle; φλέψ (phleps), a vein: from the stalking of veins 7 and 8 (R_5 and R_4) on the forewing.

860 **silacella** (Hübner, 1796) – *silaceus*, ochreous: from the pale ochreous coloration of the forewing.

Telephila Meyrick, 1923 – τηλέφιλον (tēlephilon), 'properly "far-away love" or "love-in-absence", the leaf of some plant used as a charm by lovers to try whether their love was returned: it was laid on one hand or arm and struck with the other, and a loud crack was a favourable omen' (Liddell & Scott, 1869). There is no entomological application, but the name and its meaning must have appealed to Meyrick.

861 **schmidtiellus** (Heyden, 1848) – i.h.o. A. Schmidt (d. 1899), a German entomologist, or one of several others of this name.

DICHOMERINAE (862)

Dichomeris Hübner, 1825 – δίχα, in composition διχο- (dikha, dikho-), in two; μέρος (meros), a part: from the pattern of the forewings, which are divided longitudinally or transversely by a subcostal streak or median fascia.

862 **marginella** (Fabricius, 1781) – *margo, marginis*, an edge: from the white costal streak on the forewing.

863 **juniperella** (Linnaeus, 1761) – *Juniper communis*, juniper: the larval foodplant.

864 **ustalella** (Fabricius, 1794) – *ustulatus*, scorched: from the ferruginous brown streaks, described by Fabricius as 'scorched', towards the apex of the forewing. Authors emended the spelling to *ustulella*, but this is not permissible.

865 **fasciella** (Hübner, 1796) – *fascia*, a band: from the dark brown fascia on the forewing.

Brachmia Hübner, 1825 – i.h.o. N. J. Brahm (late 18th-early 19th century), a German entomologist and lawyer who lived at Mainz.

866 **blandella** (Fabricius, 1798) – *blandus*, pleasant, charming: from the attractive appearance of the moth.

= **gerronella** (Zeller, 1850) – γέρρον (gerron), an object made of wickerwork: from a supposed reticulate pattern on the forewing.

867 **inornatella** (Douglas, 1850) – *inornatus*, plain, unadorned: from the forewing which is unmarked except for the stigmata.

868 **rufescens** (Haworth, 1828) – *rufescens*, reddish: from the ground colour of the forewing.

869 **lutatella** (Herrich-Schäffer, 1854) – *lutatus*, bedaubed with mud: from the reddish-grey-ochreous ground colour of the forewing.

SYMMOCINAE (872)

Oegoconia Stainton, 1854 – probably a typographical error, though Stainton abode by the original spelling. Two amendments have been suggested (see Kloet & Hincks, 1972: 24). (1) *Oecogenia* – οἰκογενής (oikogenēs), born in the house; (2) *Oecogonia* – οἶκος (oikos), a house; γωνία (gonia), a corner, an angle: 'in houses' (Stainton, 1859); specimens may have been found in dark corners indoors. The second spelling, *Oecogonia*, involving only the transposition of the 'c' and the 'g', is more likely to have been what Stainton intended.

870 **quadripuncta** (Haworth, 1828) – *quadri-*, four; *punctum*, a spot: from the four pale spots or fasciae on the fuscous forewing; alternatively the moth could be described as yellowish with fuscous markings.

871 **deauratella** (Herrich-Schäffer, 1854) – *deauratus*, gilded: from the yellow ground colour of the forewing.

871a **caradjai** Popescu-Gorj & Căpuşe, 1965 – i.h.o. A. von Caradja (1861–1955), a German entomologist.

Symmoca Hübner, 1825 – συν- (sun-), with, altogether (in composition); μωκός (mōkos), a mimic: perhaps from close resemblance between the two species placed by Hübner in the genus.

872 **signatella** Herrich-Schäffer, 1854 – *signatus*, clearly marked, distinctively marked: from the pattern on the forewing.

BLASTOBASIDAE (873)

Blastobasis Zeller, 1855 – βλαστός (blastos), a shoot; βάσις (basis), a step, a pedestal on which something stands: from the conspicuous pecten on the scape of the antenna.

873 **lignea** Walsingham, 1894 – *ligneus*, of wood: from the fuscous forewing (Macleod), which may suggest a fragment of dead wood. There is no reference to the early stages which were unknown to Walsingham.

874 **decolorella** (Wollaston, 1858) – *decolor*, faded: from the pale ochreous forewing.

875 **phycidella** (Zeller, 1839) – *phycis, phycidis* (φυκίς), a fish living amongst sea-weed that changes its colour at various seasons of the year: perhaps from the ground colour of the forewing being very variable.

Auximobasis Walsingham, 1892 – αὔξιμος (auximos), promoting growth, well-grown: βάσις (basis), a pedestal: collateral with *Blastobasis* above and given for the same reason.

876 **normalis** Meyrick, 1918 – *normalis*, made according to the square: perhaps from the roughly quadrate positioning of the stigmata and dark markings on the forewing.

877 see below 722

MOMPHIDAE (880)

BATRACHEDRINAE (878)

Batrachedra Herrich-Schäffer, 1853 – βάτραχος (batrakos), a frog; ἕδρα (hedra), a seat: from the frog-like posture of the moth with its anterior raised.

878 **praeangusta** (Haworth, 1828) – *prae-*, in composition, to a high degree; *angustus*, narrow: from the markedly elongate forewing.

879 **pinicolella** (Zeller, 1839) – *Pinus*, the pine genus; *colo*, to inhabit: from the larval foodplant.

MOMPHINAE (880)

Mompha Hübner, 1825 – μομφή (momphē), blame: this appears to be the only possible derivation, but is inappropriate for a genus containing beautiful and brightly coloured moths; however, the species included by Hübner were, with the exception of 885, those of more sober coloration. The name is most probably fanciful without any entomological application; if Hübner really wished to allocate blame it would probably have been upon himself.

880 **langiella** (Hübner, 1796) – i.h.o. H. G. Lang (18th century), a German jewel-cutter and amateur entomologist who supplied Hübner with the type specimen for figuring.
= **epilobiella** (Roemer, 1794) *nec* ([Denis & Schiffermüller], 1775) – see 893.

881 **terminella** (Humphreys & Westwood, 1845) – *termen*, *terminis*, the end, the tip: 'the terminal joint of the palpi pale fuscous' (Pickard *et al.*, Westwood being one of the '*al.*', though he ascribes the name to Dale). The interpretation 'from silver metallic mark on termen of forewings' (Macleod) is therefore incompatible with contemporary opinion of Dale's intention.

882 **locupletella** ([Denis & Schiffermüller], 1775) – *locuples*, *locupletis*, rich, opulent: from the rich coloration of the forewing which is orange with black, white and silver markings.

883 **raschkiella** (Zeller, 1839) – i.h.o. J. G. Raschke (1763–1815), a German entomologist.

884 **miscella** ([Denis & Schiffermüller], 1775) – *misco*, to mingle: from the coloration of the forewing in which grey and ferruginous are blended.

885 **conturbatella** (Hübner, 1819) – *conturbatus*, confused, mixed: from the coloration of the forewing in which fuscous, dull ferruginous and silvery scales are mingled.

886 **ochraceella** (Curtis, 1839) – *ochra*, yellow earth: from the ochraceous coloration of the forewing.

887 **lacteella** (Stephens, 1834) – *lacteus*, milky: from the whitish thorax and basal blotch on the forewing.

888 **propinquella** (Stainton, 1851) – *propinquus*, close, resembling: from close similarity to the preceding species.

889 **divisella** Herrich-Schäffer, 1854 – *divisus*, divided: from the pattern of the forewing which has the dorsal area paler than the rest.

890 **subdivisella** Bradley, 1951 – *sub-*, near; *divisella*, the last species. Its specific status was not fully confirmed until the discovery of the life history in 1988.

891 **nodicolella** Fuchs, 1902 – *nodus*, a knot, nob, gall; *colo*, to inhabit: from the habits of the larva which feeds in a gall in a stem of rosebay willowherb (*Epilobium angustifolium*).

892 **subbistrigella** (Haworth, 1828) – *sub-*, somewhat; *bis*, twice; *striga*, a streak: from the two rather obscure pale fasciae on the forewing. Haworth had already used the name *bistrigella* (128) and required a second name with a similar meaning.

893 **epilobiella** ([Denis & Schiffermüller], 1775) – *Epilobium*, the willowherb genus: *E. hirsutum*, the great willowherb, is the larval foodplant.
= **fulvescens** (Haworth, 1828) – *fulvus*, tawny, *fulvescens*, inclining to be tawny: from the coloration of the forewing.

COSMOPTERIGIDAE (894)

COSMOPTERIGINAE (894)

Cosmopterix Hübner, 1825 – κόσμος (kosmos), an ornament; πτέρυξ (pterux), a wing: from the bright colours and striking beauty of many of the species.

894 **zieglerella** (Hübner, 1810) – i.h.o. Dr Ziegler, a Viennese medical practitioner who used to send specimens to Hübner for figuring.

895 **schmidiella** Frey, 1856 – i.h.o. either A. Schmidt (d. 1899), a German entomologist, or one of the several other entomologists of the same name.

896 **orichalcea** Stainton, 1861 – *orichalceus*, brazen, formed from *orichalcum*, brass: from the brassy metallic markings on the forewing.

= **drurella, druryella** sensu auctt. – Fabricius bestowed this name on a gelechiid (746, q.v.) but it was misapplied to this species.

897 **lienigiella** Lienig & Zeller, 1846 – i.h.o. Madam Lienig (d. 1855), a Latvian entomologist and close friend of Zeller who collaborated with her in a list of Latvian Lepidoptera.

Limnaecia Stainton, 1851 – λίμνη (limnē), a marshy lake; οἰκέω (oikeō), to dwell: from the habitat. Meyrick (1888) emended the spelling to 'Limnoecia' which may have been what Stainton had intended.

898 **phragmitella** Stainton, 1851 – *Phragmites australis*, the common reed: the real foodplant is reedmace (bulrush) (*Typha latifolia*).

ANTEQUERINAE

From the American genus *Antequera* Clarke, 1941. This is a puzzling name and an explanation was sought from Dr J. F. G. Clarke himself, via Dr R. W. Hodges. After the lapse of half a century, the former was understandably uncertain, but on reflection he thought the explanation was as follows, *ante*, preceding; *quera*, question, doubt: the intent was to question the preceding generic association, which was 666 *Semioscopis*, a genus of the Oecophoridae. '*Quera*' could be formed from *quaero*, to question and might be a relatinization of 'query', itself derived from the imperative of *quaero*. Dr Clarke added that perhaps he should have used '*querula*'; *querulus* means complaining, from the root of *queror*, to lament. We shall never know for certain whether Dr Clarke was questioning or deploring an earlier error in classification, but this is irrelevant to the other genera in the subfamily. Incidentally, there are towns named Antequera in Spain and Paraguay.

Pancalia Stephens, 1829 – πάγκαλος (pagkalos), entirely beautiful: from the attractive appearance of the moths.

899 **leuwenhoekella** (Linnaeus, 1761) – i.h.o. Antony van Leeuwenhoek (1632–1723), a Dutch scientist who developed the microscope and included insects in his studies.

900 **latreillella** Curtis, 1830 – i.h.o. P. A. Latreille (1762–1833), the distinguished French entomologist and systematist.

Euclemensia Grote, 1874 – εὖ (eu), well, but here a prefix of little meaning other than to express distinction; *Clemensia* Packard, 1874, an American genus of Arctiidae named i.h.o. J. B. Clemens (1829–67), an American entomologist. The next species has been variously placed in more than one family but never in the Arctiidae, although its bright colours may have made it seem like a miniature tiger moth.

901 **woodiella** (Curtis, 1830) – i.h.o. R. Wood, a Manchester collector who supplied Curtis with a specimen for figuring; the species had, in fact, been discovered by another Manchester collector named Cribb who was angered at the slight.

BLASTODACNINAE (905)

Glyphipteryx Curtis 1827 – γλυφίς (gluphis), a notch; πτέρυξ (pterux), a wing: the wings are not notched and the name is indistinguishable in pronunciation from *Glyphipterix* Hübner (see 391–397). It is therefore likely that *Glyphipteryx* will be deemed to be preoccupied and the following name brought back into use.

= **Chrysoclista** Stainton, 1854 – χρυσός (khrusos), gold; κλύστος (klustos), washed: from the golden-orange markings on the forewing.

902 **lathamella** (Fletcher, 1936) – i.h.o. Dr John Latham, a British ornithologist with a secondary interest in entomology whose collection contained the then unique specimen (now lost) entitled *bimaculella* by Haworth (1828). This name was doubly preoccupied and a replacement was necessary.

903 **linneella** (Clerck, 1759) – i.h.o. Carl von Linné (Linnaeus) (1707–78). Clerck was one of his pupils and was returning a compliment since Linnaeus had already named a species (263) in Clerck's honour.

Spuleria Hofmann, 1898 – i.h.o. A. Spuler (1869–1937), the German entomological author.

904 **flavicaput** (Haworth, 1828) – *flavus*, yellow; *caput*, the head: descriptive of the adult.

Blastodacna Wocke, 1876 – βλαστός (blastos), a shoot; δάκνω (daknō), to bite: from the fact that some of the species have larvae that mine the shoots of their foodplant.

905 **hellerella** (Duponchel, 1838) – i.h.o. Professor J. F. Heller (b. 1813), a Viennese entomologist.

906 **atra** (Haworth, 1828) – *ater*, black: from the dark fuscous ground colour.

Dystebenna Spuler, 1910 – δυς- (dus-), a prefix reversing meaning; *Tebenna* Herrich-Schäffer, *nec* Billberg (τήβεννα (tēbenna), a robe of state: from the bright colours of the species (e.g. 883 *Mompha raschkiella*) that Herrich-Schäffer had included): Spuler wished to remove the next species from *Tebenna* but at the same time to indicate affinity.

907 **stephensi** (Stainton, 1849) – i.h.o. J. F. Stephens (1792–1852), the British entomological writer.

CHRYSOPELEIINAE

From the Nearctic genus *Chrysopeleia* Chambers, 1874 – χρυσός (khrusos), gold; πέλεια (peleia), the wood-pigeon, its dove-grey colour.

Sorhagenia Spuler, 1910 – i.h.o. L. F. Sorhagen (1836–1914), a German entomologist.

908 **rhamniella** (Zeller, 1839) – *Rhamnus catharticus*, buckthorn: the larval foodplant.

909 **lophyrella** (Douglas, 1846) – λόφιον (lophion), a small crest: from the numerous scale-tufts on the forewing.

910 **janiszewskae** Reidl, 1962 – i.h.o. Professor J. Janiszewska of Wroclaw (Breslau), Poland. The dedicatee is a lady.

SCYTHRIDIDAE (911)

Scythris Hübner, 1825 – σκυθρός (skuthros), sullen: from the sober coloration of most of the species.

911 **grandipennis** (Haworth, 1828) – *grandis*, large; *pennae* (pl.), a wing: from the large size of the moth compared with most of its congeners.

912 **fuscoaenea** (Haworth, 1828) – *fuscus*, dusky; *aëneus*, brassy: from the forewing which is greenish bronze with a coppery tinge towards the apex.

913 **fallacella** (Schläger, 1847) – *fallax*, *fallacis*, deceitful: probably from deceptive resemblance to another species. 'From moth's habit of hiding in grass' (Macleod); an unconvincing guess. Such behaviour is too normal to be the source of a name; the moth flies freely in afternoon sunshine.

914 **crassiuscula** (Herrich-Schäffer, 1854) – *crassius*, comparative neut. of *crassus*, fat, i.e. fatter, fattish; *-culus*, dim. suffix: from the adult which is stout in proportion to its small wingspan.

= **fletcherella** Meyrick, 1928 – i.h.o. W. H. B. Fletcher (1853–1941) or T. Bainbrigge Fletcher (1878–1950), either of whom could have been associated with the moth. The former is more likely, since Meyrick, who ascribes the name to Durrant, records the moth from Sussex, where this Fletcher lived and collected, but not from Gloucestershire, the home of the latter.

915 **picaepennis** (Haworth, 1828) – *pica*, a magpie; *pennae* (pl.), a wing: Haworth describes an irregular white fascia and gives as the English name 'pye-feather', although the dark bronzy wing usually has only a slight irroration of white scales.

f. **vagabundella** (Herrich-Schäffer, 1855) – *vagabundus*, a vagrant, a wanderer: probably from the behaviour of the adult. The name is given to the distinctive form occurring in western Scotland and Ireland which has a purplish blue gloss and no white irroration.

916 **siccella** (Zeller, 1839) – *siccus*, dry: probably because the grizzling of white scales on the forewing gives an impression of withered old age. Zeller himself offers no explanation.

917 **empetrella** Karsholt & Nielsen, 1976 – *Empetrum*, the crowberry genus: in Britain, however, the species has been associated with heather (*Erica* and *Calluna*).

= **variella** (Stephens, 1834) *nec* ([Denis & Schiffermüller], 1775) – *varius*, varying: from the ground colour which ranges from grey to dark fuscous.

918 **limbella** (Fabricius, 1775) – *limbus*, an edge: the white subdorsal spots are sometimes connected by white scaling along the dorsal margin.
= **quadriguttella** (Thunberg, 1794) – *quadri-*, four; *gutta*, a spot: there are three diffuse whitish spots on the forewing, the similarly coloured thorax perhaps supplying the fourth spot.
919 **cicadella** (Zeller, 1839) – *cicada*, the cicada or, loosely, a grasshopper: from the short, darting flight which resembles that of a grasshopper.
920 **potentillella** (Zeller, 1847) (*potentillae* misspelling) – *Potentilla*, the cinquefoil genus: however, the larva feeds on sorrel (*Rumex* spp.).
920a **inspersella** (Hübner, 1817) – *inspersus*, sprinkled: from the irroration of white scales on the dark fuscous forewing.
920b **sinensis** (Felder & Rogenhofer, 1875) – *Sinensis*, Chinese: from the type locality, the species having been named from specimens collected on a world voyage undertaken for scientific purposes by the frigate *Novara*.

TORTRICOIDEA (1033)

TORTRICIDAE (1033)

COCHYLINAE (962)

Trachysmia Guenée, 1845 – τραχύς (trakhus), rough; τραχυσμός (trakhusmos), a roughness: from the raised scale-tufts on the forewings of some species in the genus.
= **Hysterosia** Stephens, 1852 – ὕστερος (husteros), behind, posterior: from the posterior crest on the thorax.
921 **inopiana** (Haworth, 1811) – *inops*, poor, destitute: from the poverty of distinct markings on the forewing; Haworth's English name for this species was 'the plain drab'.
922 **schreibersiana** (Frölich, 1828) – i.h.o. K. F. A. von Schreibers (1775–1852), a German entomologist.
923 **sodaliana** (Haworth, 1811) – *sodalis*, a companion: from a stated resemblance to 1197 *Eucosma campoliliana* (*nigromaculana* Haworth) and 1048 *Acleris variegana* (*griseana* Haworth).

Hysterophora Obraztsov, 1944 – ὕστερος (husteros), behind, posterior; φορέω (phoreō), to bear: suggesting affinity with 921 *Hysterosia*, q.v.
924 **maculosana** (Haworth, 1811) – *maculosus*, spotted: from the forewing pattern.

Phtheochroa Stephens, 1829 – φθέω (phtheō), to fade; χρόα (khroa), colour, especially of the skin: Stephens set up this genus to accommodate 925 *rugosana* and wrote 'a very remarkable pecularity also belongs to this genus, which is that a portion of the colouring of the wings fades almost immediately upon the death of the animal, some fine rufous marks fading on that event taking place'. 'From appearance of forewings' (Macleod).
925 **rugosana** (Hübner, 1799) – *rugosus*, wrinkled: from the roughened appearance caused by the scale-tufts on the forewing.

Piercea Filipjev, 1940 – i.h.o. F. N. Pierce (1861–1943), the British entomologist and one of the authors of *The Genitalia of the British Tortricidae*.
926 see below 932
927 **minimana** (Caradja, 1916) – *minimus*, smallest: British specimens, however, are not noticeably smaller than their congeners.
928 **permixtana** ([Denis & Schiffermüller], 1775) – *per-*, intensifying prefix; *mixtus*, mingled: probably referring to the suffused forewing markings involving several colours.
929 **vectisana** (Humphreys & Westwood, 1845) – *Vectis*, the Latin name for the Isle of Wight where the species was first recognized.
930 **alismana** (Ragonot, 1883) – from *Alisma plantago-aquatica*, water-plantain, which is the foodplant.
931 **luridana** (Gregson, 1870) – *luridus*, pale yellow: from the ground colour of the forewing.

Phalonidia Le Marchand, 1933 – the genus *Phalonia* Hübner 1825; εἶδος (eidos), form: indicating affinity with *Aethes* Billberg, of which *Phalonia* is a junior synonym; φάλος (phalos), part of the helmet, perhaps the base of the plume: from the thoracic crest present in some species of the genus.

932 **affinitana** (Douglas, 1846) – *affinis*, akin: from close relationship with and similarity to 929 *Piercea vectisana*.

926 **manniana** (Fischer von Röslerstamm, 1839) – i.h.o. J. J. Mann (1804–89), a Viennese entomologist.

933 **gilvicomana** (Zeller, 1847) – *gilvus*, pale yellow; *coma*, hair of the head: from the yellow coloration of the head and forewing.

934 **curvistrigana** (Stainton, 1859) – *curvus*, curved; *striga*, a furrow, a stripe: from the curved fascia on the forewing.

Cochylimorpha Razowski, 1959 – the genus 962 *Cochylis*, q.v.; μορφή (morphē), form: a genus resembling *Cochylis*.

= **Stenodes** Guenée, 1845 *nec* Dujardin, 1844 – στενώδης (stenōdēs), somewhat narrow: from the rather narrow forewing.

935 **alternana** (Stephens, 1834) – *alternus*, alternate: from the alternation of yellowish ground colour and ochreous brown fasciae.

936 **straminea** (Haworth, 1811) – *stramineus*, straw-coloured: from the ground colour of the forewing.

Agapeta Hübner, 1822 – ἀγαπητός (agapētos), beloved, desirable: from the attractive appearance of the moths.

937 **hamana** (Linnaeus, 1758) – *hamus*, a hook: Linnaeus described the posterior streak on the forewing as '*hamata*', hook-shaped.

938 **zoegana** (Linnaeus, 1767) – i.h.o. J. Zoega, a Swiss entomologist who accompanied Fabricius to Uppsala where they became pupils of Linnaeus. Zoega captured the species at Hammarby, Linnaeus' residence (Pickard *et al.*). Earlier Clerck (1759) had figured both 937 and 938 under the name *hamana*.

Aethes Billberg, 1820 – ἀήθης (aēthēs), unusual, strange: perhaps because a yellowish ground colour is unusual in the Tortricoidea, this being a character common to the five moths included by Billberg in his genus.

939 **tesserana** ([Denis & Schiffermüller], 1775) – *tessera*, a square, a die: from the tessellate forewing pattern.

940 **rutilana** (Hübner, 1817) – *rutilus*, red: from the colour of the forewing pattern.

941 **hartmanniana** (Clerck, 1759) – i.h.o. Professor P. I. Hartmann (1727–91), a German physician and botanist.

942 **piercei** Obraztsov, 1952 – i.h.o. F. N. Pierce, see 927 (genus).

943 **margarotana** (Duponchel, 1836) – *margarita*, a pearl, μάργαρος (margaros), the pearl-oyster: from the shining white markings forming part of the forewing pattern.

944 **williana** (Brahm, 1791) – perhaps in honour of a friend or entomologist; Brahm gives no explanation. The absence of the letter 'w' in Latin indicates that the source is a proper name.

945 **cnicana** (Westwood, 1854) – from *Cnicus*, a synonym of *Cirsium*, the genus of the foodplants.

946 **rubigana** (Treitschke, 1830) – *rubigo, rubiginis* (*robigo*), rust: from the ferruginous forewing markings.

947 **smeathmanniana** (Fabricius, 1781) – i.h.o. H. Smeathman (1750–87), a British entomologist who studied termites in Sierra Leone.

948 **margaritana** (Haworth, 1811) – *margarita*, a pearl: from the pearly white ground colour of the forewing.

949 **dilucidana** (Stephens, 1852) – *dilucidus*, clear, bright, distinct: from the simple yet striking forewing pattern.

950 **francillana** (Fabricius, 1794) – i.h.o. J. Francillon (1744–1816), a British entomologist.

951 **beatricella** (Walsingham, 1898) – i.h.o. the Hon. Mrs Beatrice Carpenter, who reared specimens about 1880.

Commophila Hübner, 1825 – κομμός (kommos), decoration; φίλος (philos), dear, loving: from a reaction to the attractive wing pattern of the next species.

952 **aeneana** (Hübner, 1800) – *aëneus*, brazen: from the submetallic spots on the forewing.

Eugnosta Hübner, 1825 – εὔγνωστος (eugnōstos), easy to discern: from the distinctive wing-pattern.

953 **lathoniana** (Hübner, 1800) – Leto or Latona, Greek goddess and mother of Apollo and Artemis by Zeus: no entomological application.

Eupoecilia Stephens, 1829 – εὐποίκιλος (eupoikilos), variegated: from the forewing pattern.

954 **angustana** (Hübner, 1799) – *angustus*, narrow: from the shape of the forewing, which is relatively narrow with the costa unarched.

subsp. **thuleana** Vaughan, 1880 – the subspecies occurring in '*Ultima Thule*' (Shetland): see 14 *Hepialus humuli*.

955 **ambiguella** (Hübner, 1796) – *ambiguus*, doubtful: Hübner also called it in German 'the doubtful moth'. The termination '*-ella*' shows that he described it as a tineid *sensu lato*, and he may well have been doubtful whether he was right to do so. 'From resemblance to an Indian species' (Macleod), conjecture presented as fact, but entirely without foundation, there being no hint of this in Hübner's text.

Cochylidia Obraztsov, 1956 – the genus 962 *Cochylis*, q.v.; εἶδος (eidos), form: indicating generic affinity.

956 **implicitana** (Wocke, 1856) – *implicitus*, embraced: from its having been formerly included as a single species with 957 *C. heydeniana*.

957 **heydeniana** (Herrich-Schäffer, 1851) – i.h.o. Senator H. G. von Heyden (1793–1866), a German entomologist and friend of Stainton.

958 **subroseana** (Haworth, 1811) – *sub-*, rather; *roseus*, rosy: from the forewing which is rosy, but less so than that of 962 *Cochylis roseana*, named in the same work by Haworth.

959 **rupicola** (Curtis, 1834) – *rupes*, a rock; *colo*, to inhabit: the larva feeds on hemp-agrimony (*Eupatorium cannabinum*), which often grows on damp cliffs.

Falseuncaria Obraztsov & Swatschek, 1958 – *falsus*, false; *uncus*, the posterior extension of the tergum of abdominal segment 9 in the male. 'The great development of the uncus and general modification demands a separate genus' (Pierce & Metcalfe, 1922). Obraztsov & Swatschek complied with this suggestion.

960 **ruficiliana** (Haworth, 1811) – *rufus*, red; *cilium*, an eyelash, *cilia* (pl.), a fringe: from the cilia of the forewing which are, however, yellowish ochreous rather than red.

961 **degreyana** (McLachan, 1869) – i.h.o. the Hon. Thomas de Grey, later Lord Walsingham (1843–1919), the British microlepidopterist.

Cochylis Treitschke, 1829 – χογχύλη (khogkhulē), a conch, a shell: the link with the moths is probably one of colour. Sodoffsky (1837) emended the name *Conchylis*, which would have been a better transliteration of the Greek.

962 **roseana** (Haworth, 1811) – *roseus*, rosy: from the pink flush on the forewing; cf. 958.

963 **flaviciliana** (Westwood, 1854) – *flavus*, yellow; *cilium*, see 960: from the yellow cilia of the forewing.

964 **dubitana** (Hübner, 1799) – *dubius*, doubtful, *dubitare*, to doubt: from uncertainty whether it was a distinct species.

965 **hybridella** (Hübner, 1813) – *hibrida* (*hybrida*), a hybrid: in appearance like a cross between related species.

966 **atricapitana** (Stephens, 1852) – *ater*, black; *caput*, *capitis*, the head: from the black head which is a distinguishing feature of the adult.

967 **pallidana** Zeller, 1847 – *pallidus*, pale: from the forewing which is dull white with pale greyish markings.

968 **nana** (Haworth, 1811) – *nanus*, a dwarf: this was the smallest British cochyline when Haworth named it; 929 *Piercea vectisana* is smaller but was named 34 years later.

TORTRICINAE (1033)

Pandemis Hübner, 1825 – πάνδημος (pandēmos), belonging to the people, common; also an epithet of Aphrodite, the goddess of love. In 1825 Hübner established two other tortricid genera with the termination *-demis*, 986 *Syndemis* and 1113 *Eudemis*, his diagnosis emphasizing in each case the conspicuous diagonal fascia extending across the forewing. Plato postulated that there were two manifestations of Aphrodite, Aphrodite Urania, the goddess of heavenly love, the pure love between souls, whence came our phrase 'Platonic love'; and Aphrodite Pandemos, the goddess of the baser carnal love practised by the common people. The diagonal fascia across the moths' wings may have suggested to Hübner the bend sinister or 'fesse' (a word derived from *fascia*) on a heraldic shield, this being a mark of illegitimacy; this may have put him in mind of the goddess who presided over such activity, Pandemos being her appropriate title in such a context. *Syndemis* and *Eudemis* are names expressing affinity (see p. 35). I have no evidence to support this flight of fancy.

969 **corylana** (Fabricius, 1794) – *Corylus avellana*, the hazel: a larval foodplant.

970 **cerasana** (Hübner, 1786) – *Prunus cerasus*, dwarf cherry: a larval foodplant.

971 **cinnamomeana** (Treitschke, 1830) – κιννάμωμον (kinnamōmon), cinnamon: from the ground colour of the forewing.

972 **heparana** ([Denis & Schiffermüller], 1775) – ἧπαρ (hēpar), the liver: from the ground colour of the forewing.

973 **dumetana** (Treitschke, 1835) – *dumetum*, a thicket: in Britain, however, its habitats are fenland and downland.

Argyrotaenia Stephens, 1852 – ἄργυρος (arguros), a silver; ταινία (tainia), a band: from the silvery fasciae on the forewing of the next species.

974 **ljungiana** (Thunberg, 1797) – i.h.o. S. I. Ljung, also spelt Liung and Ljungh (1757–1828), a Swedish entomologist.

= **pulchellana** (Haworth, 1811) – *pulchellus*, dim., *pulcher*, beautiful: from the attractive pattern of the forewing.

Homona Walker, 1863 – probably a meaningless neologism.

975 **menciana** Walker, 1863 – meaning not traced; Walker spells the name with a capital letter, so it is probably derived from a person or place.

Archips Hübner, 1822 – ἀρχι- (arkhi-), chief; ἴψ (ips), a worm or larva that eats vine-buds; the larvae of several of the member species are minor pests of fruit-trees. Hübner originally proposed this name in his *Tentamen* [1806], giving *A. oporana* (now *podana*, see below) as the example; this is indeed a pest of fruit-trees.

976 **oporana** (Linnaeus, 1758) – ὀπώρα (opōra), late summer, the season when fruit ripens, sometimes the fruit itself; '*habitat in* Malo': this name and description were meant to apply to the next species, *A. podana*, and Linnaeus on the next page of *Systema Naturae* calls the present species *piceana*, giving an accurate description of its larva and feeding habits on pine. Unfortunately an example of his pine-feeding *piceana* stands as the type specimen of his fruit-feeding *oporana* and, as the type specimen takes precedence, this moth bears a name intended for another species; cf. 688 *heracliana*.

977 **podana** (Scopoli, 1763) – i.h.o. N. Poda von Neuhaus (1723–98), Professor of Physics at Graz and an entomologist.

978 **betulana** (Hübner, 1787) – *Betula*, birch: a Continental foodplant, though in Britain the larva has been found only on bog-myrtle (*Myrica gale*).

979 **crataegana** (Hübner, 1799) – *Crataegus*, hawthorn: a likely enough foodplant for this polyphagous species.

980 **xylosteana** (Linnaeus, 1758) – *Lonicera xylosteum*, fly honeysuckle: an occasional foodplant and the one listed by Linnaeus.

981 **rosana** (Linnaeus, 1758) – *Rosa*, rose: a foodplant and listed as such by Linnaeus.

981a **argyrospila** (Walker, 1863) – ἄργυρος (arguros), silver; σπίλος (spilos), a spot: descriptive of the forewing.

Choristoneura Lederer, 1859 – χωριστός (khōristos), separated; νεῦρον (neuron), a nerve, a wing-vein: from vein 2 (Cu_2) of the forewing, which is interrupted beyond the cell.

982 **diversana** (Hübner, 1817) – *diversus*, different: a distinct species, possibly from 981 *Archips rosana* or 986 *Syndemis musculana* with which it was placed by Hübner.

983 **hebenstreitella** (Müller, 1764) – i.h.o. J. E. Hebenstreit (1703–51), a German entomologist.

984 **lafauryana** (Ragonot, 1875) – i.h.o. M. Lafaury (19th century), a French entomologist who discovered the species at Dax in south-western France.

Cacoecimorpha Obraztsov, 1954 – the genus *Cacoecia* Hübner, 1825, a junior synonym of 976

Archips, from κακός (kakos), bad; οἰκία (oikia), a house, the 'bad houses' being the larval spinnings on fruit-trees, causing damage; and μορφή (morphē), shape, form: a genus resembling *Cacoecia*. Macleod unjustifiably wished to emend the name to *Caecoecia*, 'from Lat. *caecus*, hidden, and Gr. *oikeo*, dwell, with reference to moth's habit of resting on background resembling itself'; the moths have no such habit and the forewing is not cryptically patterned.

985 **pronubana** (Hübner, 1799) – having affinity with 2107 *Noctua pronuba*, q.v., in that both species have yellow hindwings.

Syndemis Hübner, 1825 – probably syncopated from σύν- (sun-), with, near, and 968 *Pandemis*, q.v., both genera being characterized by the diagonal fascia on the forewing.

986 **musculana** (Hübner, 1799) – *musculus*, dim., a small mouse: from the greyish suffusion on the forewing. '*Musculus*, muscle, from female's stoutness' (Macleod); possible but less likely; cf. 2378 *musculosa*.

subsp. **musculinana** (Kennel, 1899) – a spelling variant (not dim.) of *musculana*: the subspecies occurring in the Outer Hebrides, Orkney and Shetland Islands.

Ptychomoloides Obraztsov, 1954 – the genus 1000 *Ptycholoma*, q.v.; εἶδος (eidos), form: a genus having affinity with *Ptycholoma*.

987 **aeriferanus** (Herrich-Schäffer, 1851) – *aerifer*, bearing brass, an epithet of the worshippers of Bacchus who bore brazen cymbals: from the golden yellow ground colour of the forewing.

Aphelia Hübner, 1825 – ἀφέλεια (apheleia), plainness: from the almost unicolorous forewings of the species in the genus.

988 **viburnana** ([Denis & Schiffermüller], 1775) – *Viburnum lantana*, the wayfaring-tree: recorded as a foodplant on the Continent but not in Britain.

989 **paleana** (Hübner, 1793) – *palea*, chaff: from the whitish ochreous forewing.

990 **unitana** (Hübner, 1799) – *unitas*, uniformity: from the unicolorous forewing; cf. 642 *unitella*.

Clepsis Guenée, 1845 – κλέπτω (kleptō), to steal, to conceal, in composition κλεψ- (kleps-), often implying furtiveness: probably from the habits of the larvae which secrete themselves in spinnings. 'Perhaps because it "stole" species from other genera' (Macleod): Guenée never named a genus on such grounds.

991 **senecionana** (Hübner, 1819) – *Senecio*, the ragwort genus: not, however, recorded as a foodplant.

992 **rurinana** (Linnaeus, 1758) – formed from *rus, ruris*, the country, farmland: from the habitat, though one shared with many other species.

993 **spectrana** (Treitschke, 1830) – *spectrum*, an image, a spectre: possibly referring to the pallor of the often whitish ochreous forewing.

994 **consimilana** (Hübner, 1817) – *consimilis*, entirely similar: possibly from the often almost unicolorous forewing; there is no obvious similarity to another species.

995 **trileucana** (Doubleday, 1847) – τρι- (tri-), three; λευκός (leukos), white: an American species, presumably with three white fasciae.

996 **melaleucanus** (Walker, 1863) – μέλας (melas), black; λευκός (leukos), white: an American species, presumably with a black and white pattern.

Epichoristodes Diakonoff, 1960 – the genus *Epichorista* Meyrick, 1909, not represented in Britain and named, like 982 *Choristoneura* (q.v.), from the wing venation; εἶδος, ὠδ- (eidos, ōd-), form: a genus resembling *Epichorista*.

997 **acerbella** (Walker, 1864) – *acerbus*, bitter; Walker described this American species as a *Depressaria* and gave no reason for the name.

Epiphyas Turner, 1927 – ἐπί (epi), upon; φυάς (phuas), a shoot: from the larval feeding habits.

998 **postvittana** (Walker, 1863) – *post*, behind; *vitta*, a band: from the pattern of the forewing, in which the area beyond the median fascia is darker.

Adoxophyes Meyrick, 1881 – ἄδοξος (adoxos), ignoble; φυή (phuē), a growth, nature: from the ill-repute of 999 *A. orana* as a pest species. 'Strange, . . . from double brood of moths' (Macleod); he had confused the meaning of ἄδοξος with παράδοξος (paradoxos), and for some reason thought a second brood strange.

999 **orana** (Fischer von Röslerstamm, 1834) – *ora*, the coast, a margin: Fischer von Röslerstamm attributes the name to von Tischer and gives no explanation. The type localities are not coastal and no prominent marginal marking is described. 'Greek *hora*, autumn, from the second season of moth's appearance' (Macleod): ὥρα when applied to a season, means summer and not autumn, and authors seldom derived specific names from Greek sources; Germans did not drop their aitches as a Frenchman or Stephens might have done. A more convincing guess is that the name is an aphaeretic form of 976 *oporana* (q.v.), which was intended for 977 *podana* and used for that species in 1834; both species feed on apple and are regarded as pests.

Ptycholoma Stephens, 1829 – πτύξ, πτυκός (ptux, ptukos), a fold; λῶμα (lōma), a border, a margin: from the strong costal fold on the male forewing.

1000 **lecheana** (Linnaeus, 1758) – i.h.o. J. Leche (mid-18th century), a Finnish entomologist.

Lozotaeniodes Obraztsov, 1954 – the genus 1002 *Lozotaenia*, q.v.: εἶδος, ὠδ- (eidos, ōd-), form: a genus resembling *Lozotaenia*.

1001 **formosanus** (Geyer, 1830) – *formosus*, beautiful: in approbation.

Lozotaenia Stephens, 1829 – λοξός (loxos), oblique; ταινιά (tainia), a band: from the slanting fascia on the forewing of 1002 *L. fosterana* and the twenty-seven other species placed by Stephens in the genus; cf. 1087 *Orthotaenia*. The name is malformed for '*Loxotaenia*', Stephens having confused the Greek letters xi (ξ) and zeta (ζ) both here and in other names compounded with λοξός not cited in this work.

1002 **forsterana** (Fabricius, 1781) – i.h.o. J. R. Forster (1729–98), a friend of Fabricius; see 394 *forsterella*.

1003 **subocellana** (Stephens, 1834) – *sub-*, somewhat; *ocellus*, a small eye: this extinct species, of which no specimen survives, must have had an ill-defined ocellus.

Paramesia Stephens, 1829 – παράμεσος (paramesos), next to the middle: from the median fascia, usually reduced to a costal spot.

1004 **gnomana** (Clerck, 1759) – γνώμων (gnōmōn), the gnomon of a sundial, a carpenter's square: from the shape of the central fascia which in Clerck's figure is reduced to two striae joined at right angles.

Periclepsis Bradley, 1977 – περι- (peri-), near to; the genus 991 *Clepsis*: a related genus.

1005 **cinctana** ([Denis & Schiffermüller], 1775) – *cinctus*, girt: from the dark median fascia on each forewing; these fasciae unite when the wings are folded, giving the appearance of a belt.

Epagoge Hübner, 1825 – ἐπαγωγή (epagōgē), properly a leading up of troops, an invasion, but it can also mean leading on, enticing, alluring: Hübner gives no clue; he may have looked in his Greek-German lexicon for a word meaning 'attractive' and found this one.

1006 **grotiana** (Fabricius, 1781) – probably i.h.o. Hugo Grotius (1583–1645), the distinguished Dutch jurist and national hero. Like Grotius, Fabricius had studied at Leiden University. When he named a species after a contemporary, Fabricius usually gave a brief biographical note; here there is none, indicating that the dedicatee needed no introduction. Grotius had no connection with entomology. 'After Canadian entomologist, Grote (19th century)' (Macleod): oh dear! A. R. Grote (1841–1903) was born 60 years after the name was bestowed and was German, though he did spend much of his working life in Canada and the U.S.A.

Capua Stephens, 1834 – Capua, the chief town of Campania in Italy: Stephens confirms this derivation, there being no entomological significance. Ochsenheimer was largely responsible for setting the fashion of naming genera after towns and islands; see p. 33.

1007 **vulgana** (Frölich, 1828) – *vulgus*, the common people: a name suggesting a plebeian, undistinguished-looking moth, which it is. Earlier attempts by Hübner to name it were of no avail because he muddled his figures.

Philedone Hübner, 1825 – φιλήδονος (philēdonos), fond of pleasure: the moths fly by day and may be supposed to revel in the sunshine.

1008 **gerningana** ([Denis & Schiffermüller], 1775) – i.h.o. J. C. Gerning (18th century), a German entomologist who lived at Frankfurt and was a member of the leading entomological societies; later he used to send specimens to Hübner for him to figure and to lend him books from his extensive library.

Philedonides Obraztsov, 1954 – the genus 1008 *Philedone*, q.v.; εἶδος (eidos), form: indicating affinity between the genera.

1009 **lunana** (Thunberg, 1784) – *luna*, the moon: *'fasciis duabus fuscis lunatis'*, with two crescentic fuscous fasciae. The fasciae are more strongly expressed in the female, the subterminal being markedly crescentic.

Ditula Stephens, 1829 – δίτυλος (ditulos), with two humps: from the bifid posterior crest on the thorax of the adult.

1010 **angustiorana** (Haworth, 1811) – *angustior*, more narrow, rather narrow: from the shape of the forewing which is narrow compared with those of the somewhat similarly marked *Archips* spp. (976–981).

Pseudargyrotoza Obraztsov, 1954 – ψεῦδος (pseudos), a falsehood; *Argyrotoza* Stephens, 1829, synonym of 1035 *Croesia* – ἀργυρότοξος (argurotoxos), bearer of the silver bow, an epithet of Apollo: indicating similarity between the genera, but deceptive because it is in appearance rather than structure. The confusion by Stephens of the letters ζ and ξ (cf. 1002 *Lozotainia*) led to the amendment *Argyrotoxa* Agassiz, 1846. Adults in both genera have submetallic silver scales on the forewing.

1011 **conwagana** (Fabricius, 1775) – i.h.o. Conway, a British entomologist (18th century). He collected with Fabricius ('an indefatigable companion') in the woods round London, but Oxford is the type locality of this species. The name may be a typographical error for *'conwajana'*, 'j', the consonantal 'i', being the usual substitute for 'y' which does not occur in the Latin alphabet; cf. 345 *rajella*.

Sparganothis Hübner, 1825 – σπαργανοθείς (sparganotheis), swathed: either from the fasciae of raised scales on the forewing or from the spinnings in which the larvae involve themselves. Hübner placed eight species, including 1012 *pilleriana*, in the genus and it is likely that he knew the life history of some of them.

1012 **pilleriana** ([Denis & Schiffermüller], 1775) – i.h.o. M. Piller (18th century), a Viennese entomologist, Jesuit priest and professor; in 1794 he was *'Praepositus in Hungaria'*. He collaborated with Professor L. Mitterpacher (see 1120).

1013 }
1014 } see below 1062.

Eulia Hübner, 1825 – εὖ (eu), well; λεῖος (leios), smooth: from the forewing pattern of 1015 *ministrana*, the only species included in the genus by Hübner, in which the markings merge smoothly without clear definition.

1015 **ministrana** (Linnaeus, 1758) – *minister*, a servant: possible because the coloration suggested a footman's livery to Linnaeus, but perhaps without entomological application.

Cnephasia Curtis, 1826 – κνέφας (knephas), darkness: from the sombre grey and fuscous forewings of the species.

1016 **longana** (Haworth, 1811) – *longus*, long: from the relatively elongate forewing.

1017 **gueneana** (Duponchel, 1836) – i.h.o. A. Guenée (1809–80), the distinguished French entomologist and jurist; cf. 1284 and 2354.

1018 **communana** (Herrich-Schäffer, 1851) – *communis*, shared in common: widespread, but, strictly, not common in the sense plentiful.

1019 **conspersana** Douglas, 1846 – *conspersus*, besprinkled: from the grey irroration on the whitish forewing: 'sprinkled with griseous atoms'.

1020 **stephensiana** (Doubleday, 1849) – i.h.o. J. F. Stephens (1792–1852), the British entomologist and author; cf. 592 (genus).

f. **octomaculana** Curtis, 1850 – *octo*, eight; *macula*, a spot: from the conspicuous dark spots, four on each forewing.

1021 **asseclana** ([Denis & Schiffermüller], 1775) – *assecla*, an attendant, a follower, a hanger-on: perhaps in the sense of an addition to a group of homogeneous species.

= **interjectana** (Haworth, 1811) – *interjectus*, interposed, added: Haworth was understandably puzzled by the identity of *Pyralis cretana* of Fabricius; he first described 1020 *C. stephensiana* as *Tortrix asinana*, and then 'added' this species, saying that if the first was not, this one certainly was the *cretana* of Fabricius. For Pyralis *usu* Fabricius see p. 29.

1022 **pasiuana** (Hübner, 1799) – authors have recognized that this is a misprint, and Hübner himself (1822) emended it to *pascuana* – *pascuus*, belonging to pasture: from the habitat. Others have suggested *passivana* (*pasivana*) – *passivus*, found everywhere. However, according to I.C.Z.N. rules, the original spelling must stand.

1023 **genitalana** Pierce & Metcalfe, 1915 – *genitalia*, the genital organs: it was recognized as a distinct species only after the examination of the genitalia.

1024 **incertana** (Treitschke, 1835) – *incertus*, uncertain: referring either to the variability of the forewing pattern, or (Macleod) to uncertainty about its status as a distinct species.

Tortricodes Guenée, 1845 – the genus 1033 *Tortrix*, q.v.; εἶδος, ὠδ – (eidos, ōd-), form: a genus resembling *Tortrix*; Guenée placed the genus in the Tineae *sensu lato* although recognizing some tortricoid characters.

1025 **alternella** ([Denis & Schiffermüller], 1775) – *alternus*, alternate: from the forewing pattern in which there is an alternation of greyish white and ochreous brown. The termination '*-ella*' shows that Schiffermüller also thought the species was a tineid.

Exapate Hübner, 1825 – ἐξαπάτη (exapatē), gross deceit: the deceit for Hübner lay in a tineid looking like a tortricid; in fact, it is the other way round.

1026 **congelatella** (Clerck, 1759) – *con-*, with; *gelu*, frost; *congelatus*, frozen: the moth appears with the first frosts from late October to December. The termination '*-ella*' shows that Clerck had supposed the species to be a tineid.

Neosphaleroptera Réal 1953 – νέος (neos), new; the genus *Sphaleroptera* Guenée, 1845, a synonym of 1016 *Cnephasia*, from σφαλερός (sphaleros), delusive, uncertain; πτερόν (pteron), a wing: Guenée stated that he was not sure that his genus was distinct, but the name refers rather to the difficulty of species determination.

1027 **nubilana** (Hübner, 1799) – *nubilus*, cloudy, overcast: from the fuscous wings.

Eana Billberg, 1820 – ἑανός (heanos), a fine robe, especially one that is white: the seven species included by Billberg all have white or whitish ground colour.

1028 **argentana** (Clerck, 1759) – *argentum*, silver: from the silvery white wings.

1029 **osseana** (Scopoli, 1763) – *osseus*, of bone: from the pale greyish ochreous forewing.

1030 **incanana** (Stephens, 1852) – *incanus*, hoary: from the greyish irroration on the forewing.

1031 **penziana** (Thunberg, 1791) – i.h.o. Consiliar D. Pentz (18th century), who took the type specimen in Sweden. The first reference is in a dissertation by Thunberg's pupil P. E. Becklin who is often incorrectly cited as co-author (Karsholt & Nielsen, 1986).

subsp. **colquhounana** (Barrett, 1884) – i.h.o. Dr H. Colquhoun (19th century), a medical practitioner living in Glasgow who discovered the subspecies.

Aleimma Hübner, 1825 – ἄλειμμα (aleimma), unguent, oil: the forewing markings of the next species tend to be diffused and so to appear greasy or smudged; Hübner describes its forewings as having a soft grey pattern.

1032 **loeflingiana** (Linnaeus, 1758) – i.h.o. P. Löfling (1727–56), a Swedish botanist. He was a pupil of Linnaeus and recorded in Spain and America.

Tortrix Linnaeus, 1758 – *tortor*, fem. *tortrix*, a torturer, a twister: '*Tortricum larvae contorquent & filo connectunt folia, quae vorant & intra quae se recipiunt*' (the larvae of the Tortrices twist and spin leaves which they eat and within which they secrete themselves). Linnaeus' family names for moths are feminine and for this reason he chose the feminine form of the name; see p. 37.

1033 **viridana** (Linnaeus, 1758) – *viridis*, green: from the colour of the forewing.

Spatalistis Meyrick, 1907 – σπαταλιστής (spatalistēs), a profligate, σπατάλη (spatalē), luxury, an ornament, often of gold: the next species has only submetallic markings but Indo-Malayan members of the genus are more brightly coloured.

1034 **bifasciana** (Hübner, 1787) – *bi-*, two; *fascia*, a band: continental specimens tend to have two distinct fasciae but they usually merge in British examples.

Croesia Hübner, 1825 – Croesus, King of Lydia, 560–546 B.C., famed for his wealth in gold: some of the species in the genus have their ground colour golden; cf. 151 *croesella*.

1035 **bergmanniana** (Linnaeus, 1758) – i.h.o. T. O. Bergmann (d. 1784), a Swedish entomologist.

1036 **forsskaleana** (Linnaeus, 1758) – i.h.o. P. Forsskahl or Forsskål (1732–63), a Swedish botanist and pupil of Linnaeus. He appears to have been responsible for the vernacular name 'death's-head hawk-moth' (1973, q.v.).

1037 **holmiana** (Linnaeus, 1758) – Holmia, Stockholm, the capital of Sweden.

Acleris Hübner, 1825 – ἄκληρος (aklēros), unallotted: perhaps a genus set up to accommodate species not allocated to another; Hübner placed nine species in this genus.

1038 **laterana** (Fabricius, 1794) – *later*, a brick: from the brownish red, reticulate forewing pattern, suggestive of a brick wall.

= **latifasciana** (Haworth, 1811) – *latus*, broad; *fascia*, a band: from the broad fascia on the forewing of some forms.

1039 **comariana** (Lienig & Zeller, 1846) – *Comarium* (now *Potentilla*) *palustris*, marsh cinquefoil, a larval foodplant.

1040 **caledoniana** (Stephens, 1852) – Caledonia, Scotland, where it was first taken.

1041 **sparsana** ([Denis & Schiffermüller], 1775) – *sparsus*, bespattered: from the forewing pattern, which is indicated mainly by irroration.

1042 **rhombana** ([Denis & Schiffermüller], 1775) – ῥόμβος, Lat. *rhombus*, a rhomb: from the costal blotch on the forewing which is often so shaped.

1043 **aspersana** (Hübner, 1817) – *aspersus*, sprinkled: from the reticulation and striation of the forewing in some forms.

1044 **ferrugana** ([Denis & Schiffermüller], 1775) – *ferrugo*, the colour of iron-rust: from the colour of the forewing.

1045 **notana** (Donovan, 1806) – *nota*, a mark, a spot: the name was bestowed on the form that has the forewing coarsely irrorate with black dots.

= **tripunctana** (Hübner, 1799) – *tri-*, three; *punctum*, a spot: the name was bestowed on the form with the costal blotch broken into three dark spots.

1046 **shepherdana** (Stephens, 1852) – i.h.o. Edwin Shepherd (d. *c.*1883), one-time secretary of the Entomological Society of London, who first took the species.

1047 **schalleriana** (Linnaeus, 1761) – i.h.o. J. G. Schaller (1734–1813), a German entomologist.

1048 **variegana** ([Denis & Schiffermüller], 1775) – *variegatus*, particoloured, variegated: from the variegated forewing pattern of some forms, not from the variability of the species as supposed by Macleod; in the 18th century the varieties were regarded as distinct species.

1049 **permutana** (Duponchel, 1836) – *permuta*, to change, to vary: 'très remarquable par la variété de ses colours', the main contrast being between the whitish basal and reddish distal halves of the forewing. '*Permuto*, to change, being hardly more than variation of *A. variegana*' (Macleod): explanation entirely unsupported by Duponchel's text.

1050 **boscana** (Fabricius, 1794) – i.h.o. L. A. G. Bosc (1759–1828), a French entomologist.

1051 **logiana** (Clerck. 1759) – *logium*, the stage from which actors spoke their lines (λόγος (logos), a word), a platform: from the raised scales on the forewing.

1052 **umbrana** (Hübner, 1799) – *umbra*, a shade: from the heavy black streak extending from the base to the apex of the forewing.

1053 **hastiana** (Linnaeus, 1758) – i.h.o. Reinhart Hast, a Finnish pupil of Linnaeus. '*Hast (Reinh.) fenno, natus ad Insectorum Historiam, at in flore aetatis periit egregius discipulus*' (Reinhart Hast, a Finn, a born student of the natural history of insects and an outstanding pupil, who died, however, in the flower of his youth'): a poignant tribute and one unparallelled in the entomological pages of *Systema Naturae*. Linnaeus also attributes the discovery of the larva on *Salix* to Hast.

1054 **cristana** ([Denis & Schiffermüller], 1775) – *crista*, a tuft, a crest: from the huge scale-tuft on the forewing.

1055 **hyemana** (Haworth, 1811) – *hiems* (*hyems*), winter: the species overwinters as an adult.

1056 **lipsiana** ([Denis & Schiffermüller], 1775) – *Lipsia*, Leipzig, given by Schiffermüller as the type locality.

1057 **rufana** ([Denis & Schiffermüller], 1775) – *rufus*, red: from the reddish brown scaling on the forewing of most forms.

1058 **lorquiniana** (Duponchel, 1835) – i.h.o. P. J. M. Lorquin (1797–1873), the French entomologist who submitted the type material to Duponchel via Lefebvre.

1059 **abietana** (Hübner, 1822) – *Picea abies* (Norway spruce) and *Abies* spp. (silver firs) are foodplants.

1060 **maccana** (Treitschke, 1835) – *maccus*, a buffoon: perhaps because the contrasting colours in some forms suggest motley (Macleod), or because some of its forms mimic other species.

1061 **literana** (Linnaeus, 1758) – *litera*, a letter of the alphabet: from the black markings on the forewing which are 'reduced and fractured into sharply defined hieroglyphic-like shapes' (Bradley *et al.*, 1973).

1062 **emargana** ([Denis & Schiffermüller], 1775) – *e-*, out of; *margo*, an edge, a margin: from the strongly emarginate costa of the forewing.

CHLIDANOTINAE

From the oriental genus *Chlidanota*, Meyrick, 1906 – χλιδανός (klidanos), delicate: in approbation.

Olindia Guenée, 1845 – perhaps suggested by Olinda, a town in South America; Guenée stated that the name had no entomological significance.

1013 **schumacherana** (Fabricius, 1787) – i.h.o. C. F. Schumacher (1757 – 1830), a German entomologist.

Isotrias Meyrick, 1895 – ἴσος (isos), equal; τριάς (trias), a group of three, a triad: from the three more or less equal fasciae on the forewing.

1014 **rectifasciana** (Haworth, 1811) – *rectus*, straight; *fascia*, a band: from the median fascia which is almost directly transverse, not oblique as in many tortricids.

OLETHREUTINAE (1068)

Celypha Hübner, 1825 – κέλυφος (keluphos), a husk, a pod: Hübner gives no explanation; his diagnosis describes the wings as having dark reticulation and some pods are similarly marked.

1063 **striana** ([Denis & Schiffermüller], 1775) – *stria*, a furrow, a streak: from the strigulate forewing pattern.

1064 **rosaceana** (Schläger, 1847) – *rosaceus*, rosy: from the ground colour of the forewing.

1065 **rufana** (Scopoli, 1763) – *rufus*, red: from the ground colour of the forewing.

1066 **woodiana** (Barrett, 1882) – i.h.o. Dr J. H. Wood (1841–1914), who discovered the species in Herefordshire in 1878.

1067 **cespitana** (Hübner, 1817) – *caespes* (*cespes*), *caespitis*, a turf, a sod: from the habitat, not the foodplant; 'not really its habitat' (Macleod) – Hübner, not Macleod, was right.

1067a **rurestrana** (Duponchel, 1843) – *rurestris*, pertaining to the country, rural: from the habitat, one common to many other species.

Olethreutes Hübner, 1822 – ὄλεθρος (olethros), destruction: probably from the damage to young plants caused by common and polyphagous species such as 1076 *O. lacunana*.

1068 **rivulana** (Scopoli, 1763) – *rivulus*, a small stream: from the wavy strigulae (not the 'band', i.e. fascia, as stated by Macleod) on the forewing.

1069 **aurofasciana** (Haworth, 1811) – *aurum*, gold; *fascia*, band: from the deep yellow ground colour traversed by fuscous fasciae; Haworth did not, as Macleod supposed, get the colour of the fasciae wrong.

1070 **mygindiana** ([Denis & Schiffermüller], 1775) – i.h.o. a Danish student named Mygind who was sent down from the University of Hafnia for being too critical of the establishment (*'ob disputationem nimis acerbam'*). He died in 1787 at Vienna where Schiffermüller no doubt had met him and approved of his views. 'Mygind, a town in Denmark' (Macleod).

1071 **arbutella** (Linnaeus, 1758) – *Arbutus unedo*, the strawberry tree: from the similarity in colour between the blood-red forewings of the moth (*'alis sanguineis'*) and the fruit of the tree. 'Not really its foodplant' (Macleod): Linnaeus never suggested that it was, since he stated correctly that the larva fed on *Arctostaphylos uva-ursi*, bearberry.

1072 **metallicana** (Hübner, 1799) – *metallicus*, metallic: from the silvery metallic edging to the blackish brown markings. 'From glossy forewings' (Macleod): they are not glossy.

1073 **schulziana** (Fabricius, 1777) – i.h.o. Dr J. D. Schulz (18th century), a German entomologist who lived in Hamburg.

1074 **palustrana** (Lienig & Zeller, 1846) – *paluster*, marshy: the habitat in Britain is heathland, both dry and damp, where the larva feeds on mosses.

1075 **olivana** (Treitschke, 1830) – *oliva*, an olive: from the olive-ochreous markings on the forewing.

1076 **lacunana** ([Denis & Schiffermüller], 1775) – *lacuna*, a hole, cleft or gap: from the streak of ground colour which interrupts the darker median fascia above its middle.

1077 **obsoletana** (Zetterstedt, 1840) – *obsoletus*, old, worn-out: from a common drab, almost unicolorous, form in which the moth appears to be rubbed after having been long on the wing.

1078 **doubledayana** (Barrett, 1872) – i.h.o. Henry Doubleday (1809–75), the British entomologist and author of *A synonymic list of British Lepidoptera* (1849).

1079 **bifasciana** (Haworth, 1811) – *bi-*, two; *fascia*, a band: from the two fasciae on the forewing.

1080 **arcuella** (Clerck, 1759) – *arcus*, a bow: from the arched fasciae, especially the postmedian, of the forewing. Described in Tinea.

Pristerognatha Obraztsov, 1960 – πριστήρ, πριστῆρος (pristēr, pristēros), a saw; γνάθος (gnathos), the jaw: from the labial palpi, which are enlarged with scales.

1081 **penthinana** (Guenée, 1845) – from the genus *Penthina* Treitschke, 1830, now a synonym of 1082 *Hedya*. *Penthina* was named from πένθος (penthos), sorrow, because of the black and white coloration of the species in the genus, suggestive of funereal clothes. This species had the characters, but not the typical coloration of the genus. 'From black and white markings of forewings' (Macleod): this completely misses the point. This is *not* a black and white moth, but the name draws attention to affinity with others that are so coloured.

Hedya Hübner, 1825 – ἡδύς (hēdus), sweet, pleasing: from the attractive coloration of the moths.

1082 **pruniana** (Hübner, 1799) – *Prunus spinosa*, blackthorn: the larval foodplant.

1083 **dimidioalbana** (Retzius, 1783) – *dimidius*, through the middle, half; *albus*, white: from the pattern of the forewing, in which the basal half (roughly) is dark and the distal half white.

= **nubiferana** (Haworth, 1811) – *nubifer*, cloud-bearing: from the somewhat cloudy plumbeous markings in the distal half of the forewing.

1084 **ochroleucana** (Frölich, 1828) – ὠχρός (ōkhros), pale yellow, sallow: λευκός (leukos), white: from the ochreous-white distal half of the forewing.

1085 **atropunctana** (Zetterstedt, 1840) – *ater*, black; *punctum*, a small spot: from the sprinkling of black spots in the markings of the forewing.

1086 **salicella** (Linnaeus, 1758) – *Salix*, the willow genus, that of the larval foodplants. Although Linnaeus placed it in Tinea because it rests with involute wings, he wrote '*habitat in* Salice. *Tortrix*', i.e. in a spinning.

Orthotaenia Stephens, 1829 – ὀρθός (orthos), straight; ταινία (tainia), a band, a fascia: from the direct fascia on the forewing; cf. 1002 *Lozotaenia*.

1087 **undulana** ([Denis & Schiffermüller], 1775) – *undulatus*, marked as with waves: from the sometimes sinuous margins of the median fascia and, more especially, the wavy strigulae in the pale areas of the forewing.

Pseudosciaphila Obraztsov, 1966 – ψευδο- (pseudo-), false; the genus *Sciaphila* Treitschke, 1829 (junior synonym of 1016 *Cnephasia*), σκιά (skia), a shadow; φιλέω (phileō), to love: from the dusky colours the moths 'like' to assume. Pierce & Metcalfe (1922: 15) had incorrectly placed the next species in *Sciaphila*.

1088 **branderiana** (Linnaeus, 1758) – probably i.h.o. E. Brander, the captor of the first specimens of 2300 *Mormo maura*, 2054 *Utetheisa pulchella*, etc. Linnaeus used the Tortrices to commemorate his former pupils; see p. 20. 'After G. Brander, London antiquary of Swedish family (d. 1787)' (Macleod): very unlikely, since that Brander was not an entomologist.

Apotomis Hübner, 1825 – ἀποτομή (apotomē), a cutting off, a break: from the clearcut separation in some of the species of the dark basal and pale distal halves of the forewing, or from the abbreviated median fascia, e.g. of the next species.

1089 **semifasciana** (Haworth, 1811) – *semi-*, half; *fascia*, a band: from the form of the central fascia on the forewing, which ends abruptly in the disc and is not continued to the dorsum.

1090 **infida** (Heinrich, 1926) – *infidus*, not to be trusted: Heinrich states that it is a very variable species and as such it is deceptive.

1091 **lineana** ([Denis & Schiffermüller], 1775) – *lineus*, flaxen: from the ground colour of the forewing, which is whitish with a pale ochreous tinge.

1092 **turbidana** (Hübner, 1825) – *turbidus*, confused, muddy: from the rather complicated forewing pattern with dusky clouding.

1093 **betuletana** (Haworth, 1811) – *betuletum*, a birch thicket (not classical Latin but coined on the analogy of *quercetum*, *salictum*, etc.): from the foodplant.

1094 **capreana** (Hübner, 1817) – *Salix caprea*, goat-willow: a larval foodplant.

1095 **sororculana** (Zetterstedt, 1840) – *sororcula*, a little sister: from its similarity to and smaller size than other *Apotomis* species.

1096 **sauciana** (Frölich, 1828) – *saucius*, wounded, 'pinked': from the pinkish flush which sometimes suffuses the white ground colour. 'From penetration of basal margin of forewings by sharp white mark like hook' (Macleod); this is gibberish; the outer margin of the median fascia is indented, not wounded.

subsp. **grevillana** (Curtis, 1835) – i.h.o. Dr R. K. Greville who discovered the subspecies in Sutherland in July; Curtis does not give the year.

Endothenia Stephens, 1852 – ἔνδοθεν (endothen), inside, within: from the feeding habits of the larvae within stems, roots and seedheads.

1097 **gentianaeana** (Hübner, 1799) – *Gentiana*, the gentian genus: not the correct foodplant.

1098 **oblongana** (Haworth, 1811) – *oblongus*, rather long: from the somewhat elongate forewing.

= **sellana** (Frölich, 1828) – *sella*, a chair or stool: perhaps from the raised scales at the tornus and along the inner margin of the hindwing underside. 'From appearance of markings on forewings' (Macleod): meaningless.

1099 **marginana** (Haworth, 1811) – *margo*, *marginis*, a border: from the male hindwing, which is white with a broad fuscous terminal suffusion.

1100 **pullana** (Haworth, 1811) – *pullus*, dark-coloured, blackish grey: from the sombre coloration of the insect.

= **fuligana** sensu auctt. – *fuligo*, soot: from the dark-coloured forewing.

1101 **ustulana** (Haworth, 1811) – *ustulo*, to scorch: from the forewing which is blackened as if by scorching.

1102 **nigricostana** (Haworth, 1811) – *niger*, black; *costa*, a rib, the anterior margin of an insect's wing: probably from the veins of the forewing which are obscurely black-lined, especially in the ocellar region. The 'dark base of costa' cited by Macleod is imaginary; the median fascia is restricted to a blackish brown bar on the costa but this is not a prominent character.

1103 **ericetana** (Humphreys & Westwood, 1845) – *ericetum*, heathland (formed like *betuletum* in 1093 above): the insect is not, however, associated with this habitat.

= **trifoliana** (Herrich-Schaffer, 1851) – *Trifolium*, the clover genus: the larva, however, feeds on marsh woundwort (*Stachys palustris*). This name was introduced by Schnack (1985) in the mistaken belief that it was senior to *ericetana*, the date of which had been given incorrectly in Kloet & Hincks (1972) (Emmet, 1987b).

1104 **quadrimaculana** (Haworth, 1811) – *quadri-*, four; *macula*, a spot: from the forewing pattern which is often broken into four discrete spots.

Lobesia Guenée, 1854 – λώβησις (lōbēsis), maltreatment, ruin: possibly a reference to 1107 *L. botrana* which Guenée included in his genus and which is an important pest of vineyards. However, not all Guenée's generic names have entomological applications.

1105 **occidentis** Falkovitsh, 1970 – *occidens*, western: perhaps from its distribution being more westerly than that of *L. euphorbiana* (Freyer) with which it had previously been confused.

1106 **reliquana** (Hübner, 1825) – *reliquus*, remaining: perhaps a species that was left over after all the others had been classified; its appearance is baffling and at one time it was supposed to be a cochyline.

1107 **botrana** ([Denis & Schiffermüller], 1775) – βότρυς (botrus), Lat. *botrus*, a bunch of grapes: the larva is a pest of the grapevine.

1108 **abscisana** (Doubleday, 1849) – *abscisus*, cut off, steep, abrupt: probably from the sharply defined margins of the basal patch and median fascia.

1109 **littoralis** (Humphreys & Westwood, 1845) – *litoralis, littoralis*, pertaining to the sea-shore: from the moth's principal habitat.

Bactra Stephens, 1834 – perhaps from βάκτρον (baktron), a staff, a club, with reference to the maxillary palpus which Stephens describes as 'rather large and thickly enveloped in scales; or from Bactra, later called Balkh and now Vazirabad, a town in northern Afghanistan; it was the capital of Bactria, a province in the empire of Alexander the Great. If the latter explanation, which is given by Pickard *et al.* and Macleod, is correct, it is a 'geographical' name without entomological significance: cf. 1007 *Capua* and see p. 33. Stephens himself offers no explanation.

1110 **furfurana** (Haworth, 1811) – *furfur*, bran: from the ochreous brown markings of the forewing.

1111 **lancealana** (Hübner, 1799) – *lanceola*, a small lance: from the narrow, lanceolate forewing or, less probably, the spear-like longitudinal streak present in some specimens. Although *lancealana* is a typographical error for *lanceolana*, the emendation used in earlier textbooks is not permissible.

1112 **robustana** (Christoph, 1872) – *robustus*, robust: from having a stouter build than other *Bactra* species.

Eudemis Hübner, 1825 – probably formed by syncopation from εὖ (eu), well, and the genus 969 *Pandemis*, q.v. Both genera are named in the same work and both have a conspicuous diagonal fascia on the forewing, a character stressed by Hübner in his diagnosis. The 'εὖ' possibly gives credit for the richer coloration.

1113 **profundana** ([Denis & Schiffermüller], 1775) – *profundus*, deep: from the breadth of the forewing.

1114 **porphyrana** (Hübner, 1799) – πορφύρα (porphura), purple dye: from the forewing markings, though these are plumbeous and reddish brown rather than purple.

Ancylis Hübner, 1825 – ἀγκυλίς (agkulis), a hook, a barb: from the falcate forewing; our word 'angle' is derived indirectly from ἀγκυλίς.

1115 **achatana** ([Denis & Schiffermüller], 1775) – *achates*, agate: from the ferruginous markings of the forewing.

1116 **comptana** (Frölich, 1828) – *comptus*, adorned: in praise of the moth.

1117 **unguicella** (Linnaeus, 1758) – *unguis*, the nail of a finger, a claw: from the falcate forewing.

1118 **uncella** ([Denis & Schiffermüller], 1775) – *uncus*, a hook: from the falcate forewing.

1119 **geminana** (Donovan, 1806) – *geminus*, twin, double: from the two dorsally directed projections from the longitudinal streak on the forewing.

= **biarcuana** (Stephens, 1834) – *bi-*, two; *arcus*, a bow: from the doubly sinuate longitudinal streak.

1119a **diminutana** (Haworth, 1811) – *deminutus, diminutus*, diminutive: from this species being distinctly smaller than 1119 *A. geminana*.

1119b **subarcuana** (Douglas, 1847) – *sub-*, somewhat; *arcus*, a bow, *arcuatus*, bowed: from the undulations of the longitudinal streak being less strongly pronounced than in the last two species.

= **inornatana** (Herrich-Schäffer, 1851) – *inornatus*, unadorned: from the paler, less rich coloration of this species compared with the last two.

1120 **mitterbacheriana** ([Denis & Schiffermüller], 1775) – i.h.o. L. Mitterpacher (d. 1814), Professor of Natural History at Pesth. He collaborated with Piller (see 1012 *pilleriana*).

1121 **upupana** (Treitschke, 1835) – *upupa*, a hoopoe: there is no obvious connection between the bird and the insect; there is certainly none in 'the colour of moth's wings' as supposed by Macleod. See 489 genus.

1122 **obtusana** (Haworth, 1811) – *obtusus*, blunt: from the apex of the forewing which is less strongly falcate than in related species.

1123 **laetana** (Fabricius, 1775) – *laetus*, joyful, pleasing: a very pretty moth which understandably inspires pleasure.

= **lactana** (Fabricius, 1781) – *lac, lactis*, milk: from the whitish ground colour. It is possible that *laetana* was a misprint and that Fabricius was here seeking to correct an error in his *Systema Entomologiae* in which that name had been published.

1124 **tineana** (Hübner, 1799) – Tinea, the Linnaean family comprising moths that rest with convolute wings (see pp. 21–22). I have never seen a living example of this rare species, but it probably wraps its wings round its body.

1125 **unculana** (Haworth, 1811) – *unculus*, a small hook: from the falcate apex of the forewing.

1126 **badiana** ([Denis & Schiffermüller], 1775) – *badius*, chestnut-coloured (of horses); from the markings of the forewing, though these seem chocolate brown rather than chestnut. Cf. 2130 *baja* of the same authors.

1127 **paludana** (Barrett, 1871) – *palus, paludis*, a marsh: from the fenland habitat.

1128 **myrtillana** (Treitschke, 1830) – *Vaccinium myrtillus*, bilberry: the foodplant.

1129 **apicella** ([Denis & Schiffermüller], 1775) – *apex, apicis*, a tip: from the falcate apex of the forewing.

Epinotia Hübner, 1825 – ἐπί (epi), upon; νῶτον (nōton), the back: what is on the back is not clear; Hübner's diagnosis mentions the white spot in the terminal cilia, which could be regarded as on 'the back' when the wings are folded. Some of the species Hübner includes in his genus have white dorsal blotches which would be the obvious explanation – had he mentioned them.

1130 **pygmaeana** (Hübner, 1799) – *Pygmaei*, the pygmies: from the small size of the moth.

1131 **subsequana** (Haworth, 1811) – *subsequens*, following: in *Lepidoptera Britannica* this species follows 1254 *Cydia strobilella* which, according to Haworth, it closely resembles. 'From resemblance to *Hemimene sequana*, both species having been originally included in *Tortrix*' (Macleod): Haworth makes no mention of *sequana*, nor does it resemble this species.

1132 **subocellana** (Donovan, 1806) – *sub-*, somewhat; *ocellus*, a small eye: from the somewhat obscure ocellar marking.

1133 **bilunana** (Haworth, 1811) – *bi-*, twice; *luna*, moon, *lunatus*, crescent-shaped: Haworth describes two contiguous crescentic marks on the dorsum, one black and one grey.

1134 **ramella** (Linnaeus, 1758) – *ramus*, a branch: '*habitat intra quercus ramos*' (it lives inside the branches of oak); Linnaeus was half right: it lives inside catkins and twigs, but of birch. This species, 1138 and 1142 were described in Tinea because of the way in which the wings are folded when the moth is at rest.

1135 **demarniana** (Fischer von Röslerstamm, 1839) – i.h.o. Dr Demarné of Neustrelitz, a German entomologist who died in his 39th year.

1136 **immundana** (Fischer von Röslerstamm, 1839) – *immundus*, strictly unclean, but since *mundus* means clean, neat, ornamented, Fischer von Röslerstamm had the meaning 'plain' in mind: the moth is sober-coloured.

1137 **tetraquetrana** (Haworth, 1811) – *tetraquetrus* (late Lat.), four-angled: from a subquadrate dark area between the basal patch and median fascia on the forewing.

1138 **nisella** (Clerck, 1759) – *nisus*, a sparrow-hawk: from a fancied resemblance in colour; I have consulted Clerck's figure and the comparison is not unreasonable. Described in Tinea (see 1134).

1138a **cinereana** (Haworth, 1811) – *cinereus*, ash-coloured: descriptive of the forewing.

1139 **tenerana** ([Denis & Schiffermüller], 1775) – *tener*, soft, delicate: in praise of the moth.

1140 **nigricana** (Herrich-Schäffer, 1851) – *nigricans*, blackish: from the dark-coloured forewing.

1141 **nemorivaga** (Tengström, 1848) – *nemus, nemoris*, a wood with glades; *vagus*, wandering: from the habitat, though in Britain this is mainly a moorland species.

1142 **tedella** (Clerck, 1759) – *taeda, teda*, a resinous pine: the larval foodplants are Norway spruce and other Pinaceae. Described in Tinea (see 1134).

1143 **fraternana** (Haworth, 1811) – *fraternus*, related as a brother, akin: Haworth refers to its relationship with 1131 *E. subsequana*.

1144 **signatana** (Douglas, 1845) – *signatus*, marked: from the conspicuous sublunate dark mark in the subapical area of the forewing.

1145 **nanana** (Treitschke, 1835) – *nanus*, a dwarf: the smallest British member of the genus.

1146 **rubiginosana** (Herrich-Schäffer, 1851) – *robiginosus, rubiginosus*, rusty: from the ferruginous distal half of the forewing.

1147 **cruciana** (Linnaeus, 1761) – *crux, crucis*, a cross: the median and subterminal fasciae meet at a right angle near the tornus of each forewing, so forming a cross when the wings are folded.

1148 **mercuriana** (Frölich, 1828) – Mercurius, the god or planet Mercury: from the markings on the forewing which are reminiscent of the symbol of the planet (☿); cf. 1344 *mercurella*.

1149 **crenana** (Hübner, 1817) – *crena* (lat Lat.), a notch: in one form of the adult the dorsal half of the wing is mainly white, the costal half blackish fuscous, and there are indentations and projections where the colours meet.

1150 **abbreviana** (Fabricius, 1794) – *abbreviatus*, shortened: *'striga postica abbreviata fusca'*, the fuscous postmedian fascia is cut short, i.e. it does not reach the dorsum.

= **trimaculana** (Donovan, 1806) – *tri-*, three; *macula*, a spot: the basal blotch, median fascia and subapical markings are sometimes sharply separated by the pale areas in between.

1151 **trigonella** (Linnaeus, 1758) – *trigonus*, triangular: Linnaeus describes a double triangular white spot formed when the wings are joined at rest.

= **stroemiana** (Fabricius, 1781) – i.h.o. H. Ström (1726–98), a Norwegian entomologist.

1152 **maculana** (Fabricius, 1775) – *macula*, a spot: there are several dark spots, the most conspicuous that on the dorsum representing the basal patch.

1153 **sordidana** (Hübner, 1824) – *sordidus*, unclean, muddy: the forewing is greyish with the pattern often obsolescent, so looking shabby.

1154 **caprana** (Fabricius, 1798) – *Salix caprea*, goat-willow: the larval foodplant.

1155 **brunnichana** (Linnaeus, 1767) – i.h.o. M. T. Brünnich (18th century), a Danish entomologist; cf. 592.

1156 **solandriana** (Linnaeus, 1758) – i.h.o. D. C. Solander (1738–82), a Swede who settled in England and became curator of the Natural History Department of the British Museum. He sailed, together with Joseph Banks, in the *Endeavour* on Captain Cook's expedition of 1768–71.

Crocidosema Zeller, 1847 – κροκύς, κροκύδος (krokus, krokudos), the spelling κροκίς, κροκίδος which occurs in some manuscripts and was used by Zeller being incorrect (Liddell & Scott, 1869), the nap of woollen cloth, a piece of wool; σῆμα (sēma), a mark: from the dense pecten near the base of the cubital vein on the hindwing upperside, not the underside as stated by Macleod.

1157 **plebejana** Zeller, 1847 – *plebeius*, belonging to the common people, common: from the widespread distribution in warm, temperate regions; it is rare in England which is right on the edge of its range.

Rhopobota Lederer, 1859 – ῥώψ (rhōps), a shrub; βόσκω, βοτ- (boskō, bot-), to eat: most, but not all, of the British species feed on shrubs.

= **Griselda** Heinrich, 1923 – probably from a lady's name. Heinrich seems to have been a ladies' man; at any rate, he liked to form generic names from those of women. Cf. 1214 *Petrova*.

1158 **ustomaculana** (Curtis, 1831) – *ustus*, burnt, scorched; *macula*, a spot: from the wing base which Curtis describes as 'cinereous' (ashen), not the dorsal blotch which is 'silvery grey'.

1159 **naevana** (Hübner, 1817) – *naevus*, a mole on the skin: from a large patch of dark scales in the disc of the male hindwing, more prominent on the underside. 'From markings on forewings' (Macleod).

= **unipunctana** (Haworth, 1811) *nec* (Donovan, 1805) – *unus*, one; *punctum*, a small spot: from a black spot, often prominent, on the distal margin of the median fascia of the forewing.

1160 see below 1162

1161 **stagnana** ([Denis & Schiffermüller], 1775) – *stagnum* a fen: this is one of the moth's habitats where it feeds on devil's-bit scabious (*Succisa pratensis*), but it is also found in dry areas on small scabious (*Scabiosa columbaria*).

1162 **myrtillana** (Humphreys & Westwood, 1845) – *Vaccinium myrtillus*, bilberry: the larval foodplant.

Acroclita Lederer, 1859 – ἄκρον (akron), a point, an apex; κλιτύς (klitus), a slope: explained by Lederer as referring to the tip of the antenna, which tapers abruptly. 'From sinuate outer margin of forewings of some species' (Macleod); this is a character of 1159 *R. naevana*, described in *Tortrix* by Hübner and not included in this genus by Lederer, who makes no mention of the outer margin in his diagnosis.

1160 **subsequana** (Herrich-Schäffer, 1851) – *subsequens*, next, following, imitating: Herrich-

Schäffer described two very similar 'species' under the names *subsequana* and *consequana;* now they are synonymized under the former name.

Zeiraphera Treitschke, 1829 – ζειρά (zeira), a loose garment; φέρω (pherō), to carry: perhaps referring to the larval spinnings.

1163 **ratzeburgiana** (Ratzeburg, 1840) – i.h.o. J. T. C. Ratzeburg (1801–71), a German entomologist. It is not etiquette to name a species in your own honour and Ratzeburg did not do so. Saxesen was the actual nomenclator (Lienig & Zeller, 1846), but Ratzeburg was the first writer to use the name and therefore in accordance with I.C.Z.N. rules he becomes the author.

1164 **rufimitrana** (Herrich-Schäffer, 1851) – *rufus,* red; *mitra,* an eastern head-dress, a mitre: from the ferruginous head of the adult.

1165 **isertana** (Fabricius, 1794) – i.h.o. Dr P. E. Isert (1756–87), a German physician who emigrated to Denmark. He practised medicine and collected insects in Guinea and died there at the age of 31 on his second expedition. It was largely through his influence that Denmark became the first state to prohibit the slave trade. Authors misunderstood the name and emended it to *insertana.*

1166 **diniana** (Guenée, 1845) – Digne, Lat. *Dinia,* the type locality in the French Basses Alpes, where it was discovered by D. Donzel.

Gypsonoma Meyrick, 1895 – γύψος (gupsos), chalk, gypsum; νωμάω (nōmaō), to distribute: from the extensive white areas on the forewings of most species. The name is a rough Greek rendering of 1169 *dealbana.*

1167 **aceriana** (Duponchel, 1843) – *Acer,* the maple genus; the larva, however, feeds on poplar (*Populus* spp.).

1168 **sociana** (Haworth, 1811) – *socius,* an associate, an ally: Haworth described on the same page two forms, *sociana* and *comitana* (*comes, comitis,* a comrade), saying that they might constitute two allied species; they are now synonymized. '*Socius,* ally, from resemblance to *Tortrix comitana, G. sociana* having originally been included in *Tortrix*' (Macleod): he seems to have misunderstood what he was told at the British Museum (Natural History) and supposed that this *comitana* was another species already in existence, e.g. *comitana* [Denis & Schiffermüller], a synonym of 1142 *Epinotia tedella.*

1169 **dealbana** (Frölich, 1828) – *dealbo,* to whitewash: from the white ground colour.

1170 **oppressana** (Treitschke, 1835) – *oppressus,* pressed down, concealed: possibly from the adult's habit of resting flush on a trunk of poplar, where it is concealed by its cryptic pattern.

1171 **minutana** (Hübner, 1799) – *minutus,* very small: though quite a small species, its wingspan is about average for the genus.

1172 **nitidulana** (Lienig & Zeller, 1846) – *nitidulus,* rather bright, handsome (dim. of *nitidus*): the moth is silvery grey with a pretty pattern.

Gibberifera Obraztsov, 1946 – *gibber,* a hump on the back; *fero,* to carry: from the male genitalia in which the sacculus ends in a strongly developed projection.

1173 **simplana** (Fischer von Röslerstamm, 1836) – *simplus,* simple: from the uncomplicated pattern of the forewing which is white with most of the usual tortricid markings obsolete.

Epiblema Hübner, 1825 – ἐπίβλημα (epiblēma), a covering, used either of a cloak thrown over the shoulders or a tapestry draped on walls: Hübner's diagnosis offers no clue. Some of the moths have the thorax and base of the forewing dark, the distal area pale, and this may have suggested a cloak thrown over the shoulders when the wings are folded. Some vernacular names, 'cloaked carpet' (1793 *Euphyia biangulata*), 'short-cloaked moth' (2077 *Nola cucullatella*) and 'cloaked minor' (2341 *Mesoligia furuncula*) are based on this character. See also 1736 *cuculata.*

1174 **cynosbatella** (Linnaeus, 1758) – κυνόσβατος (kunosbatos), the dog-rose: from the larval foodplant. Described as a tineid (termination, *-ella*) because of the way in which the wings are folded at rest.

1175 **uddmanniana** (Linnaeus, 1758) – i.h.o. I. Uddmann, a Finnish entomologist and the author of *Nova Insectorum Species* (1753).

1176 **trimaculana** (Haworth, 1811) – *tri-,* three; *macula,* a spot: from the dark spots, often three in number, in the ocellar region of the forewing.

1177 **rosaecolana** (Doubleday, 1850) – *Rosa*, the rose genus; *colo*, to inhabit: from the larval foodplant.

1178 **roborana** ([Denis & Schiffermüller], 1775) – *robur*, *roboris*, oak: in Britain it feeds on *Rosa* spp. but it has been recorded on oak, as well as other foodplants, on the Continent.

1179 **incarnatana** (Hübner, 1800) – *in-*, intensive prefix; *carnatus*, fleshy, flesh-coloured: from the pinkish flush on the forewing; cf. 'incarnadine'.

1180 **tetragonana** (Stephens, 1834) – *tetragonum*, a quadrangle: from the shape of the pale dorsal blotch on the forewing.

1181 **grandaevana** (Lienig & Zeller, 1846) – *grandaevus*, of great age: the authors stress the relatively huge size of the species and their name seems whimsically to suggest that it is larger because it is older. 'From larval stage, which extends from September to May' (Macleod); had he consulted sources, he would have found that the larva was unknown and that the authors speculate on the possible foodplant.

1182 **turbidana** (Treitschke, 1835) – *turbidus*, confused, muddy: from the indistinctly marked, dull greyish brown forewing.

1183 **foenella** (Linnaeus, 1758) – Pickard *et al.* and Macleod both derive the name from *foenum*, hay, the latter, who misspells it '*foenus*' (which means interest from a loan), adding 'from colour of forewing', showing that he had looked neither at the moth nor at Linnaeus, who described the ground colour as fuscous. Eight of the nine species on the same page are named from their habitat or pabula and the meaning is probably that the moth occurs in hayfields; cf. 1294 *pascuella* and 1300 *pratella*. For the termination *-ella* see 1174.

1184 **scutulana** ([Denis & Schiffermüller], 1775) – either *scutula*, a diamond-, rhomb- or lozenge-shaped figure, or *scutulum*, a small shield: in either case, the name refers to the dorsal blotch on the forewing.

1184a **cirsiana** (Zeller, 1843) – *cirsion*, *cirsium*, a thistle: from the foodplant.

1185 **cnicicolana** (Zeller, 1847) – *cnicus*, the safflower (*Carthamus tinctorius*): in Britain the foodplant is common fleabane (*Pulicaria dysenterica*).

1186 **sticticana** (Fabricius, 1794) – στικτός (stiktos), spotted: from the conspicuous subterminal series of black spots.

= **farfarae** (Fletcher, 1938) – *Tussilago farfara*, colt's-foot: the foodplant.

1187 **costipunctana** (Haworth, 1811) – *costa*, a rib or, as here, the anterior margin of an insect's wing; *punctum*, a spot: from the black and white strigulae on the distal half of the costa.

Pelochrista Lederer, 1859 – πηλός (pēlos), clay, earth; χριστός (khristos), anointed: from the ill-defined markings of the forewing, including perhaps the roughly-scaled spot resembling a scar which is mentioned by Lederer.

1188 **caecimaculana** (Hübner, 1799) – *caecus*, blind; *macula*, a spot: from the ocellus of the forewing which lacks the metallic plumbeous edging and the black dashes present in related species. 'From black-centred "eye" on forewings' (Macleod): factually incorrect.

Eriopsela Guenée, 1845 – ἔριον (erion), wool; ψάλλω, ψηλ- (psallo, psēl-), to twitch or touch: the species Guenée placed in the genus, which included 1161 *Rhopobota stagnana*, have a whitish ground colour reminiscent of wool.

1189 **quadrana** (Hübner, 1813) – *quadra*, a square: from the subquadrate markings on the dorsum of the forewing.

Eucosma Hübner, 1823 – εὔκοσμος (eukosmos), graceful, well-adorned: an artist's judgement on the species he was figuring.

1190 **aspidiscana** (Hübner, 1817) – ἀσπίς (aspis), a shield; δίσκος (diskos), a quoit, a round plate: from the prominent ocellus. The Greek word ἀσπιδίσκος (aspidiskos) also means the boss of a shield, whence 'from crest on thorax' (Macleod); the thorax is *not* crested.

1191 **catoptrana** (Rebel, 1903) – κάτοπτρον (katoptron), a mirror, in classical times made of polished metal: from the metallic-edged ocellus. *Catoptria* Guenée, 1845, *nec* Hübner, 1825 is a synonym of 1190 *Eucosma*.

= **heringiana** (Jäckh, 1953) – i.h.o. E. M. Hering (1893–1967), the German microlepidopterist.

1192 **conterminana** (Herrich-Schäffer, 1851) – *con-*, prefix, here without special meaning; *termen*, the end, the outer margin of a wing: Herrich-Schäffer describes the terminal area as being paler than the ground colour. 'Bordering on, from resemblance to *E. subsequana*' (Macleod):

one can only suppose that he muddled *consequana* Herrich-Schäffer (a synonym of 1160 *Acroclita subsequana* (Herrich-Schäffer), q.v.) with *conterminana* Herrich-Schäffer.

1193 **tripoliana** (Barrett, 1880) – *Aster tripolium*, sea-aster: the larval foodplant.

1194 **aemulana** (Schläger, 1849) – *aemulus*, rivalling, comparable to: from resemblance to other members of the genus, but not *latiorana* Herrich-Schäffer, 1851, as suggested by Macleod, since this is an anachronism; *latiorana* is a junior synonym of *aemulana*.

1195 **lacteana** (Treitschke, 1835) – *lac*, *lactis*, milk: from the ground colour.

= **maritima** (Humphreys & Westwood, 1845) – *maritimus*, pertaining to the sea: from the coastal habitat.

1196 **metzneriana** (Treitschke, 1830) – i.h.o. Herr Metzner (d. 1861), a senior civil servant and enthusiastic collector who lived at Frankfurt-on-Oder; Treitschke attributes the name to Herr von Kuhlwein. See also 723, genus, and 726.

1197 **campoliliana** ([Denis & Schiffermüller], 1775) – *campus*, a field; *lilium*, a lily: 'Consider the lilies of the field, how they grow; they toil not, neither do they spin: And yet I say unto you, That even Solomon in all his glory was not arrayed like one of these' (Matthew, 6: 28–29); an evocative name for a beautiful tortricid ('beautiful marbled bell' (Heslop, 1964)).

1198 **pauperana** (Duponchel, 1843) – *pauper*, poor, miserable: an undeserved reproach for a not unattractive moth, although Haworth seems to have shared Duponchel's view, for he gave it the English name 'the spotted drab'.

1199 **pupillana** (Clerck, 1759) – *pupilla*, the pupil of the eye: from the strongly marked ocellus on the forewing.

1200 **hohenwartiana** ([Denis & Schiffermüller], 1775) – i.h.o. S. von Hohenwarth (Hochenwarth) (1745–1822), a Theresian professor and later Bishop of Linz. He collected in the Vienna district and wrote on the entomology of the Alps: he named, for instance, 166 *Zygaena exulans*, q.v.

1200a **fulvana** (Stephens, 1834) – *fulvus*, tawny: from the ground colour which is a distinguishing character from 1200 *E. hohenwartiana*.

1201 **cana** (Haworth, 1811) – *canus*, hoary, grey: from the colour of the streaks on the forewing; a syncopated form of *canana*.

1202 **obumbratana** (Lienig & Zeller, 1846) – *obumbratus*, overshadowed, darkened: from the dusky suffusion occupying the costal half of the forewing.

Foveifera Obraztsov, 1956 – *fovea*, a pit; *fero*, to carry: from the presence of a scale-filled depression on the forewing. A Latin, not Greek, derivation.

1203 **torridana** (Lederer, 1859) – *toridus*, hot: from the habitat of this south European species which has a doubtful claim to a place on the British list.

Thiodia Hübner, 1825 – θειώδης (theiōdēs), brimstone-like: from the yellowish ground colour of the next species.

1204 **citrana** (Hübner, 1799) – *citrus*, a citron-tree, including the lemon: from the yellowish ground colour.

Spilonota Stephens, 1829 – σπίλον (spilos), a spot; νῶτον (nōton), the back: from the dorsal blotch of the forewing which stands out prominently on the 'back' of the moth when its wings are folded at rest. Although Stephens included seventeen species in his genus, 1205 *S. ocellana* was not one of them, and the derivation of Macleod 'from basal patch on forewings' with reference to that species cannot be correct.

1205 **ocellana** ([Denis & Schiffermüller], 1775) – *ocellus*, a small eye: from the forewing ocellus which is exceptionally large, reaching the subapical area.

1205a **laricana** (Heinemann, 1863) – *Larix decidua*, European larch: the larval foodplant.

Clavigesta Obraztsov, 1946 – *clavus*, the purple stripe on the Roman *tunica*, broad for senators and narrow for knights; *gero*, in composition *gest-*, to bear: from the forewing fasciae which have a variable purplish flush.

1206 **sylvestrana** (Curtis, 1850) – *Pinus sylvestris*, Scots pine: a possible foodplant, although maritime pine (*P. pinastri*) and stone pine (*P. pinea*) are preferred; Curtis may also have been punning on I. Sylvestre, an 18th century French entomologist and friend of Fabricius.

1207 **purdeyi** (Durrant, 1911) – i.h.o. Captain W. Purdey (1844–1922), the British entomologist who took the type specimens at Folkestone, Kent in July, 1911.

Blastesthia Obraztsov, 1960 – βλαστός (blastos), a shoot; ἐσθίω (esthiō), to eat: from the larval feeding habits.

1208 **posticana** (Zetterstedt, 1840) – *posticus*, posterior: probably from the purplish or ferruginous flush in the distal (posterior) area of the forewing.

1209 **turionella** (Linnaeus, 1758) – *turio*, a shoot: '*habitat intra* Pini *turiones*', it lives in the shoots of pine. Described in Tinea because of the way in which the wings are folded.

Rhyacionia Hübner, 1825 – ῥυάκιον (rhuakion), dim. of ῥύαξ (rhuax), a stream: from the stream-like markings ('rivulets') on the forewing.

1210 **buoliana** ([Denis & Schiffermüller], 1775) – i.h.o. Baron von Buol (18th century), a Viennese entomologist who supplied Schiffermüller with the first specimens.

1211 **pinicolana** (Doubleday, 1849) – *pinus*, a pine-tree; *colo*, to inhabit: from the larval foodplant.

1212 **pinivorana** (Lienig & Zeller, 1846) – *pinus*, a pine-tree; *voro*, to devour: from the larval foodplant.

1213 **logaea** Durrant, 1911 – λογαῖος (logaios), picked out: Durrant 'picked out' the type material from amongst fifty misidentified specimens at the BM(NH). '*Logeum*, platform, from cluster of raised scales near base of forewings' (Macleod): an apparently fictitious character not mentioned by Durrant.

= **duplana** sensu auctt. – *duplus*, double: possibly from the striations on the forewing which tend to be grouped into two fasciae. *R. logaea* and *R. duplana* are distinct species of which only the former is found in Britain.

Retinia Guenée, 1845 – ῥητίνη (rhētinē), resin: the larvae of the species included by Guenée feed in, or cause, resinous exudation. The name should have been spelt '*Rhetinia*' and probably would have been if the author had not been French; cf. 424 *Yponomeuta*, etc.

= **Petrova** Heinrich, 1923 – not explained; some of the other new generic names given by Heinrich in the same paper are those of ladies; see 1158 *Griselda*.

1214 **resinella** (Linnaeus, 1758) – *resina*, resin: '*habitat intra resinae glomerem exsudantem e vulnere ramorum* Pini', it lives in a ball of resin exuding from a wound in the branches of *Pinus*. Described in Tinea; cf. 1209.

Cryptophlebia Walsingham, 1899 – κρυπτός (kruptos), hidden; φλέψ, φλεβός (phleps, phlebos), a vein: from the pocket of specialized black scales overlaid with whitish scales in the anal angle of the hindwing.

1215 **leucotreta** (Meyrick, 1913) – λευκός (leukos), white: τρητός (trētos), perforated with a hole: see generic description above, the 'hole' being the depressed pocket.

Enarmonia Hübner, 1826 – ἐναρμόνιος (enarmonios), in harmony: in appreciation of the beauty of the next species.

1216 **formosana** (Scopoli, 1763) – *formosus*, beautiful: in appreciation.

Eucosmomorpha Obraztsov, 1951 – the genus 1190 *Eucosma*, q.v.; μορφή (morphē), form: a genus having resemblance to *Eucosma*.

1217 **albersana** (Hübner, 1813) – i.h.o. J. A. Albers (1772–1821), a medical practitioner and entomologist who lived at Bremen.

Selania Stephens, 1834 – σελάνα (selana), a Doric variant of σελήνη (selēnē), the moon: from the crescentic dorsal blotch of the forewing. The Attic spelling was preoccupied by 1917 *Selenia* Hübner, 1823.

1218 **leplastriana** (Curtis, 1831) – i.h.o. Mr Leplastrier (Le Plastrier), a professional collector living at Dover; he or Curtis found the type specimen nearby.

Lathronympha Meyrick, 1926 – λαθρόνυμφος (lathronymphos), secretly married: 'an inconspicuous but peculiar form' (Meyrick, 1928), whatever that may mean, since it appears to apply to the adult which is common and easily observed (the genus is monotypic). It is better to derive from λάθρη (lathrē), secretly, and νύμφη (nymphē), a nymph or an insect in its immature stage, and to apply the name to the larva which feeds in a tight spinning.

1219 **strigana** (Fabricius, 1775) – *striga*, a swath, a furrow, a band: either from the costal strigulae or from the dark bars in the ocellus, or from both.

Collicularia Obraztsov, 1960 – *colliculus*, a little hill: from two filaments on the ductus bursae of the female genitalia which Obraztsov calls 'colliculi'.

1220 **microgrammana** (Guenée, 1845) – μικρός (micros), small; γράμμα (gramma), something drawn, a letter of the alphabet: from the fine striations across the forewing, the short costal strigulae and the small line surrounding the ocellus, all of which are mentioned by Guenée.

Strophedra Herrich-Schäffer, 1854 – στρόφος (strophos), a twisted cord; ἕδρα (hedra), a seat: from the larval feeding place, constructed by fastening one leaf of the foodplant above another by means of a series of short thongs of woven silk.

1221 **weirana** (Douglas, 1850) – i.h.o. J. J. Weir (1822–94), a British entomologist (cf. 678 *weirella*).

1222 **nitidana** (Fabricius, 1794) – *nitidus*, shining: from the glossy forewing.

Pammene Hübner, 1825 – παν- (pan-), all, the whole; μήνη (mēnē), the moon: from the circular ocellus of many species, suggesting a full moon.

1223 **splendidulana** (Guenée, 1845) – *splendidulus*, dim. formed from *splendidus*, shining, so 'bright little . . .': from the submetallic gloss on the forewing and the small size of the moth, both mentioned by Guenée.

1224 **luedersiana** (Sorhagen, 1885) – i.h.o. L. Lüders: I have been unable to trace him further but it is reasonable to surmise that he was the collector who discovered the species.

1225 **obscurana** (Stephens, 1834) – *obscurus*, obscure: from the indistinct pattern on the forewing.

1226 **agnotana** Rebel, 1914 – *ad-* (*ag-*), in addition; *notus*, known: from its being, when named, a recent addition to the genus.

1227 **inquilina** Fletcher, 1938 – *inquilinus*, a dweller in a house which is not his own, a lodger: from the feeding habits of the larva which lives in a gall caused by Hymenoptera.

1228 **argyrana** (Hübner, 1799) – ἄργυρος (arguros), silver: from the submetallic markings on the forewing.

1228a **ignorata** Kuznetsov, 1968 – *ignoratus*, unknown, i.e. unknown until Kuznetsov described it many years after most of its congeners.

1229 **albuginana** (Guenée, 1845) – *albugines* (pl.), scurf of the head: from the orange-ochreous irroration on the forewing.

1230 **suspectana** (Lienig & Zeller, 1846) – *suspectus*, mistrusted: the authors draw attention to its close similarity to 1271 *Cydia gallicana* (Guenée, 1845), and the name reflects their misgivings over its status as a distinct species.

1231 **spiniana** (Duponchel, 1843) – *Prunus spinosa*, blackthorn: the larval foodplant.

1232 **populana** (Fabricius, 1787) – *Populus*, the poplar genus: the larva, however, feeds on sallow (*Salix* spp.).

1233 **aurantiana** (Staudinger, 1871) – *aurum*, gold; *antiae*, the hair grown on the forehead, the forelock: from the golden brown hair-scales of the head and collar. Alternatively, from *aurantiacus* (late Lat.), orange, or from *aurans*, presumed pres. part. of *auro*, to gild (a verb found only in its past part., *auratus*), but this would mean 'gilding', not 'being gilt'.

1234 **regiana** (Zeller, 1849) – *regius*, royal, magnificent: from the rich markings of the forewing, consisting of orange-yellow and metallic blue on a deep blackish brown ground colour.

1235 **trauniana** ([Denis & Schiffermüller], 1775) – traditionally i.h.o. Field Marshal Traun (d. 1748). As far as is known, Traun was not an entomologist, but Denis and Schiffermüller are stated to have been soldiers in their youth and may have served under him in the Austrian army. The name first appears in print twenty-seven years after the death of the field marshal, so it was bestowed as a token of esteem and not in quest of patronage. Pickard *et al.* and Macleod both give this explanation and Heslop (1964) adds the vernacular name 'Traun's black-and-white piercer'. However, there is in Austria a River Traun which flows through Lake Traun and joins the River Danube just below Linz. The authors were covering the Lepidoptera of the Vienna district (interpreted broadly), and the valley of this river may have been the type locality. They named 1056 *Acleris lipsiana* (q.v.) from the type locality, Leipzig.

1236 **fasciana** (Linnaeus, 1761) – *fascia*, a band: from the pale median blotch which is continued like a fascia towards the apex of the forewing.

1236a **herrichiana** (Heinemann, 1854) – i.h.o. G. A. W. Herrich-Schäffer (1799–1874), the German entomologist and author whose 'work. . . , for the accuracy of description and the magnificence of its plates, leaves little to be desired' (Pickard *et al.*).

1237 **germmana** (Hübner, 1799) – *germanus*, related: from affinity with another species. Hübner emended the name to *germana* in 1822 and *germarana* in 1825. Pickard *et al.* and Macleod take the third spelling and explain the name as given in honour of Professor E. F. Germar (1786–1853), a distinguished German entomologist. This could have been an afterthought by Hübner in 1825 but when the first version of the name was given to a figure on a plate (Tortrices, 10) executed between 1796 and 1799, Germar would have been a precocious lad of between ten and thirteen years of age.

1238 **ochsenheimeriana** (Lienig & Zeller, 1846) – i.h.o. F. Ochsenheimer (1767–1822), the distinguished German scholar, playwright, actor and entomologist.

1239 **rhediella** (Clerck, 1759) – i.h.o. H. A. van Rheede tot Draakenstein (late 17th–early 18th century), a Dutch naturalist and the Governor of Malabar. His *Hortus Indicus Malabaricus* (1678–1703, published in parts), was one of the most celebrated pre-Linnaean books on natural history and was illustrated in all probability partly by Rheede himself. Clerck's name, therefore, was a tribute from one illustrator to another. 'After British entomologist Rheede (18th century)' (Macleod); I cannot trace this Rheede, so it is unlikely that he would have been well enough known on the Continent to merit a name from a Swedish artist in 1759.

Cydia Hübner, 1825 – κῦδος (kudos), glory, renown: a tribute from an artist to the beauty of the moths he was figuring. Our word 'kudos' has not obeyed the rules for the transliteration of Greek words given on p. 10.

= **Laspeyresia** Hübner, 1825 *nec* R.L., 1817 – i.h.o. J. H. Laspeyres (1769–1809), a German entomologist; see 2473, genus.

1240 **caecana** (Schläger, 1847) – *caecus*, blind: the ocellus of the forewing is less strongly developed than in most related species.

1241 **compositella** (Fabricius, 1775) – *compositus* well-ordered, neat: in praise of the moth's elegance; *-ella* because described in Tinea (see pp. 21–22).

1242 **internana** (Guenée, 1845) – *internus*, inward, inner: from the markings on the dorsum (the 'inner' margin), *macula interna pone medium bipartita*, a divided dorsal spot beyond the middle. 'From dark line across centre of wings' (Macleod); not the reason given by Guenée.

1243 **pallifrontana** (Lienig & Zeller, 1846) – *pallidus*, pale; *frons, frontis*, the forehead: the frons and labial palpus of the adult are greyish white.

1244 **gemmiferana** (Treitschke, 1835) – *gemma*, a precious stone; *fero*, to carry: from the metallic plumbeous striae and the silver-edged ocellus on the forewing.

1245 **janthinana** (Duponchel, 1835) – *ianthinus*, violet-blue: from the pinkish or purplish suffusion on the forewing.

1246 **tenebrosana** (Duponchel, 1843) – *tenebrosus*, shadowy, dark, gloomy: from the almost unicolorous, dark brown forewing.

1247 **funebrana** (Treitschke, 1835) – *funebris*, funereal: from the almost unicolorous, dark brown forewing.

1248 **molesta** (Busck, 1916) – *molestus*, troublesome: from the larva, which is a serious pest of peach, feeding in the fruit.

1249 **prunivorana** (Ragonot, 1879) – *Prunus*, the plum genus; *voro*, to devour: from the larva which feeds in fruits, mainly of *Prunus* spp.

1250 **lathyrana** (Hübner, 1813) – *Lathyrus*, a genus containing species of pea and vetch: these are the foodplants of the next three species but this one feeds on dyer's greenweed (*Genista tinctoria*).

1251 **jungiella** (Clerck, 1759) – perhaps i.h.o. R. C. Jung, who was described by Hübner in 1796 as a court official at Uffenheim who sent him moths to figure and describe. The name had been bestowed thirty-seven years previously, so if it is the same Jung he maintained a lifelong interest in entomology.

1252 **lunulana** ([Denis & Schiffermüller], 1775) – *lunula*, an ornament shaped like a small moon: from the crescentic dorsal blotch on the forewing.

= **dorsana** sensu auctt. – *dorsum*, the back, the dorsal area of a wing: from the pale dorsal blotch on the forewing; *dorsana* Fabricius, 1775, has not been found in Britain.

1253 **orobana** (Treitschke, 1830) – *Vicia orobus*, the wood bitter vetch or *Orobus tuberosus*, a former name of *Lathyrus montanus*, the bitter vetch: the larva, however, feeds on neither of these, but on related species.

1254 **strobilella** (Linnaeus, 1758) – *strobilus*, a pine-cone: '*habitat in strobilis* Abietis' (it lives in the cones of *Abies* (fir)). Like 1251, described in Tinea because of the way in which the wings are folded.

1255 **succedana** ([Denis & Schiffermüller], 1775) – *succedo*, to follow, *succedaneus*, that which succeeds to, follows after: perhaps just the 'next' species, no more suitable name having come to mind. In the *Schmetterlinge der Wienergegend* it follows 1041 *Acleris sparsana*.

1255a **medicaginis** Kuznetsov, 1962 – *Medicago*, the medick/lucerne genus: from the larval foodplants.

1256 **servillana** (Duponchel, 1836) – i.h.o. J. G. Audinet de Serville (1775–1858), a French entomologist who wrote books on Coleoptera and Hymenoptera.

1257 **nigricana** (Fabricius, 1794) – *nigricans*, blackish: from the rather dark forewing.

= **rusticella** (Clerck, 1759) – *rusticus*, belonging to the countryside: from the habitat; the tineid termination '*-ella*' from the involute method of folding the wings. This is the senior name, but application has been made to the I.C.Z.N. for the retention of *nigricana*, the name used in horticultural text-books for this pest species.

1258 **milleniana** Adamczewski, 1967 – *millenium*, the span of 1,000 years: i.h.o. the Polish millenium, celebrated in 1966.

= **deciduana** (Steuer, 1969) – *Larix decidua*, the European larch: the larval foodplant.

1259 **fagiglandana** (Zeller, 1841) – *Fagus sylvatica*, the beech; *glans*, a beech-nut: from the larval feeding place.

1260 **splendana** (Hübner, 1799) – *splendeo*, to be bright, magnificent: probably from its impressive size and robust build, rather than its colour and pattern.

1261 **pomonella** (Linnaeus, 1758) – Pomona, the goddess of fruit-trees; *pomum*, a fruit, often an apple, but here in fact a pear, '*habitat intra poma* Pyri', it lives in the fruits of pear. Linnaeus placed it in Tinea (termination *-ella*) because it rests with involute wings.

1262 **amplana** (Hübner, 1799) – *amplus*, large: presumably from the size of the moth which has a very slender claim to a place on the British list.

1263 **inquinatana** (Hübner, 1799) – *inquinatus*, polluted: presumably from the dirty appearance of this species which has no valid claim to a place on the British list; cf. 1703 *inquinata* and 1306 *inquinatella*.

1264 **leguminana** (Lienig & Zeller, 1846) – *legumen*, pulse, any leguminous plant: Lienig & Zeller describe this species as being close to 1253 *C. orobana*, *C. dorsana* Fabricius and 1251 *C. jungiana* (*jungiella* (Clerck)), all of which feed on Leguminosae (Papilionaceae), but in this instance they were wrong, since the larva feeds in decaying bark, especially of elm.

1265 **cognatana** (Barrett, 1874) – *cognatus* related: from affinity with other pine-feeding members of the genus, e.g. 1268 *coniferana*.

1266 **pactolana** (Zeller, 1840) – *pactus*, promised, betrothed: probably with the idea of affinity with related species; cf. 1265 *C. cognatana*.

1266a **illutana** (Herrich-Schäffer, 1851) – *illotus*, *illutus*, unwashed, smeared: from the obscure forewing pattern; cf. 1263 *inquinatana*.

1267 **cosmophorana** (Treitschke, 1835) – κόσμος (kosmos), an ornament (cf. cosmetic); φορέω (phoreō), to bear, to wear: from the metallic blue striae on the forewing.

1268 **coniferana** (Ratzeburg, 1840) – *conifer*, cone-bearing, a coniferous tree: from the habits of the larva, which feeds under the bark of Scots pine (*Pinus sylvestris*) and other conifers.

1269 **conicolana** (Heylaerts, 1874) – *conus*, a fir-cone; *colo*, to inhabit: from the larval feeding place inside the cones of Scots pine (*Pinus sylvestris*) and other conifers.

1270 **corollana** (Hübner, 1823) – *corolla*, a small wreath: from the large ocellus of the forewing which encloses several metallic-edged black striae.

1271 **gallicana** (Guenée, 1845) – *Gallicus*, Gallic, French: i.h.o. Guenée's native country, in which the type locality is situated.

1272 **aurana** (Fabricius, 1775) – *aurum*, gold: from the golden yellow blotches on the forewing.

Dichrorampha Guenée, 1845 – δίχροος (dikhroos), two-coloured; ῥαμφή (rhamphē), a hooked knife: from the coloration of the labial palpus, '*hoc nomine, quod ore auribusque suavius* *efficere frustra tentavi, palpos fere semper basi croceos margineque fuscos revocare volui'* (in this name, which I unsuccessfully tried to make less of a mouthful and easier on the ear, I wanted to draw attention to the palpi, which are nearly always basally yellow and laterally fuscous). It is a very apt name.

1273 **petiverella** (Linnaeus, 1758) – i.h.o. James Petiver (1660–1718), apothecary, naturalist and collector who may be regarded as the father of British entomology. Described in Tinea because of the way in which the wings are folded.

1274 **alpinana** (Treitschke, 1830) – *alpinus*, pertaining to the Alps: Treitschke received the type specimen from von Tischer (see 123, genus), but does not state the type locality.

1275 **flavidorsana** Knaggs, 1867 – *flavus*, yellow; *dorsum*, the back: from the yellow dorsal blotch.

1276 **plumbagana** (Treitschke, 1830) – *plumbago*, black lead, graphite: from the silver-plumbeous striae on the forewing.

1277 **senectana** (Guenée, 1845) – *senectus*, aged, old-looking: from the grey scales on the underside of the wings, *'subtus omnes canae, sericeae, viridi-politae'* (below the wings are entirely grey, silky, with green reflections); *canus* means hoary, *cani* are the grey hairs of advancing age. Macleod, who had not consulted Guenée, thought the name referred to the upperside of the wings and to be unsuitable.

1278 **sequana** (Hübner, 1799) – *sequor*, to follow: probably just the 'next' species for description and figuring.

1279 **acuminatana** (Lienig & Zeller, 1846) – *acuminatus*, sharpened: from the acute apex of the forewing which distinguishes this species from its close relatives.

1280 **consortana** Stephens, 1852 – *consors, consortis*, a colleague: from the similarity between this species and its close relatives.

1281 **simpliciana** (Haworth, 1811) – *simplex*, plain, simple: from the forewing which lacks distinctive markings.

1282 **sylvicolana** Heinemann, 1863 – *silva, sylva*, a wood; *colo*, to inhabit: from the habitat; it frequents damp ground often, but not necessarily, in woods. In Britain it was first found in Epping Forest.

1283 **montanana** (Duponchel, 1843) – *montanus*, pertaining to mountains: from the habitat; this is reflected in Britain where the species has a northern distribution, but lowland as well as upland.

1284 **gueneeana** Obraztsov, 1953 – i.h.o. A. Guenée (1809–80), French lawyer, lepidopterist and entomological author. The use of his name is appropriate since he was the nomenclator of the generic name 1273 *Dichrorampha*.

1285 **plumbana** (Scopoli, 1763) – *plumbum*, lead: from the plumbeous submetallic striae on the forewing.

1286 **sedatana** Busck, 1906 – *sedatus*, staid, sedate: from the sober coloration of the forewing.

1287 **aeratana** (Pierce & Metcalfe, 1915) – *aeratus*, coppery, brazen: a name analogous to 1276 *plumbagana* and 1285 *plumbana* and like them referring to the submetallic striae on the forewing, though these are plumbeous rather than coppery.

ALUCITOIDEA (1288)

ALUCITIDAE (1288)

Alucita Linnaeus, 1758 – *alucita*, a gnat: Linnaeus originally applied the name to the Alucitae + Pterophoridae, which he regarded as a single family. The term *'alucita'* is imprecise and includes crane-flies (Tipulidae) which rest with their wings extended rather like the Pterophoridae. When Geoffroy's alternative name Pterophorus was adopted for the plumes, Alucita was liberated for other use and was applied by Fabricius and Latreille to a large section of feathery-winged Tineae such as *Coleophora* and *Argyresthia*; later, however, it fell into disuse. The 'many-plumed' family was first separated from Pterophorus by Latreille (1796) under the name Orneodes, which remained current well into the present century (e.g. Meyrick, 1928). When it was realized that the Linnaean name should be revived, it was assigned to the 'wrong' half of the original family, i.e. the half less like crane-flies, and Orneodes was reduced to synonymy.

= **Orneodes** Latreille, 1796 – ὄρνις (ornis), a bird; εἶδος, ὠδ- (eidos, ōd-), form: from the feather-like lobes (plumes) into which the wings are divided.

1288 **hexadactyla** Linnaeus, 1758 – ἕξ (hex), six; δάκτυλος (daktulos), a finger: *'alis patentibus fissis: singulis sexpartitis'*, with wings extended at rest and divided, each wing into six divisions. The moth is no. 6 of Linnaeus' plumes, no doubt because of the sexpartite division of the wings. Some innumerate Englishman called it 'the twenty-plume moth'. See also 1510 Pterophoridae.

PYRALOIDEA (1416)

PYRALIDAE (1416)

CRAMBINAE (1294)

In his definition of Tinea, Linnaeus wrote *'alis convolutis fere in cylindrum'* (with wings wrapped round to body so as almost to form a cylinder). Since most Crambinae adopt the involute wing posture when at rest (see MBGBI 7(2), Pl. B, fig. 2), they were described by Linnaeus and his immediate successors in Tinea and their names given the conventional termination *-ella*. In due course Fabricius (1798) recognized that they were misplaced and set up Crambus as a family to accommodate them (see p. 29).

Euchromius Guenée, 1845 – εὖ (eu), well; χρῶμα (khrōma), the colour of the skin, colour in general: from the bright colour of the species included by Guenée in the genus, of which 1289 *E. ocellea* was not one.

1289 **ocellea** (Haworth, 1811) – *ocellus*, a little eye: from the ocellar region, which is bounded proximally by ochreous yellow and distally by golden metallic spots.

Chilo Zincken, 1817 – χεῖλος (kheilos), a lip: from the elongate labial palpus (*labium*, a lip).

1290 **phragmitella** (Hübner, 1805) – *Phragmites australis* (formerly *Arundo phragmites*), the common reed: the larval foodplant.

Acigona Hübner, 1825 – ἀκίς (akis), a point, a dart; γωνία (gōnia), an angle: from the white costal streak on the forewing, which narrows to a point at the apical angle.

1291 **cicatricella** (Hübner, 1824) – *cicatrix*, a scar: from the conspicuous white costal streak on the forewing of the female, suggestive of a weal.

Calamotropha Zeller, 1863 – κάλαμος (kalamos), a reed; τροφή (trophē), food: from the larval foodplant of the next species.

1292 **paludella** (Hübner, 1824) – *palus, paludis*, a marsh: from the habitat.

Chrysoteuchia Hübner, 1825 – χρυσός (khrusos), gold; τεύχω (teukhō), to make, to furnish with: from the strongly golden metallic cilia of the forewing.

1293 **culmella** (Linnaeus, 1758) – *culmus*, a stalk, a stem: from the habitat (*'habitat in pascuis'*, it lives in pastures); probably also from the moth's habit of resting head-downwards on grass-stems. The larva feeds at the base of grass-stems but the life history was not known to Linnaeus.

= **hortuella** (Hübner, 1796) – *hortus*, a garden: from the habitat.

Crambus Fabricius, 1798 – κράμβος (krambos), dry, parched: from the yellowish brown coloration of the forewing of many species, suggestive of hay or withered vegetation; possibly also from the mainly dry grassland habitat. Macleod derives from κράμβη (krambē), a cabbage, which would be absurd unless one regards it as a Fabrician pun; however, the spelling would then have been 'Crambe'. See also Crambinae above. Haworth (1811) reapplied the name to the 2476 Hypeninae.

1294 **pascuella** (Linnaeus, 1758) – *pascuum*, a pasture: from the habitat (*'habitat in pascuis'*, it lives in pastures).

1295 **leocoschalis** (Hampson, 1898) – λευκός (leukos), white; σχάλις (skhalis), a forked stick: from the notch projecting from the longitudinal streak on the forewing, the character which serves to separate this species from the preceding one.

1296 **silvella** (Hübner, 1813) – *silva*, a wood: from the habitat, though this is more commonly boggy moorland.

1297 **uliginosellus** Zeller, 1850 – *uliginosus*, marshy: from the habitat.

1298 **ericella** (Hübner, 1813) – *Erica*, the heath genus: from the heathland habitat.

1299 **hamella** (Thunberg, 1788) – *hamus*, a hook, a fish-hook: from the central white streak on the forewing, which appears to be barbed.

1300 **pratella** (Linnaeus, 1758) – *pratum*, a meadow: from the habitat.
= **dumetella** (Hübner, 1813) – *dumetum*, a thicket: from the supposed habitat.

1301 **lathoniellus** (Zincken, 1817) – Λητώ, Λαθώ, Lat. Leto, Latona, the mother of Apollo and Artemis; cf. 953 *lathoniana* and 1603 *lathonia*.
= **nemorella** (Hübner, 1813), *nec* (Thunberg, 1788) – *nemus, nemoris*, a woodland glade: from the habitat.

1302 **perlella** (Scopoli, 1763) – *perla* (late Lat.), a pearl: from the glossy, whitish forewing.

Agriphila Hübner, 1825 – ἀγρός (agros), a field; φιλέω (phileō), to love: from the habitat.

1303 **selasella** (Hübner, 1813) – σέλας (selas), brightness: from the shining white longitudinal streak on the forewing.

1304 **straminella** ([Denis & Schiffermüller], 1775) – *stramen, straminis*, straw: from the ground colour of the forewing.

1305 **tristella** ([Denis & Schiffermüller], 1775) – *tristis*, sad, mournful: from the heavy fuscous shading sometimes present on the forewing; according to Guenée, the name was bestowed on a rare aberration.

1306 **inquinatella** ([Denis & Schiffermüller], 1775) – *inquinatus*, polluted, stained: from the fuscous irroration on the forewing.

1307 **latistria** (Haworth, 1811) – *latus*, broad; *stria*, a furrow, a stripe: from the broad white longitudinal stripe on the forewing.

1308 **poliellus** (Treitschke, 1832) – πολιός (polios), grey: from the ground colour which is greyer than in related species.

1309 **geniculea** (Haworth, 1811) – *geniculum*, a little knee: from the two geniculate fuscous fasciae on the forewing.

Catoptria Hübner, 1825 – κάτοπτρον (katoptron), a mirror: from the glossy pearl markings on some of the species.

1310 **permutatella** (Herrich-Schäffer, 1848) – *permutatus*, altered: this species was formerly confused with *C. myella* (Hübner, 1796). When Herrich-Schäffer recognized that there were two species, he had to change the name of the specimens which were not *C. myella*.

1311 **osthelderi** (de Lattin, 1950) – i.h.o. L. Osthelder (1877–1954), an Austrian entomologist.

1312 **speculalis** Hübner, 1825 – *speculum*, a mirror: cf. 1310 *Catoptria* above, a genus set up to accommodate this species.

1313 **pinella** (Linnaeus, 1758) – *pinus*, a pine-tree: '*habitat in pinetis*', it lives in pine-woods; in 1761 Linnaeus sought to change the name to *pinetella*, but according to the I.C.Z.N. rules the original spelling must stand. Linnaeus never gave pine as the foodplant, as supposed by Macleod.

1314 **margaritella** ([Denis & Schiffermüller], 1775) – *margarita*, a pearl: from the pearly white longitudinal stripe on the forewing.

1315 **furcatellus** (Zetterstedt, 1840) – *furcatus*, forked: from the white stripe on the forewing, which is branched on its dorsal margin.

1316 **falsella** ([Denis & Schiffermüller], 1775) – *falsus*, false, deceptive: because it is 'a poor imitation of [1294 *Crambus*] *pascuella*'. The explanation given by Macleod 'by contrast with similar species *C. verellus* ("true")' is an anachronism by forty-two years and ignores the reason the authors give for their name.

1317 **verellus** (Zincken, 1817) – *verus*, true: the name was given as an antonym to *falsellus* above, indicating both affinity and distinction.

1318 **lythargyrella** (Hübner, 1796) – *lithargyrus*, the oxide of lead, removed in refining silver: from the submetallic forewing, which is ochreous yellow, often with a greyish tinge.

Chrysocrambus Bleszynski, 1957 – χρυσός (khrusos), gold; the genus 1294 *Crambus*: a genus resembling *Crambus*, but with the terminal cilia more strongly golden metallic.

1319 **linetella** (Fabricius, 1781) – *linea*, a linen thread, a line, a net: from the forewing pattern of brown lines between the veins, crossed by two rust-coloured fasciae.

1320 **craterella** (Scopoli, 1763) – *crates*, a hurdle: Scopoli describes the forewing as having eight longitudinal and two transverse streaks.

Thisanotia Hübner, 1825 – θίς (this), a sand-hill, a beach; ἄνω (anō), on the upper part: from the habitat.

1321 **chrysonuchella** (Scopoli, 1763) – χρυσός (khrusos), gold; νύχιος (nukhios), pertaining to the night: from the golden metallic cilia, which contrast with the dusky longitudinal streaks on the forewing. Macleod derived from ὄνυξ (onux), a claw, but admitted that the legs were not golden.

Pediasia Hübner, 1825 – πεδιάσιος (pediasios), variant of πεδιακός (pediakos), of the plain: from the habitat.

1322 **fascelinella** (Hübner, 1813) – *fascelinus*, dim. coined from *fascia*, a band: from the two narrow, indistinct fasciae on the forewing.

1323 **contaminella** (Hübner, 1796) – *contaminatus*, defiled: from the 'dirty' fuscous irroration on the forewing.

1324 **aridella** (Thunberg, 1788) – *aridus*, dry: although a coastal species, it frequents sand-hills and the drier parts of salt-marshes. However, since the type series also contains specimens of 1293 *Chrysoteuchia culmella* and 1306 *Agriphila inquinatella* (Karsholt & Nielsen, 1986), the name may well be derived from the habitat of one of those species.

Platytes Guenée, 1845 – πλατύτης (platutēs), breadth: the species in the genus tend to have broader and less elongate forewings than other Crambinae.

1325 **alpinella** (Hübner, 1813) – *alpinus*, alpine, montane: from the habitat. In Britain this is almost exclusively coastal, but on the Continent it occurs inland as well and so potentially on mountains.

1326 **cerussella** ([Denis & Schiffermüller], 1775) – *cerussa*, white lead: from the ground colour of the female's forewing; that of the male is brown.

Ancylolomia Hübner, 1825 – ἀγκυλίς (agkulis), a bend; λῶμα (lōma), a fringe: from the excavate termen of the forewing. Macleod, who derived the name impossibly from ὅμοιος (homoios), like, gives 'from the rounded wings', the antithesis of the meaning intended by Hübner and a character existing only in his imagination.

1327 **tentaculella** (Hübner, 1796) – *tentaculum* (late Lat.), a claw, a tentacle: from the strongly developed labial palpus. 'From hairy antennae' (Macleod).

SCHOENOBIINAE (1328)

Schoenobius Duponchel, 1836 – σχοῖνος (skhoinos), a rush, a reed; βιόω (bioō), to live: from the larval foodplant and the habitat.

1328 **gigantella** ([Denis & Schiffermüller], 1775) – *giganteus*, huge: this species has the biggest wingspan of all the British Microlepidoptera.

Donacaula Meyrick, 1890 – δόναξ (donax), a reed; αὐλή (aulē), a courtyard, a dwelling-place: from the habitat of 1330 *D. mucronellus*, the only species in the genus as at first constituted, not its foodplant which was then unknown. The genus differs from 1328 *Schoenobius* only in a detail of wing venation and the name has a similar meaning.

1329 **forficella** (Thunberg, 1794) – *forfex*, a pair of shears; *-ella*, because Thunberg described it in Tinea: from the way in which the dorsal margins of the forewings overlap when the moth is at rest, like the blades of a pair of scissors (see Goater, 1986, frontispiece, fig. 3). 'From marking of forewings' (Macleod): what marking? See also 1356 *forficalis*.

1330 **mucronellus** ([Denis & Schiffermüller], 1775) – *mucro, mucronis*, a sharp point: from the acute apex of the forewing.

1331 see below 1355

SCOPARIINAE (1332)

Scoparia Haworth, 1811 – *scopae*, twigs, a besom: from the maxillary palps which Haworth describes as having a terminal tuft resembling a paint-brush. Macleod invents scale-tufts on the wings and Spuler derives the name from σκώψ (skōps), an owl!

1332 **subfusca** Haworth, 1811 – *subfuscus*, brownish: from the ground colour of the forewing. At Oxford, undergraduates have to wear 'subfusc' clothes when they sit examinations.

= **cembrella** sensu auctt. – *Pinus cembra*, the arolla pine or Swiss stone-pine: from the adult's habit of resting on tree-trunks, including those of this tree.

1333 **pyralella** ([Denis & Schiffermüller], 1775) – see 1416 *Pyralis*.

= **arundinata** (Thunberg, 1792) – *arundo, arundinis*, a reed: from one of the moth's habitats.

= **dubitalis** (Hübner, 1796) – *dubitare*, to doubt: from difficulty in determination or (Macleod) from great variability.

1334 **ambigualis** (Treitschke, 1829) – *ambiguus*, uncertain: from difficulty in determination. Macleod's explanation is anachronistic.

1334a **basistrigalis** Knaggs, 1866 – *basis*, base; *striga*, a furrow, a line: from the black strigula at the base of the forewing.

1335 **ancipitella** (de la Harpe, 1855) – *anceps*, doubtful: from difficulty in determination.

= **ulmella** Knaggs, 1867 – *ulmus*, an elm-tree: from the habits of the adult, which often rests on elm-trunks.

1336 }
1337 } see below 1338

Dipleurina Chapman, 1912 – δίς (dis), two; πλευρά (pleura), a rib: from the male genitalia, in which the sacculus projects from the valva.

1338 **lacustrata** (Panzer, 1804) – *lacus*, a lake, *lacuster* (late Lat.), pertaining to a lake: from one of the moth's habitats.

= **crataegella** (Hübner, 1796) – *Crataegus*, hawthorn: the larva feeds on moss, often that growing on tree-trunks, including those of hawthorn. The name, however, may refer to the subfamily habit of resting on tree-trunks.

Eudonia Billberg, 1820 – εὕδω (heudō), to rest: from the adult's habit of resting on tree-trunks or rocks. 1344 was the only species included by Billberg in the genus.

1336 **pallida** (Curtis, 1827) – *pallidus*, pale: from the pallid ground colour.

1337 **alpina** (Curtis, 1850) – *alpinus*, Alpine, montane: the species occurs on the mountains of Scotland, generally above 1000m.

1339 **murana** (Curtis, 1827) – *murus*, a wall: the larva feeds on mosses growing on walls and rocks, and the adult rests on the same substrate.

1340 **truncicolella** (Stainton, 1849) – *truncus*, a tree-trunk; *colo*, to inhabit: from the adult's habit of resting on tree-trunks by day.

1341 **lineola** (Curtis, 1827) – *linea*, a linen thread, a fine line: from the two narrow pale fasciae on the forewing.

1342 **angustea** (Curtis, 1827) – *angustus*, narrow: from the forewing which is narrower than those of related species.

1343 **delunella** (Stainton, 1846) – *de-*, lacking; *luna*, the moon: from the dark costal blotch which conceals the crescent-shaped mark present in related species.

= **vandaliella** (Herrich-Schäffer, 1851) – perhaps from the Vandals, an ancient German tribe; Herrich-Schäffer gives no explanation.

= **resinella** sensu auctt. – *resina*, resin: from the habit of the adult of resting on the trunks of conifers which exude resin.

1344 **mercurella** (Linnaeus, 1758) – Mercurius, the god or the planet Mercury: from the symbol of the planet, depicted thus ☿ by Linnaeus in his description of the forewing markings.

NYMPHULINAE (1350)

Elophila Hübner, 1822 – ἕλος (elos), a water-meadow; φίλος (philos), loving: from the habitat.

1345 **nymphaeata** (Linnaeus, 1758) – *Nymphaea*, the water-lily genus: '*habitat in Nymphaea*', it lives amongst water-lilies.

1346 **difflualis** (Snellen, 1882) – *diffluus*, flowing apart: from the divergent markings on the forewing.

= **enixalis** Swinhoe, 1885 – *enixus*, strenuous: application obscure.

1347 **melagynalis** Agassiz, 1978 – μέλας (melas), dark, black; γυνή (gynē), a woman: from the female, which is much darker than the male.

1347a **manilensis** Hampson, 1917 – Manila, in the Philippine Islands, is the type locality.

1348 }
1349 } see below 1350

Nymphula Schrank, 1802 – νύμφη (numphē), a nymph, especially a Naiad or water nymph: from the larval habit of feeding below the surface of the water.

1350 **stagnata** (Donovan, 1806) – *stagnum*, a pond: from the habitat.

Parapoynx Hübner, 1825 – παρά (para), beside; πῶυγξ (pōugx), a species of heron: a lacustrine genus with members which 'live beside the heron'. I am indebted to the late Duke of Newcastle for this explanation. Most entomologists have wished to correct the spelling to *Paraponyx*, then deriving the second half from ὄνυξ (onux), a banded chalcedony, referring to lines ('china-marks') on the wings. The emendment gives an excellent and likely enough sense, but is both unnecessary and impermissible.

1348 **stratiotata** (Linnaeus, 1758) – *Stratiotes aloides*, the water-soldier: '*habitat in* Stratiote aloide *sub aquis*', it lives on water-soldier beneath the surface of the water. In Britain other aquatic plants are more frequently used.

1349 **obscuralis** (Grote, 1881) – *obscurus*, dark, obscure: from the dark ground colour.

1350 see above 1348

1351 **diminutalis** Snellen, 1880 – *deminutus*, *diminutus*, diminished, small: probably because the male is much smaller than the female, the species itself not being unduly small.

1351a **fluctuosalis** Zeller, 1852 – *fluctuosus*, pertaining to waves, with wave-like marks: either or both meanings are possible, referring to the aquatic habitat and/or a narrow, curved orange band on the forewing.

1351b **crisonalis** (Walker, 1859) – probably, like a number of Walker's names, without meaning.
= **stagnalis** sensu Agassiz, 1981 – *stagnum*, a pond: from the habitat.

Oligostigma Guenée, 1854 – ὀλίγος (oligos), few; στίγμα (stigma), a spot: from the sparse pattern on a member species.

1352 **angulipennis** Hampson, 1891 – *angulus*, an angle; *pennae*, feathers, a wing: from a small notch in the termen of the hindwing.

1353 **bilinealis** Snellen, 1876 – *bi-*, two; *linea*, a line: the forewing pattern has a series of more or less parallel lines, two of which, being white, are more prominent.

1353a **polydectalis** Walker, 1859 – πολυδέκτης (poludektēs), the All-receiver, an epithet of Hades: probably without entomological application.

Cataclysta Hübner, 1825 – κατακλύζω (katakluzō), to flood: from the larval habitat beneath the surface of water.

1354 **lemnata** (Linnaeus, 1758) – *Lemna*, the duckweed genus: '*habitat in* Lemna, *ex qua collecta nidum construit*', it lives on duckweed, from collected fragments of which it makes a nest – an accurate description.

Synclita Lederer, 1863 – συγκλίνω (sugklino), to incline together, to converge: from the converging forewing fasciae of some members of this tropical genus.

1355 **obliteralis** (Walker, 1859) – *obliteratus*, rubbed out: Walker describes the forewing markings as very indistinct.

ACENTROPINAE (1331)

Acentria Stephens, 1829 – ἀ, alpha privative; κέντρον (kentron), a spur: from the virtual absence of the tibial spurs, which are vestigial.
= **Acentropus** Curtis, 1834 – ἀ, alpha privative; κέντρον (kentron), a spur; πούς, ποδός (pous, podos), the foot: the same meaning as *Acentria* above.

1331 **ephemerella** ([Denis & Schiffermüller], 1775) – ἐφήμερος (ephēmeros), living but a day: hardly an exaggeration, the average life span of the adult being two days.
= **nivea** (Olivier, 1791) – *niveus*, snowy: from the white forewing.

EVERGESTINAE (1356)

Evergestis Hübner, 1825 – εὐεργής (euergēs), well-wrought; ἐσθής (esthēs), a garment: from the goodly vestiture of the adults. Macleod derives from εὐεργέτης (euergetēs), a benefactor, without attempting an explanation.

1356 **forficalis** (Linnaeus, 1758) – *forfex, forficis*, a pair of scissors or shears: a name based on the way in which the wings are folded in repose. Linnaeus defined the Pyrales thus (Pl. III), '*alis* *conniventibus in figuram deltoideam forficatam'*, words carefully chosen which may be translated literally 'with wings winking into the shape of a scissored deltoid'. A butterfly closes its wings with a hinged movement like a door, but a pyrale brings them together on a single plane like the sashes of a window or the lids when the eyes are closed. Shears or scissors close in a similar manner but to cut effectively the blades must meet fully or overlap. A deltoid is a triangular figure, shaped like the capital Greek delta (Δ). *E. forficalis*, therefore, has a name suggesting that it is the quintessential pyrale. When its wings are folded, they do not overlap like those of 1329 *Donacaula forficella*, q.v., but form a tectiform deltoid, meeting each other over the abdomen and concealing it. As defined by Linnaeus, both the Geometrae and Pyrales rest with their wings extended over the substrate, but in the former they stop short at the abdomen, leaving it exposed; the wings do not close fully like scissors. The way the wings are held at rest is not always easy to observe and as a result species were sometimes placed in the 'wrong' family with the 'wrong' termination; of the following species, Scopoli thought his a pyrale, Hufnagel his a geometer. In consequence, Fabricius (1775) placed the two families together in Phalaena, which he divided into three sections, the Geometrae pectinicornes and Geometrae seticornes of Linnaeus, and a third group headed *'Alis forficatis'* comprising the pyrales and hypenines with 'scissored' wings and the termination *-alis*. *'Forfex*, shears, from appearance of markings on forewings' (Macleod).

1356a **limbata** (Linnaeus, 1767) – *limbus*, a border; from the dark terminal area of the wings.

1357 **extimalis** (Scopoli, 1763) – *extimus*, *extremus*, both superlatives of *exter*, outermost; *-alis*, the pyrale termination: from the conspicuously dark termen and terminal cilia.

1358 **pallidata** (Hufnagel, 1767) – *pallidus*, pale; *-ata*, geometrid termination: from the pale yellowish ground colour. Compare the terminations of the last two species with my remarks on classification given under 1356 *E. forficalis* above.

ODONTIINAE

From the genus *Odontia* Duponchel, 1832 – ὀδούς, ὀδόντος (odous, odontos), a tooth: see next species.

Cynaeda Hübner, 1825 – κύων, κύνος (kuōn, kunos), a dog; εἶδος (eidos), a form: from the dog-tooth pattern on the forewing of the next species. *'Kinaidos*, peach, from colour of forewings' (Macleod); κίναιδος (kinaidos), a lecherous fellow (Liddell & Scott, 1869).

1359 **dentalis** ([Denis & Schiffermüller], 1775) – *dens*, *dentis*, a tooth: from the strongly dentate fascia on the forewing.

Metaxmeste Hübner, 1825 – μέταξα (metaxa), raw silk; μεστός (mestos), filled with: from the glossy wings.

1359a **phrygialis** (Hübner, 1796) – *phrygius*, embroidered, as worked by the Phrygians: perhaps from the wing pattern.

GLAPHYRIINAE

From the genus *Glaphyria* Hübner, 1823 – γλαφυρός (glaphuros), polished, well-finished: the next species, the only one of this tropical subfamily to have been taken in Britain, has glossy hindwings.

Hellula Guenée, 1858 – Guenée gives no clue to the meaning; like some of his other generic names, it may be 'sans étymologie'.

1360 **undalis** (Fabricius, 1781) – *unda*, a wave: from the sinuate fasciae on the forewing.

PYRAUSTINAE (1361)

Pyrausta Schrank, 1802 – πυραύστης (puraustēs), a moth that gets singed in the candle (Aeschylus, Aristotle), from πῦρ (pūr), fire and αὔω (auō), to kindle or burn. The word was used for a moth in general and also in a metaphorical sense; 'it is variously stated to be applicable to the short-lived, to those destroyed for the sake of some small pleasure, or those who die ignobly through their own folly' (Beavis, 1988). *Pyrausta* or *pyraustes* according to Pliny was an unknown insect supposed to live in fire. It is unfortunate that the name is now applied to a genus of day-flying moths that are only rarely attracted to light.

1361 **aurata** (Scopoli, 1763) – *auratus*, golden: from the yellowish gold markings of the forewing.

1362 **purpuralis** (Linnaeus, 1758) – *purpura*, purple: from the purple ground colour of the forewing.

1363 **ostrinalis** (Hübner, 1796) – *ostrinus*, purple (ὄστρεον (ostreon)): from the ground colour of the forewing.

1364 **sanguinalis** (Linnaeus, 1767) – *sanguineus*, blood-coloured: from the crimson fasciae.

1365 **cespitalis** ([Denis & Schiffermüller], 1775) – *caespes* (*cespes*), *caespitis*, a turf: from the dry, grassland habitat.

1366 **nigrata** (Scopoli, 1763) – *nigratus*, blackened: from the dark ground colour.

1367 **cingulata** (Linnaeus, 1758) – *cingula*, a girdle: from the white fascia on each forewing which together form a continuous band when the wings are folded.

Margaritia Stephens, 1827 – *margarita*, a pearl: from the colour of the forewing of some of the thirty species Stephens included in his genus. The name must be regarded as a generalization and not as having special reference to the next species.

1368 **sticticalis** (Linnaeus, 1761) – στίκτος (stiktos), spotted: from the pale spot situated between the darker stigmata on the forewing.

Uresiphita Hübner, 1825 – perhaps from οὔρησις (ourēsis), making water (urine); φιτύω (phituo), to bring into being (here pleonastic): from the yellow hindwing of the next species. However, Hübner included eight other species with grey hindwings in his genus as well, so Macleod's derivation from οὐρεσιφοίτης (ouresiphoitēs), mountain-haunting, may be correct, though it seems to lack application to any of the species.

1369 **polygonalis** ([Denis & Schiffermüller], 1775) – *Polygonum aviculare*, knotgrass: this is given incorrectly by the authors as the foodplant. 'Gr. *polu-*, many, *gonia*, angle; from triangular forewings' (Macleod); he had not consulted Denis & Schiffermüller and supposed that they did not know the difference between a polygon and a triangle.

= **limbalis** sensu auctt. – *limbus*, a border: from the dusky border to the yellow hindwing.

Sitochroa Hübner, 1825 – σιτόχροος (sitokhroos), of the colour of ripe wheat: from the ground colour of species in the genus.

1370 **palealis** ([Denis & Schiffermüller], 1775) – *palealis*, pertaining to chaff (*palea*): from the pale yellowish ground colour.

1371 **verticalis** (Linnaeus, 1758) – *vertex*, an eddy, the highest point: a name that has puzzled authors who have assumed that the second meaning was the one intended. Pickard *et al.* give in explanation 'the top, the highest point, the largest species in the genus'; this cannot be correct because Linnaeus included the even larger 2477 *Hypena proboscidalis* as a pyrale on the same page. Macleod wrote 'from two vertical lines across forewings'; this, too, must be wrong because the lines are not vertical and it is unlikely that the Latin *vertex* could bear that meaning. Linnaeus himself gives a clue, '*alis glabris pallidis subfasciatis: subtus fusco undatis*' (with smooth, pale, faintly fasciated wings: below with wavy fuscous markings); see Goater, 1986, text figure 7. The link is between *vertex*, an eddy and *unda* a wave and the name refers to the sinuous markings on the under surface.

Paracorsia Marion, 1959 – παρά (para), alongside, close to; *Epicorsia* Hübner, 1818 (a genus not represented in Britain) – ἐπί (epi), upon; κόρση (korsē), the side of the head, the temples: from the labial palpus (see below).

1372 **repandalis** ([Denis & Schiffermüller], 1775) – *repandus*, turned-up: from the ascending labial palpus. Duponchel gives 'froment', wheat, as the meaning, i.e. treating this name as a variant of 1373 *pandalis*; this is an anachronism, the present name being senior by fifty years.

Microstega Meyrick, 1890 – μικρός (mikros), small; στέγη (stegē), a roof, a covering: from the larval habit, unusual for the Pyralidae, of making a case from leaf fragments.

1373 **pandalis** (Hübner, 1825) – Panda, a Roman goddess identified with Ceres, the goddess of agriculture: from the corn-coloured forewing. '*Pandus*, bent, from margin of forewings' (Macleod), an apochryphal character.

1374 **hyalinalis** (Hübner, 1796) – ὕαλος (hualos), glass: from the hyaline, pale yellow forewing.

Sclerocona Meyrick, 1890 – σκληρός (sklēros), hard; κῶνος (kōnos), a pine-cone, any cone-shaped object: from the cone-shaped chitinous prominence on the frons of the adult.

1374a **acutellus** (von Eversmann, 1842) – *acutus*, sharp: from the subacute apex of the forewing.

Ostrinia Hübner, 1825 – *ostrinus* (ὄστρεον), purple: from the purplish brown ground colour of the male of the next species.

1375 **nubilalis** (Hübner, 1796) – *nubilum*, cloudy weather: from the relatively dark wing markings, especially in the male.

Eurrhypara Hübner, 1825 – εὖ (eu), well; ῥυπαρός (rhuparos), greasy: from the wings of the next species (the only one included by Hübner) which have a glossy sheen as if they were slightly oily.

1376 **hortulata** (Linnaeus, 1758) – *hortus*, dim. *hortulus*, a garden, an orchard: '*habitat in* Urtica, *hortis pomonae*' (it lives on nettle, in fruit orchards). Linnaeus placed it in the Geometrae, hence the termination *-ata*.

Perinephela Hübner, 1825 – περί (peri), around; νεφέλη (nephelē), a cloud: from the dusky, cloudy markings of the next species.

1377 **lancealis** ([Denis & Schiffermüller], 1775) – *lancea*, a light spear: from the elongate (lanceolate) forewing.

Phlyctaenia Hübner, 1825 – φλύκταινα (phluktaina), a blister: from the circular pale blotches on the wings of 1378 *P. coronata*.

1378 **coronata** (Hufnagel, 1767) – *corona*, a wreath; *-ata*, geometrid termination: from the fuscous markings which appear to enwreathe the pale blotches on the wings.
= **sambucalis** ([Denis & Schiffermüller], 1775) – *Sambucus*, the elder genus, which includes some of the foodplants.

1379 see below 1384

1380 **perlucidalis** (Hübner, 1809) – *per*, intensive prefix; *lucidus*, bright: from the relatively glossy whitish ground colour.

1381, 1382, 1383 } see below 1384, 1379

1384 **stachydalis** (Germar, 1822) – *Stachys*, the woundwort genus to which the larval foodplants belong.

Mutuuraia Munroe, 1976 – i.h.o. A. Mutuura, a contemporary Japanese entomologist with whom Munroe has collaborated in several papers on the Pyralidae.

1379 **terrealis** (Treitschke, 1829) – *terreus*, earthen: from the rather drab, fuscous wings.

Anania Hübner, 1825 – ἀνάνιος (ananios), without pain: Hübner resorts to litotes to express his pleasure in the beauty of 1381 *A. funebris*. Macleod's speculation that the name is a corruption of Anagnia, a city in Latium, might have been correct if Ochsenheimer or Stephens had been the author, but it would have been uncharacteristic of Hübner to form a generic name from that of a town without reason for doing so. There is no Hübnerian name in Appendix 3.

1381 **funebris** (Ström, 1768) – *funebris*, funereal: from the black and white coloration.
= **octomaculata** (Linnaeus, 1771) – *octo*, eight; *macula*, a spot; *-ata*, the termination Linnaeus used for species he regarded as Geometrae: from the eight white spots, two on each of the four wings.

1382 **verbascalis** ([Denis & Schiffermüller], 1775) – *Verbascum*, the mullein genus: the foodplant, however, is wood-sage (*Teucrium scorodonia*).

Psammotis Hübner, 1825 – ψαμμωτός (psammōtos), sanded: either from the yellowish ground colour or from the irroration on the wings of the next species.

1383 **pulveralis** (Hübner, 1796) – *pulvis*, *pulveris*, dust: from the fuscous irroration on the wings.

1384 see below 1380

Ebulea Doubleday, 1849 – *Sambucus ebulus*, dwarf elder, is a foodplant of 1378 *Phlyctainia coronata* (= *sambucalis*) which was the original type species of Doubleday's genus; now that species has been transferred to a senior genus and *Ebulea* survives as an unsuitable name for the genus of the next species.

1385 **crocealis** (Hübner, 1796) – *croceus*, saffron-coloured: from the ground colour.

Opsibotys Warren, 1890 – ὄψις (opsis), the look, the appearance; the genus *Botys* Latreille, 1802 which contained six of the species from 1371 to 1397, now placed in various genera:

Warren's name denotes affinity with the discarded genus. Pickard *et al.* derive *Botys* doubtfully from βῶτις (bōtis), a shepherdess.

1386 **fuscalis** ([Denis & Schiffermüller], 1775) – *fuscus*, dark-coloured: from the ground colour.

Nascia Curtis, 1835 – perhaps from Nascio, the goddess of birth. Curtis gives no explanation.

1387 **cilialis** (Hübner, 1796) – *cilia*, eyelashes, the fringe of an insect's wing: this species has the cilia white beyond a well-defined fuscous base, a conspicuous feature.

Udea Guenée, 1845 – οὖδας, οὔδεος (oudas, oudeos), the surface of the earth: the larvae feed on low-growing plants and the adults frequent rough pasture.

1388 **lutealis** (Hübner, 1809) – *lŭteus*, clay-coloured: from the ground colour. Macleod's derivation from *lūteus*, yellow, is less likely since the wings are pale ochreous white rather than rich yellow, the meaning of *lūteus*.

= **elutalis** sensu auctt. – *ex*, *e-*, (in this context) to a lesser degree; *lūteus*, yellow: i.e. yellowish, descriptive of the forewing. Alternatively, the name may be derived from *elutus*, pale, washed-out, with much the same meaning. Distinct from *U. lutealis* and not found in Britain.

1389 **fulvalis** (Hübner, 1809) – *fulvus*, reddish yellow: from the ground colour.

1390 **prunalis** ([Denis & Schiffermüller], 1775) – *Prunus spinosa*, blackthorn, is one of the many foodplants.

= **nivealis** (Fabricius, 1781) – *niveus*, snowy: relevance obscure.

1391 **decrepitalis** (Herrich-Schäffer, 1848) – *decrepitus*, worn out: from the pale grey forewing, described by Herrich-Schäffer as 'dirty white'.

1392 **olivalis** ([Denis & Schiffermüller], 1775) – *oliva*, an olive-tree: possibly a foodplant on the Continent; in Britain widely polyphagous but found mainly on herbaceous plants. 'From colour of forewings' (Macleod): unlikely, since these are greyish ochreous without any shade of olive.

1393 **uliginosalis** (Curtis, 1830) – *uliginosus*, marshy: from the habitat which is grassland close to streams in the Scottish mountains.

1394 **alpinalis** ([Denis & Schiffermüller], 1775) – *alpinus*, Alpine, montane: from the habitat.

1395 **ferrugalis** (Hübner, 1796) – *ferrugo*, the colour of iron-rust: from the ground colour of the forewing.

= **martialis** (Guenée, 1854) – Mars, the god of war and the red planet: from the reddish ground colour.

Mecyna Doubleday, 1849 – μηκύνω (mēkuno), to lengthen: from the elongate forewing and abdomen of the adults.

1396 **flavalis** ([Denis & Schiffermüller], 1775) – *flavus*, yellow: from the ground colour of the forewing.

subsp. **flaviculalis** Caradja, 1916 – *flaviculus*, yellowish: this subspecies is paler than the nominate subspecies.

1397 **asinalis** (Hübner, 1819) – *asinus*, an ass: from the grey ground colour.

Nomophila Hübner, 1825 – νομός (nomos), a pasture; φιλέω (phileō), to love: from the habitat.

1398 **noctuella** ([Denis & Schiffermüller], 1775) – *noctu*, by night: though a night-flying moth, it is also easily disturbed by day. The authors may have been puzzled over its correct systematic placing; although they assigned it to Tinea, they may have seen characters suggestive of Noctua.

1398a **nearctica** Munroe, 1973 – νέος (neos), new; ἀρκτικός (arktikos), near the Bear (constellation), northern, belonging to the Nearctic Region, i.e. that part of the Holarctic Region situated in the New World. A North American migratory species.

Dolicharthria Stephens, 1834 – δολικός (dolikos), long; ἄρθρα (arthra), limbs: from the exceptionally long legs of the next species.

1399 **punctalis** ([Denis & Schiffermüller], 1775) – *punctum*, a spot: from the prominent white discal spot on the forewing.

Antigastra Lederer, 1863 – ἀντι- (anti-), against; γαστήρ (gastēr), the belly: from the next species' habit, when at rest, of flexing its abdomen upwards; cf. 1413 *Hypsopygia*.

1400 **catalaunalis** (Duponchel, 1833) – Catalaunus, Catalonia: the Spanish district in which Barcelona, the type locality, is situated.

Maruca Walker, 1859 – unexplained and possibly a meaningless neologism, like some other of Walker's names.

1401 **testulalis** (Geyer, 1832) – *testula*, a small potsherd, sometimes used as a writing tablet: either from the chocolate-brown ground colour or because the white markings on the forewing are suggestive of lettering.

Diasemia Hübner, 1825 – διάσημος (diasēmos), clear, distinct: from the sharply defined wing pattern of the next species, the only one included by Hübner.

1402 **reticularis** (Linnaeus, 1761) – *reticulum*, a net: from the vaguely reticulate wing pattern.
= **litterata** (Scopoli, 1763) – *litera, littera*, a letter of the alphabet: because the dark wings appear to have white letters scribbled over them, the letter 'V', according to Scopoli.

Diasemiopsis Munroe, 1957 – the genus 1402 *Diasemia*, q.v.; ὄψις (opsis), the look, the appearance: a genus resembling *Diasemia*.

1403 **ramburialis** (Duponchel, 1834) – i.h.o. J. P. Rambur (1801–70), a French entomologist.

Hymenia Hübner, 1825 – 'Υμήν, Hymen, the god of marriage: apparently without entomological application.

1404 **recurvalis** (Fabricius, 1775) – *recurvus*, bent back: from the shape of the labial palpus.

Pleuroptya Meyrick, 1890 – πλευρόν (pleuron), a rib; πτύον (ptuon), a winnowing fan: perhaps from the ample wings of the next species.

1405 **ruralis** (Scopoli, 1763) – *ruralis*, pertaining to the countryside: from the habitat.

Herpetogramma Lederer, 1863 – ἑρπετόν (herpeton), a snake; γράμμα (gramma), a letter, figure or mark: from sinuous wing markings, although these are not present on the two species that have occurred as adventives in Britain.

1406 **centrostrigalis** (Stephens, 1834) – *centrum*, the centre; *striga*, a furrow, line: from the median fascia of the forewing.

1407 **aegrotalis** (Zeller, 1852) – *aegrotus*, sick: perhaps because the pinkish flush on the wings is suggestive of fever: cf. 1902 *chlorosata*.

Palpita Hübner, 1808 – *palpo*, to touch, whence palpus, a 'feeler'; from the labial palpus which has dense projecting scales on the underside of segment 2, both in the following species and 1375 *Eurrhypara hortulata*, which was the example Hübner gave in his *Tentamen* [1806], where he first proposed this generic name.

1408 **unionalis** (Hübner, 1796) – *Unio*, the pearl-mussel genus, whence *unio*, a single pearl, a union (cf. Hamlet v. ii. 286): from the translucent white ground colour. '*Unio*, singleness, from this being the only species' (Macleod): in 1796 Hübner placed it together with 130 other species in *Pyralis*. Duponchel gives the correct explanation.

Diaphania Hübner, 1818 – διά (dia), through; φαίνω, φαν- (phaino, phan-), to make to appear, to show, to shine: from the diaphanous wings.

1409 **hyalinata** (Linnaeus, 1767) – *hyalinus*, of glass: from the translucent wings.
= **lucernalis** (Hübner, 1796) – *lucerna*, a lamp: from the bright, translucent wings.

Agrotera Schrank, 1802 – ἀγρότερος (agroteros), rustic, 'Αγροτέρα, an epithet of Artemis (Diana), the huntress: a name possibly suggested by that of the next species, though it originally bore family status.

1410 **nemoralis** (Scopoli, 1763) – *nemus, nemoris*, a grove, a glade: from the habitat.

Leucinodes Guenée, 1854 – λευκός (leukos), white; εἶδος, ὠδ- (eidos, ōd-), form, appearance: from the white ground colour of some of the species.

1411 **vagans** Tutt, 1890 – *vagans*, wandering: a migrant species.

Sceliodes Guenée, 1854 – σκέλος (skelos), the leg; εἶδος, ὠδ- (eidos, ōd-), form: from the valvae ('tablier'), which are well separated and more strongly developed than in other Pyralidae, projecting from the abdomen like a man's legs. The genus when first erected was monotypic, the single member being Australian.
= **Daraba** Walker, 1859 – not explained by Walker. The genus was erected to contain the following species, the name of which is not explained either.

1412 **laisalis** Walker, 1859 – Walker gives no explanation. There are towns called Lais in Sumatra and the Philippine Islands, one of which may have prompted the name, though the type locality is 'Cape; Africa'. Walker spells the name with a capital.

PYRALINAE (1416)

Hypsopygia Hübner, 1825 – ὕψος (hupsos), height; πυγαῖος (pugaios), pertaining to the rump: from the moth's habit of flexing its abdomen upwards when it is at rest.

1413 **costalis** (Fabricius, 1775) – *costa*, a rib, the anterior margin of an insect's wing: from the two gold spots on the forewing costa.

Synaphe Hübner, 1825 – συναφή (sunaphē), union: 'perhaps from the partial union of some veins on the hindwing' (Macleod).

1414 **punctalis** (Fabricius, 1775) – *punctum*, a spot: from the small, yellowish white dots on the costa of the forewing.

= **angustalis** ([Denis & Schiffermüller], 1775) – *angustus*, narrow: from the narrow, elongate wings.

Orthopygia Ragonot, 1891 – ὀρθός (orthos), straight; πυγαῖος (pugaios), pertaining to the rump: when at rest, the moth holds its abdomen straight, not flexed upwards like 1413 *Hypsopygia*, q.v.

1415 **glaucinalis** (Linnaeus, 1758) – *glaucus*, bluish grey: *'alis glabris glaucis'*, with smooth bluish grey wings.

Pyralis Linnaeus, 1758 – πυραλίς (puralis), *pyrallis*, an unknown species of bird or winged insect which was supposed to live in fire (πῦρ) (Pliny); cf. 1361 *Pyrausta*. The name has no classical association with this family and Linnaeus may have chosen it because of the frequency with which some pyralids are attracted to light. Pyralis was one of the seven families into which Linnaeus divided the Phalaenae (moths other than hawk-moths); see Pl. III and p. 21.

1416 **lienigialis** (Zeller, 1843) – i.h.o. Madam Lienig (d. 1855), a Livonian (Latvian) entomologist who collaborated with Zeller.

1417 **farinalis** Linnaeus, 1758 – *farina*, flour: *'habitat in farina'*, it lives on flour.

1418 **manihotalis** Guenée, 1854 – Guenée offers no explanation for this name.

1419 **pictalis** (Curtis, 1834) – *pictus*, painted, bright: from the bright coloration.

Aglossa Latreille, 1796 – ἀ-, alpha privative; γλῶσσα (glōssa), the tongue: the haustellum is obsolete. The name was intended as a suprageneric taxon, equal in rank to Papilio, Sphinx, Bombyx, etc.

1420 **caprealis** (Hübner, 1809) – apparently a misprint for *cuprealis*, to which it is emended in many text-books (e.g. Meyrick, 1928). *Cupreus*, coppery: from the ferruginous irroration on the forewing.

1421 **pinguinalis** (Linnaeus, 1758) – *pinguis* (adj.), *pingue* (n.), fat: *'habitat in pinguibus, Butyro aliisque frequens, intra domos et culinas, rarius intra ventriculo humano, inter vermes pessima'*, it lives on fats, usually on butter, etc., in houses and kitchens, occasionally in the human belly, the most loathsome of worms. 'Greasy, from appearance' (Macleod).

1422 **dimidiata** (Haworth, 1809) – *dimidiatus*, halved, half: because it is sometimes not more than half of the size of the previous species which it resembles. 'From broken markings on forewings' (Macleod).

1423 **ocellalis** Lederer, 1863 – *ocellus*, a small eye: from a series of small whitish rings on the forewing.

Endotricha Zeller, 1847 – ἔνδον (endon), within; θρίξ, τριχός (thrix, trikhos), hair: from the male tegulae which are elongate and end in a tuft, the tegulae being 'inner' in relation to the wings in a set specimen.

1424 **flammealis** ([Denis & Schiffermüller], 1775) – *flammeus*, flame-coloured; from the reddish purple suffusion on the wings.

1424a **consobrinalis** Zeller, 1852 – *consobrinus*, strictly a cousin-german on the maternal side, loosely any cousin or relative: from the close relationship between this species and the last.

GALLERIINAE (1425)

Galleria Fabricius, 1798 – γαλερός (galeros), cheerful (Spuler); '*Galleria*, from the habit of the larva of forming *galleries* in honeycomb' (Pickard *et al.*). Both are probably correct. Fabricius may have wished to convey the sense suggested by Pickard *et al.* 'Gallery' is derived from the O. Fr. 'galerie' (Italian 'galleria') and not from a classical Latin or Greek word; so, following his usual punning practice, he looked for a Greek word of any meaning which sounded like it. His aim was to remove the species from Tinea in which Linnaeus had described them because the adults rest with convolute wings; see MBGBI 7(2), Pl. B, fig. 3.

1425 **mellonella** (Linnaeus, 1758) – Mellona, the goddess of bee-keeping; *mel*, *mellis*, honey: from the larval pabulum.

Achroia Hübner, 1819 – ἄχροια (akhroia), lack of colour, paleness: from that character in the next species.

1426 **grisella** (Fabricius, 1794) – *griseus* (late Lat.), grey: from the ground colour of the forewing.

Corcyra Ragonot, 1885 – *Corcyra*, the classical name for the Ionian island of Corfu. When Stainton named 1427 *cephalonica*, he predicted that a new genus would be required to accommodate it; in complying, Ragonot coined the generic name from another island in the same group as Kefalmia (see next species).

1427 **cephalonica** (Stainton, 1866) – Cephalonica, now Kefalmia, one of the Ionian Islands off the west coast of Greece. The type specimen was reared from dried currants imported from Greece, but there is no evidence that the precise source was Kefalmia.

Aphomia Hübner, 1825 – ἀφόμοιος (aphomoios), unlike: a genus for species differing sufficiently to warrant separation; they had been placed in Tinea by Linnaeus and Galleria by Fabricius. 'Resembling; perhaps from resemblance to *Melissoblaptes*' (Macleod); *Aphomia* means *not* resembling and there is an anachronism, *Aphomia* being an older name than *Melissoblaptes* by fourteen years.

1428 **sociella** (Linnaeus, 1758) – *socius*, associating together: from the social habits of the larvae.

Melissoblaptes Zeller, 1839 – μέλισσα (melissa), a bee; βλάπτω (blaptō), to harm: Zeller erected the genus to accommodate two species, one of them 1428 *Aphomia sociella*, which are pests of bees' nests; the next species has no association with bees and was not included in the genus by Zeller.

1429 **zelleri** (Joannis, 1932) – i.h.o. P. C. Zeller (1808–83), the distinguished German micro-lepidopterist.

Paralipsa Butler, 1879 – παρά (para), beside, close to; the genus *Alispa* Zeller, 1848, now synonymized with 1465 *Nephopterix*. Butler described the type species (Japanese) of his genus as a phycitine instead of a galleriine and misspelled *Alispa*; had he not also given the specific name *angustella*, his generic name would have baffled explanation. Bradley & Fletcher (1986) make the same spelling mistake as Butler in giving the synonym of *Nephopterix*.

1430 **gularis** (Zeller, 1877) – *gula*, the throat, gluttony (figurative): the larva is a voracious pest of stored products.

Arenipses Hampson, 1901 – *arena*, sand; ἴψ (ips), a worm, a larva, an insect, generally when regarded as a pest: see next species.

1431 **sabella** Hampson, 1901 – *sabulo*, coarse sand, gravel: from the habitat.

PHYCITINAE (1452 (10))

The revision of the Phycitinae adopted by Goater (1986) and Bradley & Fletcher (1986) has been followed here and since the sequence of species differs markedly from the serial numbering of Bradley & Fletcher (1978), the reader may find difficulty in locating the one he is looking for. The Log Book number for each species is therefore followed by a second number in brackets representing the new sequence. A special index to the subfamily appears on p. 288 giving the old numerical order with cross-references to the new.

Anerastia Hübner, 1825 – ἀνέραστος (anerastos), unloved: 1432 [1] *lotella* is an important pest of rye on the Continent.

1432 [1] **lotella** (Hübner, 1813) – either, *Lotus*, the bird's-foot trefoil genus, from a mistake over the foodplant (perhaps from confusion with 1441 [11] *Oncocera semirubella*); or, *lotus*, washed, neat (Pickard *et al.*), from the pale, washed-out ground colour.

Cryptoblabes Zeller, 1848 – κρυπτός (kruptos), hidden; βλάβη (blabē), damage: from the larva which feeds internally; it cannot refer to the damage done by 1434 [3] *C. gnidiella* as a pest, since this species was named nineteen years after the genus and then not included in it.

1433 [2] **bistriga** (Haworth, 1811) – *bi-*, two; *striga*, a furrow, a streak: from the two pale fasciae on the forewing.

1434 [3] **gnidiella** (Millière, 1867) – *Daphne gnidium*, a species of mezereon: a foodplant abroad; in Britain the larva is most often found in imported pomegranates.

Salebriopsis Hannemann, 1965 – *Salebria* Zeller, 1846, a genus now reduced to synonymy with 1441 [11] *Oncocera*; ὄψις (opsis), aspect, appearance: a genus closely resembling *Salebria*.

1446 [4] **albicilla** (Herrich-Schäffer, 1849) – *albus*, white; *cilia*, eyelashes, fine hairs such as those in the fringe of an insect's wing: from the head and base of the antenna in the male, which are clad in white hair-scales.

Metriostola Ragonot, 1893 – μέτριος (metrios), moderate, modest; στολή (stolē), dress: from the sober, dark coloration of 1450 [5] *M. betulae.*

1450 [5] **betulae** (Goeze, 1778) – *Betula*, birch: the larval foodplant.

Trachonitis Zeller, 1848 – τραχύς (trakhus), rough: from the scale-tufts on the forewing of the next species.

1437a [6] **cristella** (Hübner, 1796) – *cristatus*, tufted: from the scale-tufts on the forewing.

Selagia Hübner, 1825 – σελαγέω (selageō), to shine: from the highly glossy forewing of 1448 [7] *S. argyrella.*

1448 [7] **argyrella** ([Denis & Schiffermüller], 1775) – ἄργυρος (arguros), silver: from the glossy forewing.

Microthrix Ragonot, 1888 – μικρός (mikros), small; θρίξ (thrix), hair: from the small sinuation filled with hair-scales at the base of the male antenna.

1449 [8] **similella** (Zincken, 1818) – *similis*, like: from resemblance to another species, possibly 1433 [2] *Cryptoblabes bistriga*; Macleod's statement 'similar, i.e. to *Dioryctria hostilis*, which used to be included in *Nephopterix*' is an anachronism by sixteen years.

Pyla Grote, 1882 – possibly from πύλη (pule), a gate, possibly from Pylos, the name of several Greek cities, and possibly a neologism; almost certainly without entomological application.

1451 [9] **fusca** (Haworth, 1811) – *fuscus*, dusky: from the dark forewing.

Etiella Zeller, 1839 – Treitschke named the following species (q.v.) twice; Zeller adopted the junior synonym as the generic name.

1451a [9a] **zinckenella** (Treitschke, 1832) – i.h.o. Dr J. L. Zincken (early 19th century), a German entomologist; see 129 *zinckenii.*

= **etiella** (Treitschke, 1835) – perhaps from αἴτιος (aitios), blameworthy: the larva feeds on members of the Fabaceae, the bean family, and is therefore a potential pest.

Phycita Curtis, 1828 – φύκος (phukos), a sea-weed from which a red dye was prepared, φυκῖτις (phykītis), a precious stone showing this colour: from the presence of red in the pattern of the forewing.

1452 [10] **roborella** ([Denis & Schiffermüller], 1775) – *Quercus robur*, oak: the larval foodplant.

Oncocera Stephens, 1829 – ὄγκος (ogkos), a tumor, a swelling; κέρας (keras), a horn, the antenna: from the dense, projecting scales which fill the large sinuation near the base of the antenna.

1441 [11] **semirubella** (Scopoli, 1763) – *semi-*, half; *ruber*, red; from the central area of the forewing from base to termen which is pale crimson, the costal and dorsal areas being differently coloured.

Pempelia Hübner, 1825 – πεμπέλος (pempelos), an obscure epithet of aged persons, supposed by Spuler (and probably by Hübner, too) to mean greyish.

1442 [12] **palumbella** ([Denis & Schiffermüller], 1775) – *palumbes*, a wood-pigeon: from the colour of the forewing.

1443 [13] **genistella** (Duponchel, 1836) – *Genista*, the greenweed and petty whin genus, formerly more widely applied; in Britain the foodplant is gorse (*Ulex* spp.).

1444 [14] **obductella** (Zeller, 1839) – *obductus*, concealed: from the larval habit of feeding in a dense spinning.

1445 [15] **formosa** (Haworth, 1811) – *formosus*, beautiful: in appreciation.

Sciota Hulst, 1888 – σκιωτός (skiōtos), shaded: from the dusky ground colour.
= **Nephopterix** sensu auctt. – see 1465 [23], genus.

1447 [16] **hostilis** (Stephens, 1834) – *hostilis*, hostile: Stephens gives no explanation for this apparently censorious name for an attractive moth. 'By contrast with *D.* [*S.*] *adelphella* ("brother")' (Macleod): an anachronism.

1447a [16a] **adelphella** (Fischer von Röslerstamm, 1836) – ἀδελφός (adelphos), a brother: Macleod's suggestion 'from resemblance to *D.* [*S.*] *hostilis*' is chronologically possible.

Hypochalcia Hübner, 1825 – ὑπόχαλκος (hupokhalkos), containing a mixture or proportion of copper: with reference to 1457 [17] *H. ahenella*, q.v.

1457 [17] **ahenella** ([Denis & Schiffermüller], 1775) – *aëneus*, *aheneus*, of bronze: from the faint bronze gloss on the forewing.

Epischnia Hübner, 1825 – ἰσχνός (iskhnos), dry; ἐπισχναίνω (episkhnainō), to make dry: perhaps because one of the species resembled withered vegetation, not, however, 1456 [18] *E. bankesiella* which was not named for another sixty-three years.

1456 [18] **bankesiella** Richardson, 1888 – i.h.o. E. R. Bankes (1861–1929) for his services to entomology in Dorset; Richardson himself discovered the species.

Dioryctria Zeller, 1846 – διορυκτής (dioruktēs), a digger: from the larval habit, in some of the species, of boring into pine-cones.

1454 [19] **abietella** ([Denis & Schiffermüller], 1775) – *Abies*, the fir-tree genus: the larva feeds on various Pinaceae including *Abies* spp.

1454a [20] **schuetzeella** Fuchs, 1899 – i.h.o. K. J. Schütze, who reared the type material at Racklau, Germany.

1455 [21] **mutatella** Fuchs, 1903 – *mutatus*, changed, (n.) a change: Fuchs described it, not as a species, but as a variety of 1454 [19] *D. abietella*.

Pima Hulst, 1888 – probably a meaningless neologism.

1453 [22] **boisduvaliella** (Guenée, 1845) – i.h.o. Dr J. A. Boisduval (1799–1879), a French entomologist: '*ex Helvetia accepta ab amicissimo Boisduval, cuius nomine gaudet*', received from Switzerland from my very dear friend Boisduval, in whose name it rejoices.

Nephopterix Hübner, 1825 – νέφος (nephos), a cloud; πτέρυξ (pterux), a wing: from the patch of raised scales on 1465 [23] *N. angustella*, now restored to this genus.
= **Alispa** Zeller, 1848 – ἀ, alpha privative; λίσπος (lispos), smooth: the forewing is not smooth, having a patch of strongly raised dark scales.

1465 [23] **angustella** (Hübner, 1796) – *angustus*, narrow: from the somewhat elongate forewing. Guenée suspected a misprint for *augustella*, saying that the wings were not particularly narrow and that the moth flew in August; however, the name as spelt is fully acceptable and its retention obligatory. Hübner described the species in Tinea.

Pempeliella von Caradja, 1916 – a genus close to 1442 [12] *Pempelia*, q.v.

1462 [24] **diluta** (Haworth, 1811) – *dilutus*, diluted, thin, pale: from the colour of this species being supposedly paler than that of 1463 [25] *P. ornatella*.
= **dilutella** sensu auctt. – see above.

1463 [25] **ornatella** ([Denis & Schiffermüller], 1755) – *ornatus*, adorned: from the variegated forewing.

Acrobasis Zeller, 1839 – ἄκρον (akron), a point; βάσις (basis), a step, that which is stepped on, a base: from the horny tooth at the apex of the scape (the basal segment) of the antenna.

1435 [26] **tumidana** ([Denis & Schiffermüller], 1775) – *tumidus*, swelling: from the strong antemedian ridge of raised scales on the forewing. 'From male's antennae' (Macleod).

1436 [27] **repandana** (Fabricius, 1798) – *repandus*, bent backwards, turned up: probably from the labial palpus which is moderately long, curved and ascending; cf. 1372 *repandalis*.
= **tumidella** (Zincken, 1818) – *tumidus*, swelling: from the antemedian ridge of strongly raised scales on the forewing.

1437 [28] **consociella** (Hübner, 1813) – *consocius*, a companion: from the larvae which feed gregariously.

Numonia Ragonot, 1893 – probably a neologism.

1438 [29] **suavella** (Zincken, 1818) – *suavis*, agreeable: in approbation.

1439 [30] **advenella** (Zincken, 1818) – *advena*, a stranger: Zincken had found this species only at Brunswick, where he lived, but he knew it to be resident, since he described the life history. It was a 'stranger', therefore, because of its rarity, not because it was a visitor. By contrast, this is a common species in Britain, whereas the previous species is rare.

1440 [31] **marmorea** (Haworth, 1811) – *marmoreus*, marbled: from the forewing pattern.

Apomyelois Heinrich, 1956 – ἀπο- (apo-), away from, differing from; the genus 1458 [36] *Myelois*, q.v.

1486 [32] **bistriatella** (Hulst, 1887) – *bis, bi-*, two; *stria*, a furrow, a streak: from the two fasciae on the forewing, often reduced to dorsal striae.

subsp. **neophanes** (Durrant, 1915) – νεοφανής (neophanēs), just come into sight: a newly discovered species.

Ectomyelois Heinrich, 1956 – ἐκτός (ektos), outside; the genus 1458 [36] *Myelois*, q.v.: a genus similar to *Myelois*, but having larvae that feed on pith from the outside, not as miners.

1460 [33] **ceratoniae** (Zeller, 1839) – *Ceratonia*, the carob: the larva feeds in the beans.

Mussidia Ragonot, 1888 – Mussidan, a town in the Dordogne, southern France, but not the type locality of 1466 [34] *M. nigrivenella*.

1466 [34] **nigrivenella** Ragonot, 1888 – *niger*, black; *vena*, a vein: from the fuscous irroration along the veins of the forewing.

Eurhodope Hübner, 1825 – εὖ (eu), well, intensive prefix; ῥοδωπός (rhodōpos), rosy-faced: from the colour of the head; 1459 [35] *E. cirrigerella* has the head orange-ochreous but in other species, formerly included in the genus, it is rufous-tinged.

1459 [35] **cirrigerella** (Zincken, 1818) – *cirrus*, a lock of hair, a tuft, a fringe on a tunic; *gero*, to bear: from the yellow anal tuft, the yellow cilia on the hindwing, or both.

Myelois Hübner, 1825 – μυελός (muelos), marrow, pith: from the habits of the larva of 1458 [36] *M. cribrella*, which mines the pith of thistle stems.

1458 [36] **cribrella** (Hübner, 1796) – *cribrum*, a sieve: from the spots on the forewing which are suggestive of the holes in a sieve; cf. 2053 *cribraria*.

Gymnancycla Zeller, 1848 – γύμνος (gumnos), naked; ἀγκύλη (agkule), the bend of the arm, a loop or noose in a cord: here referring to the sinuation at the base of the male antenna which is not filled with scales as in most related genera.

1464 [37] **canella** ([Denis & Schiffermüller], 1775) – *canus*, grey, hoary: from the pale forewing which often has a greyish tinge.

Zophodia Hübner, 1825 – ζοφώδης (zophōdēs), dark, gloomy: not applicable to the member of the genus which has occurred in Britain.

1464a [38] **grossulariella** (Zincken, 1818) – *Ribes uva-crispa*, formerly *R. grossularia*, gooseberry: the larval foodplant.

Assara Walker, 1863 – possibly from *assarius*, roasted: from a fancied scorched appearance. Some of Walker's names, however, are meaningless neologisms.

1461 [39] **terebrella** (Zincken, 1818) – *terebra*, an instrument for boring: from the larval habit of boring into the cones of Norway spruce (*Picea abies*).

Euzophera Zeller, 1867 – εὖ (eu), well, intensive prefix; ζοφερός (zopheros), dusky: from the dark coloration of some members of the genus.

1469 [40] **cinerosella** (Zeller, 1839) – *cinerosus*, ashen: from the ground colour.

1470 [41] **pinguis** (Haworth, 1811) – *pinguis*, fat: Haworth gave this species the English name 'the tabby knothorn' and evidently associated it for some reason with 'the large tabby' (1421 *Aglossa pinguinalis*, q.v.). He described the larva correctly as a bark-feeder. Macleod's guess 'from greasy appearance of forewings' is without foundation and a reproach to a handsome moth.

1471 [42] **osseatella** (Treitschke, 1832) – *osseus*, of bone, coloured like bone: from the forewing ground colour.

1472 [43] **bigella** (Zeller, 1848) – *bis, bi-*, two; *gero*, to bear: from the two pale fasciae on the forewing. Spuler derives the name from *bigae*, a two-horsed chariot and Guenée, who misspells the name *bigaella*, appears to have had the same opinion; there is nothing, however, in Zeller's description to support this fanciful interpretation.

Nyctegretis Zeller, 1848 – νύξ (nux), night; ἐγείρω (egeiro), to arouse, awaken) from the time of flight of the moth. In this case and contrary to his usual practice, Zeller explains the meaning of his name. 'From moth's habit of sitting by night on restharrow' (Macleod): call my bluff!

1468 [44] **lineana** (Scopoli, 1786) – *linea*, a line: from the two fasciae on the forewing.
= **achatinella** (Hübner, 1824) – *achates*, agate: from the ferruginous markings on the forewing.

Ancylosis Zeller, 1839 – ἀγκύλος (agkulos), curved: from the 'rather small, curved palps'.

1467 [45] **oblitella** (Zeller, 1848) – *oblitus*, smeared, rubbed out: Zeller describes the fasciae of the forewing as dilute.

Homoeosoma Curtis, 1833 – ὅμοιος (homoios), like; σῶμα (sōma), body: from the similarity of the abdomen in both sexes.

1480 [46] **nebulella** ([Denis & Schiffermüller], 1775) – *nebula*, a mist, smoke: from the dark shade on the outer half of the costa of the forewing.

1481 [47] **sinuella** (Fabricius, 1793) – *sinus*, a curve, *sinuosus*, curved: the median and postmedian lines consist of dark spots which are often linked by dark shading so as to form zigzag fasciae. 'From arched front margin of forewings' (Macleod), a character drawn from his imagination.

1482 [48] **nimbella** (Duponchel, 1836) – *nimbus*, a rain-cloud: the name expresses affinity with 1480 [46] *H. nebulella*, although the dark shade which gave that species its name is absent in this one. Macleod is wrong again.

Phycitodes Hampson, 1917 – the genus 1452 [10] *Phycita*, q.v.; εἶδος, ὠδ- (eidos, od-), form, shape: a genus resembling *Phycita*, not, however, in colour.

1483 [49] **binaevella** (Hübner, 1813) – *bis, bi-*, two; *naevus*, a mole on the body, a spot: from the two black discal spots on the forewing.

1484 [50] **saxicola** (Vaughan, 1870) – *saxum*, a stone, a rock; *colo*, to inhabit: from the coastal habitat; there are shingle beaches at the type locality, Leigh-on-Sea, Essex.

1485 [51] **maritima** (Tengström, 1848) – *maritimus*, pertaining to the coast: from the habitat, which is mainly coastal.
= **carlinella** (Heinemann, 1865) – *Carlina vulgaris*, carline thistle, is a foodplant.

Plodia Guenée, 1845 – not explained by Guenée; either an invented word or from *plaudere* (*plodere*), to applaud, in approbation of the rich ferruginous markings of 1479 [52] *P. interpunctella*, the only species he included in the genus.

1479 [52] **interpunctella** (Hübner, 1813) – *inter*, between; *punctus*, dotted: from the fuscous irroration which occurs mainly between the median and postmedian fasciae.

Ephestia Guenée, 1845 – ἐφέστιος (ephestios), beside the hearth, domestic: three of the four species included by Guenée occur in buildings.

1473 [53] **elutella** (Hübner, 1796) – *elutus*, washed out, pale: from the ground colour.

1474 [54] **parasitella** Staudinger, 1859 – *parasitus*, a sponger, a parasite: the species is exceptional in the genus in that it does not eat food for human consumption and Staudinger reared the type specimen from the parasitic plant *Cytinus hypocistis*; in Britain, the larva feeds in old ivy (*Hedera helix*), which is an epiphyte, not a parasite.
subsp. **unicolorella** Staudinger, 1881 – *unicolor*, all of one colour: this subspecies is more unicolorous than the nominate subspecies.

1475 [55] **kuehniella** Zeller, 1879 – i.h.o. Professor Kühn, Director of the Department of Agriculture at the University of Halle, who sent adults to Zeller for determination in June, 1877.

1476 [56] **cautella** (Walker, 1863) – probably, like many of Walker's names, a meaningless neologism, but possibly from the root of *cauter*, a branding iron, since Walker describes the hindwing as cinereous (ashy). The type material consisted of a single female taken in Ceylon and placed in the collection of the British Museum (Natural History), so Macleod's explanation '*cautus*, wary, from larva's habit of living under thick web' is fabrication.

1477 [57] **figulilella** Gregson, 1871 – *ficus*, a fig: an intentional spelling variant to avoid confusion with *E. ficella* Stainton (Gregson follows Stainton in attributing that name to Douglas), a synonym of 1478 [58] *E. calidella*. Gregson reared *figulilella* from dried figs and gives characters to distinguish it from *E. ficella*. Macleod is wrong in dubbing the name as a spelling mistake.

1478 [58] **calidella** Guenée, 1845 – *calidus*, warm: Guenée gives no explanation; he did not know the life history, so the name cannot refer to breeding in heated warehouses. The name just possibly refers to the type locality, the Île d'Hyères on the Côte d'Azure, one of the warmest parts of France.

PTEROPHOROIDEA (1510)

PTEROPHORIDAE (1510)

General note on the nomenclature

1) In 29 species the specific name ends in *-dactyla* or *-dactylus*, from the Greek δάκτυλος (dactulos), a finger, with reference to the lobes into which the wings are divided. Having been given here, this explanation is not repeated under each species.

2) Linnaeus had placed the plumes in Alucita but in 1762 Geoffroy referred to them under the name Pterophorus and, the rule of priority not yet having been formulated, Pterophorus was readily accepted by contemporary writers. However, Geoffroy did not use the name in combination with a species, with the result that when rules were made the name was ascribed to Schäffer, 1766, the first author to do so. Early writers all attribute the name to Geoffroy. They regarded Alucita as available for other use and it was applied to tineids *sensu lato* with feathery cilia on their wings. Long after Latreille had separated the 'many-plumed' moths from the plumes under the name Orneodes, it was recognized that Alucita should be returned to its original Linnaean use. There were now two choices and the name went to the 'many-plumed'. Thus Pterophorus was the first name above specific level to be added to the Linnaean system.

3) Linnaeus had given the plumes names from Greek roots and Geoffroy complied in coining Pterophorus. This may have given rise to the non-Linnaean convention that generic names should be formed from the Greek. Specific names, on the other hand, are normally formed from the Latin, but this is not the case with plumes with the termination *-dactyla*: of the twenty-nine, twenty-three are wholly Greek and six are of mixed language.

4) In 1758 Linnaeus defined the Alucitae as follows: *alis digitatis fissis ad basin* (Pl. III), with wings divided in a basal direction into fingers. He named his six species as follows, *monodactyla*, *bidactyla*, *tridactyla*, *tetradactyla*, *pentadactyla* and *hexadactyla*. Only the last two can possibly refer to the actual number of 'fingers', and then on a different system of counting, *pentadactyla* having five fingers on forewing + hindwing, whereas *hexadactyla* (now 1288 *Alucita hexadactyla*) has six fingers on the forewing and six on the hindwing. It is therefore better to regard the names as forming a numbered sequence, the two that are 'right' being so only by coincidence. Later Hübner was to use yet another method of counting when he named a species *dodecadactyla*.

AGDISTINAE (1487)

Agdistis Hübner, 1825 – Agdistis, according to some classical writers, is an alternative name for the nature goddess Cybele, the 'great mother'; according to others, a male who was castrated either by his own hand or by the gods. Where the testes fell an almond tree sprouted and a nymph called Nana, after eating the fruit, bore a beautiful son called Attis. The myth symbolized death and rebirth in nature and Attis was worshipped by the Phrygians as part of their fertility rites. The adoption of a name for entomological purposes from the Phrygian as opposed to the Greek and Latin mythology is unusual. 'A fabulous monster' (Macleod).

1487 **staticis** Millière, 1875 – *Limonium* (formerly *Statice*) *binervosum*, rock sea-lavender: the larval foodplant.

1488 **bennetii** (Curtis, 1833) – i.h.o. E. Bennet, FLS, who discovered the species on the salt-marshes at Tollesbury, Essex.

PLATYPTILIINAE (1499)

Oxyptilus Zeller, 1841 – ὀξύς (oxus), sharp; πτίλον (ptilon), a downy feather, an insect's wing: having the lobes of the forewing pointed rather than squared, so being an antonym to the older name 1497 *Amblyptilia*, q.v.

= **Crombrugghia** Tutt, 1906 – i.h.o. G. E. M. de Crombrugghe de Picquendaele (late 19th–early 20th century), a Belgian entomologist.

1489 **pilosellae** (Zeller, 1841) – *Hieracium pilosella*, mouse-ear hawkweed: the larval foodplant.

1490 **parvidactylus** (Haworth, 1811) – *parvus*, small: one of the smallest plumes.

1491 **distans** (Zeller, 1847) – *distans*, standing apart, separate, i.e. from *O. laetus* which Zeller named in the same paper.

1492 **laetus** (Zeller, 1847) – *laetus*, joyful: an instance of the pathetic fallacy.

Buckleria Tutt, 1905 – i.h.o. W. Buckler (1814–84), the British entomologist and author of *The larvae of British butterflies and moths*.

1493 **paludum** (Zeller, 1839) – *palus*, gen. pl. *paludum*, a marsh: from the habitat on boggy heaths.

Capperia Tutt, 1905 – i.h.o. S. J. Capper (1825–1912), a founder member of the Lancashire & Cheshire Entomological Society. According to Macleod, 'after C. Capper, who discovered it [how does one discover a genus?] near Beachy Head in 1902'. Tutt gives the type species as '*heterodactyla*', a misidentification of the next species, with *teucrii* Jordan in synonymy, making no mention of C. Capper. *C. britanniodactyla*, under its other names, had been well known for many years and Macleod's anecdote appears to be apocryphal; neither I nor the Duke of Newcastle could trace any entomologist named C. Capper.

1494 **britanniodactyla** (Gregson, 1869) – this species was at first supposed to occur only in Britain.

Marasmarcha Meyrick, 1886 – μαρασμός (marasmos), decay; ἀρχή (arkhē), the beginning: the name appears in a paper entitled 'On the classification of the Pterophoridae'. Meyrick considered the two most primitive genera to be 1487 *Agdistis* with undivided wings, and the non-British *Mimescoptilus* Wallengren with divided wings, the two being descended from a common ancestor. The next most primitive genus from the *Mimescoptilus* stock was his *Marasmarcha*, marking an early stage (ἀρχή) in the degeneration (μαρασμός) of the entire wing into plume structure. Meyrick himself gave no explanation and what I have written is inference based on his text. Macleod thought the name was a misprint.

1495 **lunaedactyla** (Haworth, 1811) – *luna*, the moon: from the crescentic pale fascia at the base of the cleft on the forewing.

Cnaemidophorus Wallengren, 1862 – κνημιδοφόρος (knēmidophoros), wearing greaves or leg-armour: from the tufts of scales adorning the legs of the adult.

1496 **rhododactyla** ([Denis & Schiffermüller], 1775) – a pun on ῥόδον (rhodon), a rose, a rose-tree, the foodplant, and the rosy-red colour of the forewing. No doubt there is also a literary allusion to the recurrent Homeric line ἦμος δ' ἠριγένεια φάνη ῥοδοδάκτυλος Ἠώς (as soon as early-rising, rosy-fingered Dawn appeared).

Amblyptilia Hübner, 1825 – ἀμβλύς (amblus), blunt; πτίλον (ptilon), a downy feather, an insect's wing: from the blunt tips to the lobes of the forewing; cf. 1489 *Oxyptilus*.

1497 **acanthadactyla** (Hübner, 1813) – ἄκανθα (akantha), a thorn: from the scale-teeth on the dorsal lobe of the hindwing.

1498 **punctidactyla** (Haworth, 1811) – *punctum*, a dot: the forewing is striated with black and has small white costal spots.

Platyptilia Hübner, 1825 – πλατύς (platus), wide; πτίλον (ptilon), an insect's wing: from the distinct breadth of the lobes, especially those of the forewing.

1499 **tesseradactyla** (Linnaeus, 1761) – τέσσαρες (tessares), four: Linnaeus apparently miscounted, there being five lobes (see Family introduction), but he may have intended this species to take the place of *tetradactyla* Linnaeus, 1758, which is a synonym of 1510 *tridactyla*.

1500 **calodactyla** ([Denis & Schiffermüller], 1775) – καλός (kalos), beautiful: in appreciation.

1501 **gonodactyla** ([Denis & Schiffermüller], 1775) – γωνία (gōnia), an angle: from the weakly falcate apex of the forewing.

1502 **isodactylus** (Zeller, 1852) – ἴσος (isos), equal: the wing-lobes are all of relatively equal width.

1503 **ochrodactyla** ([Denis & Schiffermüller], 1775) – ὤχρα (ōkhra), ochreous yellow: from the ground colour.

1504 **pallidactyla** (Haworth, 1811) – *pallidus*, pale: from the ground colour.

Stenoptilia Hübner, 1825 – στένος (stenos), narrow; πτίλον (ptilon), an insect's wing: because the wing-lobes, and especially the dorsal lobe of the hindwing, are narrower than they are in related genera; cf. 1499 *Platyptilia*.

1505 **graphodactyla** (Treitschke, 1833) – γραφή (graphē), something drawn, a character: from the black dash on the first lobe of the forewing.

= **pneumonanthes** sensu auctt. – *Gentiana pneumonanthe*, the marsh gentian: the larval foodplant.

1506 **saxifragae** Fletcher, 1940 – *Saxifraga*, the saxifrage genus which contains the larval foodplants.

1507 **zophodactylus** (Duponchel, 1840) – ζόφος (zophos), the gloom of the Underworld: the wings are fuscous-grey with darker irroration, coloured more soberly than those of most pterophorids.

1508 **bipunctidactyla** (Scopoli, 1763) – *bis-*, *bi-*, two; *punctus*, spotted: from the two sometimes confluent dots at the base of the forewing fissure; each lobe also has two dots in the terminal cilia, but at the time when Scopoli was writing their diagnostic value had not been recognized. Recent research has shown that a species complex has been embraced by this name and that the following species should be added to the British list:

1508a **aridus** (Zeller, 1847) – *aridus*, dry: from the habitat.

= **gallobritannidactyla** Gibeau, 1985 – *Gallia*, Gaul, France; *Britannia*, Britain: from the distribution, the name being formed on the analogy of 1494 *britanniodactyla*, *q.v.*

1508b **picardi** Gibeau, 1987 – i.h.o. J. Picard, contemporary French entomologist.

1508c **scabiodactylus** (Gregson, 1869) – *Scabiosa columbaria*, small scabious: the larval foodplant.

1508d **islandicus** (Staudinger, 1857) – *Islandicus*, Icelandic: from the situation of the type locality.

= **pelidnodactyla** sensu Bradley & Fletcher, 1986 – πελιδνός (pelidnos), livid, pale, ashen: from the coloration of the forewing.

1509 **pterodactyla** (Linnaeus, 1761) – πτερόν (pteron), a feather, a wing: Linnaeus has now ceased to number the plumes consecutively; this name would be suitable for any member of the family and may have prompted the name Pterophorus, Geoffroy (see family introduction).

PTEROPHORINAE (1510)

Pterophorus Schäffer, 1766 – πτερόν (pteron), a feather, a wing; φορέω (phoreō), to carry: originally Geoffroy's name for the family in which the wings are divided into feathery lobes; see p. 25.

1510 **tridactyla** (Linnaeus, 1758) – τρι-, three: the third 'plume' described by Linnaeus (see family introduction); his *tetradactyla* is now regarded as a synonym of this species and his *tesseradactyla* (1499) may have been intended to take its place in the sequential numbering.

1511 **fuscolimbatus** (Duponchel, 1844) – *fuscus*, dusky; *limbatus*, bordered: from the contrast between the dark cilia and the yellowish ground colour of the forewing.

subsp. **phillipsi** (Huggins, 1955) – i.h.o. R. A. Phillips (1866–1945), an Irish entomologist who lived in Cork.

= **icterodactylus** Mann, 1855 – ἴκτερος (ikteros), *icterus*, a yellow bird (see 1838 *icterata*): from the yellowish ground colour.

1512 **baliodactylus** Zeller, 1841 – βαλιός (balios), spotted: from the dark costal markings.

1513 **pentadactyla** (Linnaeus, 1758) – πέντε (pente), five: this is the only pterophorid named by Linnaeus in 1758 to be assigned the correct number of lobes.

1514 **galactodactyla** ([Denis & Schiffermüller], 1775) – γάλα, γάλακτος (gala, galaktos), milk: from the white ground colour.

1515 **spilodactylus** Curtis, 1827 – σπῖλος (spīlos), a spot: from the fuscous blotches on the yellowish white ground colour.

Pselnophorus Wallengren, 1881 – ψέλιον (pselion), a bracelet; φορέω (phoreō), to carry: from the tufts of scales on the legs; cf. 1496 *Cnaemidophorus* of the same author.

1516 **heterodactyla** (Müller, 1764) – ἕτερος (heteros), other, different: signifying 'another' pterophorid at a time when only few had been named.

Adaina Tutt, 1905 – the new pterophorid genera Tutt erected in 1905 were named in honour of entomologists and this one is unlikely to be an exception. It was not R. Adkin (*Adkinia* Tutt, junior synonym of 1505 *Stenoptilia*), but perhaps a lady entomologist or friend named Ada.

1517 **microdactyla** (Hübner, 1813) – μικρός (mikros), small: one of the smallest pterophorids.

Leioptilus Wallengren, 1862 – λεῖος (leios), smooth, bald; πτιλόν (ptilon), a feather, not necessarily a wing: from the lack of scale-tufts on the legs; cf. 1496 *Cnaemidophorus* and 1516 *Pselnophorus* of the same author.

1518 **lienigianus** (Zeller, 1852) – i.h.o. Madam Lienig (d. 1855), the Latvian entomologist with whom Zeller collaborated.

1519 **carphodactyla** (Hübner, 1813) – κάρφος (karphos), straw: from the ground colour of the forewing.

1520 **osteodactylus** (Zeller, 1841) – ὀστέον (osteon), a bone: from the ground colour of the forewing.

1521 **chrysocomae** Ragonot, 1875 – χρύσος (khrusos), gold; κόμη (komē), hair of the head: from the yellowish ground colour.

= **bowesi** (Whalley, 1960) – i.h.o. A. J. L. Bowes (1913–42), a British collector killed in action in the Second World War.

1522 **tephradactyla** (Hübner, 1813) – τέφρα (tephra), ashes: from the greyish fuscous forewing irrorate with blackish scales.

Oidaematophorus Wallengren, 1862 – οἴδημα (oidēma), a swelling; φορέω (phoreō), to carry: from the scale-tufts on the legs; cf. 1496 *Cnaemidophorus*, 1516 *Pselnophorus* and 1518 *Leioptilus* of the same author.

1523 **lithodactyla** (Treitschke, 1833) – λίθος (lithos), a stone: from the ground colour of the forewing (though stones are of every colour!).

Emmelina Tutt, 1905 – on the analogy of his other names in the same paper, Tutt is likely to be honouring an entomologist named Emmel; an alternative derivation, given by Macleod, is from ἐμμελής (emmelēs), harmonious, agreeable.

1524 **monodactyla** (Linnaeus, 1758) – μονός (monos), one, single: although he defines the Alucitae as having *alis digitatis fissis ad basin* (wings cleft into fingers in a basal direction), Linnaeus says of this species *alis . . . indivisis* (wings undivided). The apparent discrepancy may be explained by the moth's attitude when at rest, the wings then being tightly rolled into a single tubular unit; see MBGBI 7(2), Pl. B, fig. 7. It is, however, best to regard the name as 'number one' of the series of six described in Alucita (see Family introduction).

HESPERIOIDEA (1529)

HESPERIIDAE (1529)

HESPERIINAE (1529)

Carterocephalus Lederer, 1852 – καρτερός (karteros), strong; κεφαλή (kephalē), the head: from the broad head and widely separated antennae.

1525 **palaemon** (Pallas, 1771) – the future name of Melicerta, son of Athamas and Ino. When Athamas went mad and sought to destroy his family, Ino, carrying the babe Melicerta in her arms, leapt from a cliff into the sea. There Poseidon (Neptune) took pity on them and metamorphozed them both into sea-deities, Melicerta under the new name Palaemon. The Romans identified him with Portumnus, the guardian of harbours. The story is told in Telemann's dramatic cantata *Ino*.

Heteropterus Dumeril, 1806 – ἕτερος (heteros), other, different; πτερόν (pteron), a wing: from the exceptionally broad wings, shaped differently from those of most skipper butterflies.

1525a **morpheus** (Pallas, 1771) – the god of dreams, who was adept at mimicking the appearance and mannerisms of human beings.

Thymelicus Hübner, 1819 – θυμελικός (thumelikos), a member of the chorus in Greek drama: the chorus were the dancers and the name reflects the lively movements of the butterflies. A variant of *Thymela*, the name Fabricius (1807) bestowed on the black skippers (Pyrginae), but now regarded as a junior synonym of *Erynnis* Schrank, 1801.

1526 **sylvestris** (Poda, 1761) – *silvestris, sylvestris*, pertaining to a wood: from the habitat, but rather inapt since the species occurs mainly on rough grassland. However, this name set a fashion and several of the others given to skippers refer to woods, hunting or woodland deities.

= **linea** (Müller, O. F., 1766) – *linea*, a line: from the male sex-brand.

1527 **lineola** (Ochsenheimer, 1808) – *lineola*, dim. of *linea*, a small line: from the male sex-brand which is shorter than that of the last species, then known as *linea*.

1528 **acteon** (Rottemburg, 1775) – Actaeon, a Greek hunter who surprised Artemis whilst she was bathing and was turned by her into a stag; thereupon he was set on and devoured by his own hounds.

Hesperia Fabricius, 1793 – one of the Hesperides, the nymphs who guarded the golden apples of Hera. Linnaeus (1758) had divided the butterflies into six 'phalanges' (Pl. II), the first four containing the larger or 'noble' species and the sixth the 'Barbari', the barbarians or foreigners which did not fit into the first five categories. Fabricius took the fifth group, the 'Plebeji parvi', the little commoners, out of Papilio and placed them in a new family, Hesperia. Linnaeus had subdivided the *Plebeji* into two groups, the Plebeji urbicolae, 'city commoners', for the skippers, and Plebeji rurales, 'country commoners', for the blues. Fabricius placed both groups in Hesperia. Schrank (1798–1803) separated them under the names Erynnis and Cupido; Latreille (1804) accepted this classification but used the Fabrician name Hesperia for the skippers and introduced a name of his own, Polyommatus, for the blues. This may not have pleased Fabricius, as later (1807) he used Hesperia and Lycaena for his two blue families. Latreille's usage of Hesperia for the skippers had priority and the name stayed with that family against the wishes of Fabricius. The application of the name has been steadily whittled down and now it is applied to a small genus with a single British representative; it is, however, *usu* Latreille, the source of the family name. Fabricius was fond of devising names that were puns or had double meanings and a link with ἕσπερα (hespera), evening, is likely, the noble butterflies constituting the 'Diurni', the species of broad daylight, and the humble smaller species those of the small light or twilight; there is, of course, no implication of crepuscular flight. The original inclusion of the blues precludes the derivation on taxonomic grounds suggested by Pickard *et al.*, 'the Hesperidae forming the connecting link between the Diurni and Nocturni'.

1529 **comma** (Linnaeus, 1758) – κόμμα (komma), the comma, from the sex-brand on the male forewing. In classical Greek κόμμα signified a 'period' or sentence in a speech, or the caesura or pause in a line of verse; later it became the name of a mark of punctuation, cf. 1526 *linea* and 1527 *lineola*. Linnaeus' description is brief and ambiguously worded. He speaks of white spots without stating that they are on the hindwing underside. In consequence early British authors like Harris (1775a; 1775b) concluded that *comma* was the name of the large skipper which (unless they were right) is not described in *Systema Naturae*. It is significant that Linnaeus makes no reference to the descriptions of the large skipper by Petiver and Ray.

Hylephila Billberg, 1820 – ὕλη (hulē), a wood; φιλέω (phileō), to love: from the habitat; cf. 1526 *sylvestris*.

1530 **phyleus** (Drury, 1773) – from or belonging to Phyle, a town in Attica. If this derivation is correct, it is a very early example of a geographical name without entomological significance, antedating those of Ochsenheimer by many years. The butterfly is a native of North America.

Ochlodes Scudder, 1872 – ὀχλώδης (okhlōdēs), turbulent, unruly; from the swift, erratic flight of the butterflies; cf. 1526 *Thymelicus* and 1532 *Erynnis*.

1531 **venata** (Bremer & Grey, 1852) – *venatus* (n.), the art of hunting, the chase: a replacement name for the preoccupied *sylvanus* Esper, 1779, chosen to preserve the imagery of woodland and venery.

subsp. **faunus** (Turati, 1905) – *faunus*, a faun, a woodland deity, either Pan himself or, more often, one of his attendants; in later literature they were often confused with the satyrs. The woodland imagery is sustained.

PYRGINAE (1534)

Erynnis Schrank, 1801 – from the Erynyes (Erynnes) or Furies who harried wrongdoers, hounding them from place to place. The name, therefore, refers to the restless movement of the butterflies, as if perpetually chased by the avenging goddesses; cf. 1526 *Thymelicus* and 1531 *Ochlodes*. Fabricius (1793) had placed the blues and skippers together in Hesperia; Schrank was the first to give them separate families. Since Erynnis included the orange skippers, Macleod's suggestion 'perhaps from dark colour' cannot be correct.

1532 **tages** (Linnaeus, 1758) – the name of a boy with the wisdom of an old man who rose suddenly from the ground and instructed the Etruscans in the art of divination. His words were preserved in the Book of Tages.

subsp. **baynesi** Huggins, 1956 – i.h.o. E. S. A. Baynes (1890–1972), the Irish lepidopterist and author of *A revised catalogue of Irish Lepidoptera*.

Carcharodus Hübner, 1819 – καρχαρόδους (karkharodous), with sharp, jagged teeth: the teeth are, presumably, the conspicuous bars in the terminal cilia.

1533 **alceae** (Esper, 1780) – *Alcea*, a genus of Malvaceae, or *Malva alcea*, a species of mallow not found in Britain: from the larval foodplants.

Pyrgus Hübner, 1819 – πύργος (purgos), a tower on a wall, a battlement: presumably from the chequered terminal cilia; cf. 1533 *Carcharodus*, named by Hübner in the same work. Macleod's derivation from Pyrgi, a town in the Peloponnesus, is unlikely; none of Hübner's generic names is certainly geographical (see Appendix 3).

1534 **malvae** (Linnaeus, 1758) – *Malva*, the mallow genus: incorrectly stated by Linnaeus to be that of the foodplants, probably through confusion with several related species, all much alike, which do feed on Malvaceae.

1535 **armoricanus** (Oberthür, 1910) – Armorica, the Latin name for Brittany where Rennes, the type locality, is situated.

PAPILIONOIDEA (1539)

PAPILIONIDAE (1539)

PARNASSIINAE (1536)

Parnassius Latreille, 1804 – Parnassus, the mountain range near Delphi at the top of which the Muses lived. The butterflies occur high up on mountains.

1536 **apollo** (Linnaeus, 1758) – Phoebus Apollo, the god of the sun and patron of the arts; Parnassus was one of his abodes. Placed by Linnaeus in the Heliconii (see 1583, subfamily and p. 16) and by Schrank in Pieris (1549, q.v.).

1537 **phoebus** (Fabricius, 1793) – φοῖβος (phoibos), radiant, an epithet of Apollo in his capacity as sun god. The name indicates affinity with the last species.

ZERINTHIINAE (1538)

Zerynthia Ochsenheimer, 1816 – Zerynthius, a title of Apollo who had a temple at the town of Zerynthus in Thrace.

= **Parnalius** Rafinesque, 1815 – probably a portmanteau word formed from the genus 1536 *Parnassius* and *alius*, other, different, and indicating affinity. The name was rejected by I.C.Z.N. Opinion No. 1134 (1979).

1538 **rumina** (Linnaeus, 1758) – Rumina, the Roman goddess of suckling, from the root of *ruma* or *rumis*, a teat; also a title of the fig-tree under which the she-wolf suckled Romulus and Remus.

PAPILIONINAE (1539)

Papilio Linnaeus, 1758 – *papilio*, a butterfly: the name was applied by Linnaeus to all the butterflies (Pl. II); Schrank (1801) restricted it to the Nymphalidae (p. 29) but Latreille (1804) reassigned it to the phalanx Equites (p. 31), and later it was still further restricted to this genus.

1539 **machaon** Linnaeus, 1758 – a doctor who served on the Greek side in the Trojan War; Aesculapius, often regarded as the god of healing, was his father and Epione (1907, genus) his mother.

subsp. **britannicus** Seitz, 1907 – the subspecies resident in Britain.

subsp. **gorganus** Fruhstorfer, 1922 – possibly from Gorgan, an Iranian town situated near the south-eastern shore of the Caspian Sea.

= **bigeneratus** Verity, 1947 – *bi-*, twice; *generatus*, begotten: from the two generations of the Continental subspecies, the British subspecies being predominantly univoltine. Verity first published the name in 1919 but without description, the name therefore ranking as *nomen nudum*.

Iphiclides Hübner, 1819 – one related to Iphicles, a half-brother of Hercules.

1540 **podalirius** (Scopoli, 1763) – the brother of Machaon (1539) and also a doctor who participated in the Trojan War.

PIERIDAE (1549)

DISMORPHIINAE

From the genus *Dismorphia* Hübner, 1816 – δις (dis), twice, doubly; μορφή (morphē), shape: from the similarity of shape between the forewing and hindwing which is closer than in most butterflies, the same shape appearing twice.

Leptidea Billberg, 1820 – λεπτός (leptos), thin, delicate; εἶδος (eidos), form, appearance: from the slender abdomen and delicate structure in general of the members of the genus.

1541 **sinapis** (Linnaeus, 1758) – *Sinapis*, a synonym of *Brassica*, a genus of the Cruciferae: *'habitat in* Brassica *& affinibus'*, it lives on *Brassica* and its relatives. The larva really feeds on Papilionaceae.

subsp. **juvernica** Williams, 1946 – Juverna, a variant of Hibernia, Ireland: the subspecies occurring in Ireland.

COLIADINAE (1542)

Colias Fabricius, 1807 – the name of promontory on the east coast of Attica where there was a temple of Aphrodite (Venus); the Battle of Salamis was fought nearby. Fabricius set up two families to accommodate the present Pieridae, Colias for those that were yellow (clouded yellows and brimstones), and 1552 Pontia for those that were white (including orange-tips), giving both of them names associated with the goddess of beauty, perhaps simply because the butterflies themselves were beautiful. However, in view of his fondness for punning names and word-play, the following explanation is also possible: κολίας (kolias), a kind of tunny-fish (Westwood, 1855), with a pun on χολή, χόλος (kholē, kholos), bile, because of its yellow colour; cf. 166 *Zygaena*, named from the hammer-headed shark, with a pun on ζυγόν (zugon), a yoke.

1542 **palaeno** (Linnaeus, 1761) – the name of one of the 50 daughters of Danaus; see 1552 *daplidice* and 1630, genus. The name is in accordance with the policy Linnaeus adopted for the Danai Candidi (see p. 16).

1543 **hyale** (Linnaeus, 1758) – the name of another of the daughters of Danaus; see the last species and the cross-references there cited.

1544 **alfacariensis** Berger, 1948 – belonging to Alfacar, a town in southern Spain which was the type locality of the original specimens described by Ribbe in 1905 as belonging to a form of the last species.

= **australis** Verity, 1911 – *australis*, southern: from occurrence mainly in southern Europe. The name has been reduced to synonymy because Verity bestowed it on a 'race' which by his definition was a category ranking below the subspecies.

1545 **croceus** (Geoffroy *in* Fourcroy, 1785) – *croceus*, saffron yellow: from the ground colour. Petiver had called this species *Papilio croceus* in 1703, more than 50 years before the accepted beginning of scientific nomenclature; this may have influenced Geoffroy's choice of name.

Gonepteryx Leach, 1815 – γωνία (gōnia), an angle; πτέρυξ (pterux), a wing: from the subfalcate forewing and subcaudate hindwing.

1546 **rhamni** (Linnaeus, 1758) – *Rhamnus catharticus*, buckthorn: the foodplant.

subsp. **gravesi** Huggins, 1956 – i.h.o. P. P. Graves (1876–1953), British and Irish entomologist; the subspecies occurring in Ireland.

1547 **cleopatra** (Linnaeus, 1767) – the name of the wife of Meleager who presided over the wild boar hunt in Calydon; see 1590 *atalanta* and 1630 *plexippus*. Derivation from the more famous beautiful Egyptian queen is less likely, since Linnaeus preferred mythical to historical sources for his names.

PIERINAE (1549)

Aporia Hübner, 1819 – ἀ, alpha privative; πόρος (poros), a narrow passage, a strait, whence ἀπορία (aporia), difficulty in passing on a narrow track, or awkwardness, perplexity in general: possibly from the unexplained problem of its fluctuation in numbers (Macleod), 1548 *A. crataegi* being the only species in the genus. Through the sense of 'difficulty', ἀπορία came also the mean 'need' (cf. our phrase 'to be in difficulties'), and Westwood (1855) and Spuler with equal probability explain as 'shortage', with reference to the thin scaling of the wings. Another possibility is shortage in numbers, rarity. Knowledge of the status of *A. crataegi* in Germany in the second decade of the 19th century might point definitely to one of these explanations.

1548 **crataegi** (Linnaeus, 1758) – *Crataegus*, hawthorn: a larval foodplant.

Pieris Schrank, 1801 – one of the Muses (Pierides), who were supposed to live on Mt. Pierus, close to Mt. Olympus. This was one of the five families into which Schrank divided the butterflies. In its first conception it embraced all the swallowtail and white butterflies, the first name being 1536 *apollo*; Apollo was the patron of the Muses. Latreille (1804) separated the swallowtails and whites, applying this name to the latter with the same coverage as the present family name Pieridae.

1549 **brassicae** (Linnaeus, 1758) – *Brassica*, the cabbage genus: from the foodplants.

1550 **rapae** (Linnaeus, 1758) – *Brassica rapa*, wild turnip: a foodplant.

1551 **napi** (Linnaeus, 1758) – *Brassica napus*, rape: a foodplant.

subsp. **sabellicae** (Stephens, 1827) – Stephens stated that he had adopted the name from Petiver but had forgotten the reference. I, too, have been unable to trace it; it would appear to have been taken from a foodplant and I have also searched Ray's *Synopsis* (1724) without success.

subsp. **britannica** Verity, 1911 – the British, but more strictly the Irish, subspecies.

subsp. **thomsoni** Warren, 1968 – a northern subspecies named i.h.o. G. Thomson, a contemporary Scottish entomologist.

Pontia Fabricius, 1807 – πόντιος (pontios), of or from the sea, an epithet of Thetis (see 1668, genus), the nereids (sea-nymphs) and especially Aphrodite (Venus), who was born from the sea, as depicted in Botticelli's painting, *The birth of Venus*; it was also the name of a rocky islet off the coast of Latium. Fabricius divided the present Pieridae into two families, Colias (1542, q.v.), comprising those that were yellow, and Pontia, embracing all the whites and orange-tips; the latter name is now restricted to this small genus. There is no obvious reason for the choice of name.

1552 **daplidice** (Linnaeus, 1758) – the name of one of the daughters of Danaus (1630), King of Argos. Linnaeus placed the present Pieridae in his phalanx Danaus, in the section entitled 'Danai Candidi', white Danai; the other section, 'Danai Festivi', gaily-coloured Danai, embraced mainly the Satyrinae (Pl. II). Of the former he wrote in a footnote 'Danaorum Candidorum *nomina a filiabus Danai . . . mutuatus sum*', I have derived the names of the Danai Candidi from the daughters of Danaus. For some sisters of Daplidice, see 1542 and 1543; for the kind of girls they were, see 1629 *hyperantus*.

Anthocharis Boisduval, 1833 – ἄνθος (anthos), a flower; χάρις (kharis), grace: either because the butterflies have the grace of a flower, or because they lend grace to the flowers they frequent. Pickard *et al.*, followed by Macleod, derive the name from χαίρω (khairō), to rejoice, but then its natural formation would have been '*Anthochaera*'.

1553 **cardamines** (Linnaeus, 1758) – *Cardamine*, the bitter-cress genus, stated by Linnaeus to include the foodplants of this species.

subsp. **britannica** (Verity, 1908) – the subspecies occurring in Britain.

subsp. **hibernica** (Williams, 1916) – the subspecies occurring in Ireland.

Euchloe Hübner, 1819 – εὖ (eu), well, good; χλόη (khloē), the light green colour of spring vegetation; εὔχλοος (eukhloos) blooming: from the coloration of the hindwing underside.

1554 **simplonia** (Freyer, 1829) – there are two European species, the dappled white and the mountain dappled white. Boisduval named the latter *simplonia* in 1828, but without description, so the name ranked as *nomen nudum* and was invalid. The next year Freyer gave a description, using Boisduval's name, but of the other species; so it comes about that the dappled white, the lowland species that has reached our shores, bears the name originally intended for its relative, the mountain dappled white, and derived from what should have been the type locality, the Simplon Pass.

= **ausonia** sensu auctt. Ausonius, a Roman poet and scholar of the 3rd century A.D. *E. ausonia* (Hübner, 1804) is the name of the mountain dappled white.

LYCAENIDAE (1561)

THECLINAE (1556)

Callophrys Billberg, 1820 – κάλλος (kallos), beauty; ὀφρύς (ophrus), the eyebrow: probably from the metallic green scales on the frons between the eyes, more or less concealed by brown hair-scales in fresh specimens; the white ring round the eye is a family character and unlikely to have been chosen for the source of a generic name. 'Perhaps from "hairstreak" on underside of wings' (Macleod): this hardly resembles an eyebrow and in any case is less strongly expressed in the next species than in the other hairstreaks.

1555 **rubi** (Linnaeus, 1758) – *Rubus*, the bramble genus: Linnaeus gives *Rubus aculeatus* as the foodplant. The larva certainly does feed on *Rubus* which was the only foodplant known to the early collectors, e.g. Harris (1765a); others are now regarded as more usual and *Rubus* is omitted.

Thecla Fabricius, 1807 – the name of a virgin and martyr commemorated by the Greek Orthodox Church. Fabricius was more ready than most other early entomologists to use as name sources characters in history or literature other than the classical, e.g. 1006 *grotiana* and 1590 *Vanessa*. He was the first person to give the hairstreaks their own family under this name.

1556 **betulae** (Linnaeus, 1758) – *Betula*, birch: not the foodplant, but Linnaeus also gives blackthorn (*Prunus spinosa*), which is.

Quercusia Verity, 1943 – formed from the specific name of the next species.

1557 **quercus** (Linnaeus, 1758) – *Quercūs* (gen. sing.), of the oak genus: the correct foodplant.

Satyrium Scudder, 1876 – Σάτυρος (Saturos), a satyr, a mythical being associated with the worship of Bacchus, in art often depicted with the horns and tail of a goat. The satyrs engaged in voluptuous dances with the nymphs and this name, like 1531 *Ochlodes* Scudder, draws attention to the spritely flight of the butterflies. Another possible source is a plant called σατύριον (saturion), which was used as an aphrodisiac. Derivation from Saturium, a town in southern Italy, is unlikely, since the Latin 'u' should not be changed to a 'y'.

= **Strymonidia** Tutt, 1908 – the genus *Strymon* Hübner, in which the next two species had formerly been placed; εἶδος (eidos), form: a genus resembling but distinct from *Strymon*, itself named from the R. Strymon, now the R. Struma, which forms the boundary between Macedonia and Thrace.

1558 **w-album** (Knoch, 1782) – from the white (*albus*) 'hairstreak', shaped like the letter 'W', on the underside of the hindwing.

1559 **pruni** (Linnaeus, 1758) – *Prunus*, the blackthorn genus: correctly given by Linnaeus, although his references to earlier British authors show that at least in part he was confusing this species with the previous one.

Rapala Moore, 1881 – apparently a meaningless neologism.

1560 **schistacea** Moore, 1881 – *schistaceus*, slaty, late Latin from another adjective, *schistos* (Gr. σχιστός), split, often applied to stone like slate that can readily be cleft: from the slaty gloss on the wings.

LYCAENINAE (1561)

Lycaena Fabricius, 1807 – one of the three families (Thecla, Lycaena and Hesperia) into which Fabricius divided the hairstreaks, coppers and blues after he had separated them from the skippers (see p. 31). It has given its name to the family although it is junior to 1569 *Cupido* Schrank, 1801 and 1574 *Polyommatus* Latreille. 1804, both erected for the same purpose. Possibly from λύκαινα (lukaina), a she-wolf, but authors have jibbed at this derivation. Sodoffsky (1837) suggested Lycia, a title of Diana; Pickard *et al.* quote Sodoffsky without suggesting any alternative; Macleod, with greater probability, proposed Λυκαῖος (Lukaios), Arcadian, pointing out that several of the species bore the names of Arcadian shepherds. Yet another possibility is Λύκειον (Lukeion), the Lyceum, an Athenian gymnasium, with reference to the lively antics of the butterflies. However, scientific names need not have any entomological application and the first explanation is most probably correct, though with echoes of one or more of the others. The problem is akin to that posed by other names given by Fabricius, e.g. 166 *Zygaena*, in which he seems to be indulging wittily in a kind of rhyming slang, following a train of thought and ending up with a name related in sound but of a different meaning.

1561 **phlaeas** (Linnaeus, 1761) – spelt by Linnaeus with a diphthong which can be read either as -ae or -oe; φλέω, φλοίω (phleō, phloiō), to team, overflow, bloom, various adjectives having been formed from this root as epithets of deities such as Proserpina, Aphrodite and Bacchus; the Latin verb *floreo*, to flourish, is derived from the same root: from the bright copper colour, suggesting floral splendour. This is rather involved and Spuler preferred to derive the name from φλέγω (phlegō), to blaze up, also referring to the ground colour.

subsp. **eleus** (Fabricius, 1798) – *Eleus*, of or from Elis, a district of the Peloponnesus, but not the type locality.

subsp. **hibernica** Goodson, 1948 – the subspecies occurring in Ireland.

1562 **dispar** (Haworth, 1803) – *dispar*, unalike: from the disparity of the sexes, in their wing pattern rather than their size, as supposed by Macleod.

subsp. **batavus** (Oberthür, 1923) – *Batavus*, Dutch: the subspecies occurring in Holland.

subsp. **rutilus** Werneburg, 1864 – *rutilus*, yellowish red: from the distinctive coloration of the orange areas of the underside.

1563 **virgaureae** (Linnaeus, 1758) – *Solidago virgaurea*, goldenrod, which Linnaeus incorrectly gave as the foodplant; equally wrongly he gave a reference to Ray's description of 1561 *phlaeas*.

1564 **tityrus** (Poda, 1761) – the name of a shepherd mentioned by Virgil in the *Eclogues*; cf. 77 *tityrella*. See also the reference to Arcadia under 1561 *Lycaena*.

1565 **alciphron** (Rottemburg, 1775) – the name of a Greek author of the 2nd century A.D.

1566 **hippothoe** (Linnaeus, 1761) – ἱππόθοος (hippothoös), swift-riding, also used as a proper name; Hippothoüs was a hero who fought on the Trojan side and was slain by Ajax, Hippothoë a beautiful maiden abducted by Poseidon. Rennie (1832) used the English name 'swift copper' for this species.

POLYOMMATINAE (1574)

Lampides Hübner, 1819 – λαμπάς (lampas), a torch; εἶδος (eidos), form, appearance: from the lustre of the long, plumose androconial scales on the upperside of the male of the next species.

1567 **boeticus** (Linnaeus, 1767) – Boetica, the Roman name for a province in southern Spain: the butterfly occurs in Boetica but the type locality is Algeria.

Syntarucus Butler, 1901 – σύν (sun), with, allied to; the genus *Tarucus* Moore, the 'blue pierrots', a mainly Indo-Malaysian genus of the Lycaenidae: signifying affinity.

1568 **pirithous** (Linnaeus, 1767) – Pirithoüs, son of Ixion, King of the Lapathae in Thessaly and bosom friend of Theseus.

Cupido Schrank, 1801 – Cupid or Eros, the god of love. After Fabricius (1793) had placed the skippers and blues together in Hesperia (1529), Schrank was the first to separate them, this, therefore, being the oldest name for the present Lycaenidae (Theclinae + Lycaeninae + Polyommatinae). Schrank took as his family name the first specific name of the Plebeji of Linnaeus (1758) and it is possible that this aroused the disapproval of Latreille (1804) who rejected Cupido in favour of his own name, Polyommatus, although he was happy to accept Pieris, named by Schrank in the same volume. The rump that remains of Cupido today does scant justice to a name of historic importance.

1569 **minimus** (Fuessly, 1775) – *minimus*, smallest: the smallest of the Plebeji parvi to be named up to 1775.

Everes Hübner, 1819 – not traced. 'Greek hero' (Macleod); probably correct, but he was apt to jump to conclusions. There is a word εὐήρης (euērēs), meaning well-poised, of oars, *i.e.* with the weight inboard and outboard correctly adjusted; by metonymy this could become the name of a skilful oarsman.

1570 **argiades** (Pallas, 1771) – the species 1571 *argus, q.v.*; εἶδος (eidos), form, appearance: from supposed resemblance.

Plebejus Kluk, 1802 – *plebeius*, plebeian, belonging to the *plebs*, the Roman common people. The Plebeji were the fifth of the six phalanges into which Linnaeus divided the butterflies, a group including all the smaller species (blues and skippers); see p. 25. As with 1594 *Nymphalis* and 1630 *Danaus*, Kluk was the first to use the Linnaean name in a way that complied with future I.C.Z.N. rules for the establishment of generic names and is therefore deemed the author.

1571 **argus** (Linnaeus, 1758) – Zeus was in love with Io (1597), and to indulge himself without arousing the jealousy of his wife Hera, he turned Io into a heifer. Hera, however, learnt of this and placed Io in the care of Argus who had a hundred eyes. Not to be outdone, Zeus enlisted the aid of Hermes who lulled Argus to sleep with the sound of his flute and then cut off his head; the lascivious Zeus then had the opportunity he had sought. Hera consoled herself by transplanting the eyes of Argus into the tail of the peacock. The name, therefore, is apt and refers to the eye-spots on the underside of the butterfly's wings. Linnaeus here gives references to the descriptions and figures of 1574 *Polyommatus icarus* by Petiver, Ray and Wilkes, so it is not altogether certain to which species the name properly belongs; see also 1580 *Celastrina argiolus*.

subsp. **cretaceus** Tutt, 1909 – *cretaceus*, pertaining to chalk: the subspecies found on chalk downland.

subsp. **masseyi** Tutt, 1909 – the extinct subspecies which occurred at Witherslack: 'we are indebted entirely to Mr Massey for our examples of the species' (Tutt).

subsp. **caernensis** Thompson, 1937 – the subspecies found in Caernarvonshire.

Aricia R. L., 1817 – an ancient town of Latium where there was a sacred grove and temple of Diana. The identity of 'R. L.', the nomenclator, is believed to be Reichenbach of Leipzig.

1572 **agestis** ([Denis & Schiffermüller], 1775) – derivation obscure; Pickard *et al.* suggest a typographical error for *agrestis*, rustic, an adjective formed from *ager*, a field, and alluding to the grassy habitat. However, almost all the names bestowed on butterflies by Denis & Schiffermüller are taken from classical mythology and this one is unlikely to be an exception. It may be a corruption of Argestes (Ἀργέστης), the god of the north-west wind.

1573 **artaxerxes** (Fabricius, 1793) – the name of several Persian kings.

subsp. **salmacis** (Stephens, 1831) – Salmacis, one of the Naiads, who fell in love with a beautiful youth named Hermaphroditus. When he refused her advances, she prayed to the gods that she might be united with him for ever and her prayers were answered by the fusion of their bodies. The legend is the source of the word 'hermaphrodite'.

Polyommatus Latreille, 1804 – πολυόμματος (poluommatos), many-eyed, an epithet of Argus (1571). Originally a family name for all blue butterflies (see p. 31).

1574 **icarus** (Rottemburg, 1775) – Icarus, the son of Daedalus who provided wings for him so that he could escape from Crete. The wings were attached with wax which melted when he flew too near the sun, and he fell into the Aegean (Icarian) Sea and was drowned.

subsp. **mariscolore** (Kane, 1893) – *maris colore*, with the colour of the sea: this Irish race has the female more extensively and more brilliantly blue. Kane may have remembered that Icarus had a sea named after him.

Lysandra Hemming, 1933 – the name of an Egyptian princess, the daughter of Ptolemy I.

1575 **coridon** (Poda, 1761) – Corydon, the name of a shepherd in Virgil's *Eclogues*.

1576 **bellargus** (Rottemburg, 1775) – *bellus*, beautiful; Argus, see 1571. The English name, Adonis blue, is taken from *adonis* ([Denis & Schiffermüller], 1775), the scientific name in use until it was reduced to synonymy; Adonis was a beautiful youth beloved of Aphrodite.

Plebicula Higgins, 1969 – dim. of 1571 *Plebejus*.

1577 **dorylas** ([Denis & Schiffermüller], 1775) – the name of one of the Centaurs, who were half man and half horse.

Cyaniris Dalman, 1816 – κύανος (kuanos), dark blue; Ἶρις (Iris), the personification of the rainbow: from the iridescent violet-blue male of the next species.

1578 **semiargus** (Rottemburg, 1775) – *semi-*, half; Argus, see 1571: from the presence of fewer eye-spots on the underside than in *Plebejus argus*.

Glaucopsyche Scudder, 1872 – γλαυκός (glaukos), bluish green, grey; ψύχη (psykhē), a butterfly, since the soul in personification was represented with the wings of a butterfly: from the underside coloration which is pale grey with the basal area green.

1579 **alexis** (Poda, 1761) – the name of a shepherd in Virgil's *Eclogues*.

Celastrina Tutt, 1906 – κηλάστρα (kēlastra), a tree supposed to be the holly, the foodplant of the next species.

1580 **argiolus** (Linnaeus, 1758) – dim. of *argus* (1571), which it follows in *Systema Naturae*, *'praecedenti similis, sed minor'*, like the last species, but smaller. *C. argiolus* is smaller than 1574 *Polyommatus icarus* but larger than *Plebejus argus*, q.v.; it is also more similar to *P. icarus*.
subsp. **britanna** (Verity, 1919) – the subspecies occurring in Britain.

Maculinea Eecke, 1915 – *macula*, a spot; *linea*, a line: from the postdiscal series of somewhat elongate black spots on the forewing upperside, characteristic of the species in the genus.

1581 **arion** (Linnaeus, 1758) – a Greek poet and musician of the 7th century B.C. When he was sailing home to Corinth after winning a music competition in Sicily, the sailors decided to murder him and steal his prize money, but he obtained their permission to play for the last time. His music attracted a school of dolphins and, when he jumped overboard, one of them transported him to safety on its back.
subsp. **eutyphron** (Fruhstorfer, 1915) – derivation not traced; perhaps from a lesser-known figure in Greek mythology. Fruhstorfer gives no explanation.

RIODININAE

From the genus *Riodina* Westwood, 1851, one of a group of anagrammatic or nearly anagrammatic names formed from *Diorina* Boisduval; others are *Diorhina* Doubleday, E. and *Riodinia* Westwood. The derivation appears to be from δῖος (dios), divine, noble, remarkable; ῥίς, ῥινός (rhis, rhinos), the nose: from the long, porrect labial palpus characteristic of certain genera. The dropping of the aitch in the original name *Diorina* by Boisduval is usual with a French author.

= NEMEOBIINAE

From the genus *Nemeobius* Stephens, 1827 – νέμος (nemos), a grove; βιόω (bioō), to live: from the habitat.

Hamearis Hübner, 1819 – ἅμα (hama), at the same time as; ἔαρ (eär), the spring: from the flight season of the next species.

1582 **lucina** (Linnaeus, 1758) – Lucina, the goddess who brings light or to light, hence the goddess of childbirth. Linnaeus gives references only to Petiver and Ray and follows them in placing this species among the fritillaries; Petiver seems to have originated the vernacular name 'fritillary' and it is given as well in a footnote by Linnaeus (1758, p. 480).

NYMPHALIDAE (1594)

HELICONIINAE

The Heliconii were the second of the 'phalanges' into which Linnaeus divided the butterflies (Pl. II). They were named from Mt. Helicon, part of a range of snow-capped mountains in Boeotia, sacred to Apollo and the Muses, 1536 *Parnassius apollo* being the first species. Others bear the names of Muses and Graces.

Dryas Hübner, 1807 – Δρυάς, a Dryad or wood-nymph. This name first appears in Hübner's *Tentamen* [1806], where he proposed a series of generic names for butterflies based on the nymphs of Greek mythology. The example he gave for *Dryas* was [1608 *Argynnis*] *paphia*, a woodland butterfly.

1583 **julia** (Fabricius, 1775) – the personal name of various aristocratic Roman ladies; cf. 2057 *caja*. Both these names appear in masculine form in Caius Julius Caesar.

subsp. **delila** (Fabricius, 1775) – Delilah, Samson's lover who betrayed him to the Philistines; a glamorously attractive creature. See Judges: 16. An adventive species recorded only once in Britain.

LIMENITINAE

From the genus *Limenitis* Fabricius, 1807 – λιμενῖτις (limenītis), harbour-keeping: an epithet applied to deities who protected harbours. It was originally intended as a family name for all the white admirals (see p. 30) and is retained as the generic name for the next species by Continental systematists (Leraut, 1980; Schnack, 1985). The name need have no entomological application, but Fabricius may have remembered that the earliest specimen of the white admiral, historically speaking, was captured by Dr David Krieg at the harbour town of Leghorn; he sent it to Petiver (1703b), who accordingly called it '*Papilio Livornicus*, the Leghorn white admiral'.

Ladoga Moore, 1898 – probably a meaningless neologism, but perhaps a 'geographical' name from L. Ladoga, a large lake north-east of Leningrad.

1584 **camilla** (Linnaeus, 1764) – the name of a Volscian princess mentioned by Virgil in the *Aeneid*.

APATURINAE (1585)

Apatura Fabricius, 1807 – the name has puzzled authors and may be another of Fabricius' trick names like 166 *Zygaena*, 1561 *Lycaena*, etc. It was set up as a family name for all the emperors (see p. 30), but the next species is the one most likely to have influenced its derivation. Spuler is probably right in supposing that the main source is ἀπατάω (apataō), to deceive, from the deceptive structural colour of the male upperside; to account for the termination *-ura*, he suggests that it is compounded with οὐρά (oura), a tail, from the slightly elongate, but not fully tailed, tornus of the hindwing. Pickard *et al.*, followed by Macleod, consider that it is an adaptation of Ἀπατούρια (Apatouria), 'a surname of Venus which she obtained from a trick (ἀπάτη, apatē) that she played on some Giants'. Apaturia was also a title of Athena and a three-day festival celebrated annually in her honour at Athens was called the Apaturia. Several of these ideas may have been in Fabricius' mind, but the dominant theme seems to be that of deception.

1585 **iris** (Linnaeus, 1758) – Iris was the messenger of the gods and the personification of the rainbow, an appropriate name because of the 'iridescence' of the male upperside. Linnaeus attributes the name to J. C. Richter (1689–1751) of Leipzig, whose museum was one of the sources he studied in the preparation of *Systema Naturae*.

NYMPHALINAE (1594)

Junonia Hübner, 1819 – belonging to Juno, the wife of Jupiter.

1586 **villida** (Fabricius, 1787) – meaning not explained or traced. Fabricius bestowed the name on a specimen in the Bankes collection which had been taken on Amsterdam Is. in the Indian Ocean.

= **hampstediensis** (Jermyn, 1824) – 'Albin's Hampstead eye', the mysterious butterfly first figured by Petiver (1717). Albin appears to have believed in all good faith that he had captured it at Hampstead, but there can be no doubt that he had in fact muddled specimens of British and oriental origin.

1587 **oenone** (Linnaeus, 1758) – the wife of Paris before he carried off Helen.

Colobura Billberg, 1820 – κολοβοῦρος (kolobouros), stump-tailed: from the shape of the hindwing of the next species (see MBGBI 7(1), Pl. 23, figs 9,10).

1588 **dirce** (Linnaeus, 1758) – Dirce was taken to wife by Lycus, king of Thebes, after he had divorced Antiope (1596). Antiope's sons, Amphion and Zethus, marched against Thebes and killed Dirce by tying her to a bull to be dragged about until she perished.

Hypanartia Hübner, 1821 – perhaps ὕπο (hypo), intensifying prefix; ἀνάρσιος (anarsios), incongruous: the type species is *H. demonica* Hübner which suggests monstrosity.

1589 **lethe** (Fabricius, 1793) – Lethe, the River of Oblivion in the Underworld: there is no entomological significance.

Vanessa Fabricius, 1807 – from the Vanessa of Dean Swift's poem *Cadenus and Vanessa*, Vanessa being Swift's pet name for Esther *Van*hombrugh. Like other generic names given by Fabricius (e.g. 166 *Zygaena*, 1561 *Lycaena* and 1585 *Apatura*), it has puzzled authors and Sodoffsky (1837) emended it to *Phanessa* from the Greek words φαίνω (phaino), to shine, φανή (phanē), a torch and φανός (phanos), bright. In his quest for a suitable name for a family (p. 30) of brightly-coloured butterflies, these Greek words may well have come into Fabricius' mind and using his characteristic word-play he chose a name that echoed their sound. English literature would have been familiar to him from his frequent sojourns in this country and he may have wished to pay us a compliment by selecting a name from an English rather than a classical author.

1590 **atalanta** (Linnaeus, 1758) – the famous beauty and athlete who raced her suitors and killed them if they lost. She was eventually beaten by Milanion who threw golden apples in front of her during the race; these appealed so much to Atalanta that she had to stop to pick them up. She was the first to wound the monstrous wild boar in the hunt at Calydon (see 1630 *plexippus*).

1590a **indica** Herbst, 1794 – a species of red admiral occurring in India.

Cynthia Fabricius, 1807 – Cynthia, a mountain in the island of Delos, was the birthplace of Diana who accordingly received the title of Cynthia. Fabricius bestowed this name at the same time as 1590 *Vanessa* and since Cynthia was a popular name with 18th century English lyric poets there may in this case too be a veiled literary allusion. For its family usage see p. 30.

1591 **cardui** (Linnaeus, 1758) – *Carduus*, the thistle genus, correctly given as containing the foodplants by Linnaeus. Earlier, he had used Petiver's name *Bella Donna* (Linnaeus, 1746).

1592 **virginiensis** (Drury, 1773) – from Virginia, U.S.A.; the actual type locality is New York, but Drury gave this name because the earliest description had been made by Petiver (1704), who received a specimen from Virginia and accordingly called it *Papilio Virginiana*.

Aglais, Dalman, 1816 – ἀγλαός (aglaos), beautiful, ἀγλαία (aglaia), beauty.

1593 **urticae** (Linnaeus, 1758) – *Urtica*, the nettle genus, correctly given as containing the foodplants by Linnaeus.

Nymphalis Kluk, 1802 – νύμφη (numphē), a bride, a nymph, all too often made a bride against her will by a lascivious god. The nymphs were lesser deities associated with springs, groves, mountains, etc., good entomological habitats. The Nymphales formed the fourth of the phalanges into which Linnaeus divided the butterflies (Pl. II), Kluk being considered the first to have used the name in a generic sense.

1594 **polychloros** (Linnaeus, 1758) – πολύς (polus), many; χλωρός (khlōros), pale green or just pale: Linnaeus took the name from Aldrovandus (Ulysses Aldrovandi (1522–1605), Professor of Natural History at Bologna), who wrote (1602) of this species '*Septimus* πολύχλωρος *dici queat, propter colorum diversitatem*' (the seventh may be called *polychloros* on account of its varied colours). Pickard *et al.* suggest that Aldrovandus confused the Greek χλωρός with the Latin *color*, colour; more probably, since he was using the Greek, he confused it with χρῶμα (chrōma), colour (cf. our word 'polychromatic').

1595 **xanthomelas** ([Denis & Schiffermüller], 1775) – ξανθός (xanthos), yellow; μέλας (melas), black: from the black spots on an orange ground colour. The authors have here departed from their usual practice of naming butterflies from mythological personages and have formed a name based on colour, analogous with 1594 *polychloros* and thereby expressing affinity.

1596 **antiopa** (Linnaeus, 1758) – Antiope: after giving birth to twin sons, Amphion and Zethus, by Zeus, Antiope married Lycus, king of Thebes. He later divorced her in favour of Dirce (1588, q.v.).

Inachis Hübner, 1819 – a title of Io which she acquired from her father Inachus, the first king of Argos.

1597 **io** (Linnaeus, 1758) – Io, the daughter of Inachus, was beloved of Zeus, but because of trouble from his wife Hera, she was metamorphosed into a heifer and placed in the care of Argus (1571), who had a hundred eyes. After Zeus had engineered the death of Argus, Hera set his eyes in the tail of the peacock. In his *Fauna Suecica* (1746), Linnaeus had adopted Petiver's name *Oculus pavonis*, the peacock's eye, for this species, and the link between Io, Argus and the peacock may have influenced his choice of name.

Polygonia Hübner, 1819 – πολύς (polus), many; γωνία (gōnia), an angle: from the jagged termen of the wings.

1598 **c-album** (Linnaeus, 1758) – from the white (*albus*) c-shaped discal mark on the underside of the hindwing.

Araschnia Hübner, 1819 – ἀράχνιον (arakhnion), a spider's web: from the reticulate wing pattern of the next species.

1599 **levana** (Linnaeus, 1758) – the name of an obscure Roman goddess raised (*levo*, to lift) from the earth.

ARGYNNINAE (1603)

Boloria Moore, 1900 – βόλος (bolos), a fishing net: from the reticulate wing pattern.

1600 **selene** ([Denis & Schiffermüller], 1775) – σελήνη (selēnē), the moon; also used as a title of Artemis (Diana), the goddess of the moon.

subsp. **insularum** (Harrison, 1937) – *insularum* (gen. pl.), of the islands: the subspecies occurring in the Hebridean Islands.

1601 **euphrosyne** (Linnaeus, 1758) – the name of one of the three Graces who personified elegance and beauty; cf. 1607 *aglaja* and 1613 *athalia* (Thalia).

1602 **dia** (Linnaeus, 1767) – either Dia, an ancient name for the island of Naxos, or *dia* (*diva*), a goddess.

Argynnis Fabricius, 1807 – Argynnus, a lady beloved by Agammemnon. After her death he erected a temple in her honour where Aphrodite (Venus) was worshipped; thus Argynnis came to be used as an epithet of Aphrodite. This was Fabricius' family name for all the larger fritillaries which had been called 'Perlati' by Latreille (1804) because of the pearly markings on the underside; with his fondness for word play, Fabricius is probably punning on ἄργυρος (arguros), silver, with reference to these underside markings.

1603 **lathonia** (Linnaeus, 1758) – Λητώ or Λαθώ, Lat. Lētō or Latona, the mother of Apollo and Artemis; cf. 953 *lathoniana* and 1301 *lathoniellus*.

1604 **aphrodite** (Fabricius, 1787) – Aphrodite or Venus, the goddess of love.

1605 **niobe** (Linnaeus, 1758) – the daughter of Tantalus and wife of Amphion (see 1596). Her punishment for boasting that she had more children than Leto (1603) was that the children were all killed and Niobe herself was metamorphosed into a rock that wept unceasingly ('like Niobe, all tears', *Hamlet* I, ii, 149).

1606 **adippe** ([Denis & Schiffermüller], 1775) – in *Fauna Suecica* (Edn 2) (1761) Linnaeus called this species *cydippe* after a Nereid (a sea-nymph). This name, however, had already been allocated to another species and later (Linnaeus, 1767) he wrote '*in* Fauna Suecica Cydippe *perperam pro* Adippe *legitur*' (in *Fauna Suecica, Cydippe* is read in error for *Adippe*). Uncertainty over which name should be used was eventually ended in 1958 by a ruling of the I.C.Z.N. which suppressed the name *cydippe* and established *adippe* with Denis & Schiffermüller as the authors. *Adippe* appears to be an invented name designed to be reminiscent of *cydippe*.

subsp. **vulgoadippe** (Verity, 1929) – *vulgus*, common; the name *adippe*: the then common British subspecies.

1607 **aglaja** (Linnaeus, 1758) – Aglaia, one of the three Graces; see 1601 *euphrosyne*.

subsp. **scotica** (Watkins, 1923) – the subspecies occurring in Scotland and Ireland.

1608 **paphia** (Linnaeus, 1758) – Paphia (fem.), of Paphos, a town on the west coast of Cyprus. After her birth among the waves (see 1552 *Pontia*), Aphrodite (Venus) came ashore at Paphos which therefore became one of the principal centres for her worship and she is often referred to as Paphia, the Paphian goddess.

1609 **pandora** ([Denis & Schiffermüller], 1775) – Pandora was the first woman on earth, wrought by Hephaestus at the bidding of Zeus to bring misery upon the human race. All the gods bestowed on her two-edged gifts (πᾶν (neut.) (pan), all; δῶρον (dōron), a gift), one of them a box containing every human ill, all of which escaped when she opened it.

MELITAEINAE (1611)

Eurodryas Higgins, 1978 – Euro-, of Europe; Δρυάς (Druas), a Dryad or wood-nymph: a European genus accommodating species formerly placed in *Euphydryas* Scudder – εὐφυής (euphuēs), of goodly shape; Δρυάς.

1610 **aurinia** (Rottemburg, 1775) – the name of a prophetess revered by the ancient Germans and mentioned by Tacitus.

subsp. **hibernica** (Birchall, 1873) – the subspecies occurring in Ireland.

Melitaea Fabricius, 1807 – another of the names from Fabricius which have puzzled authors. Sodoffsky (1837) emended it to *Melinaea* which he said was a surname of Aphrodite (Venus); Pickard *et al.* derive it from Melitaea, the name of a town in Thessaly; Macleod from μελιτόεις (melitoeis), honeyed, according to him an epithet of Aphrodite; and Spuler from μελιταῖος (melitaios), of or belonging to Malta. Any one of the last three may be right. Fabricius placed the fritillaries in two families, the larger ones in Argynnis (1603, q.v.), the smaller in Melitaea. Word-play was suggested for the former name and is possible here too, an association with μέλι (meli), honey, from the butterflies' love of nectar, being intended; μελίτειον (meliteion), mead, is another possible source.

1611 **didyma** (Esper, 1779) – δίδυμος (didumos), a twin: from close resemblance to *M. phoebe* ([Denis & Schiffermüller], 1775).

1612 **cinxia** (Linnaeus, 1758) – *cinctus*, girdled: Cinxia was the title of Juno when, in her capacity as Lucina (1582), the goddess of childbirth, she unloosed the girdles of brides.

Mellicta Billberg, 1820 – μέλι (meli), honey; λίκτης (liktēs), one who licks: from the feeding habits of the butterflies, perhaps influenced by 1611 *Melitaea*.

1613 **athalia** (Rottemburg, 1775) – generally explained as from Athalia, the daughter of Omri, King of Israel, and mother of King Ahaziah. After her son had been killed, she reigned for six years until she herself suffered a similar fate. It is difficult to see why an obscure and unmeritorious queen of the Northern Kingdom (not Judah, as incorrectly stated by Macleod) should have been chosen to give her name to a butterfly; in any case, authors did not turn to the Bible for their names. Probably Rottemburg wanted to use Thalia, one of the Graces, the names of the other two having already been used for fritillaries (1601, 1607), but finding that Linnaeus had already given it to a heliconiine, he made a slight modification. No biblical connection was intended.

SATYRINAE

From the genus *Satyrus* Latreille, 1810 – Σάτυρος (Satyros), a satyr, one of the rustic deities, half man and half goat, who dwelt mainly in forests and attended on Bacchus. The genus formerly included most members of the subfamily. Latreille (1804) had already used the name in the plural (Satyri) for one of the three groups into which he had divided the Nymphales (see p. 31).

Pararge Hübner, 1819 – παρά (para), beside, close to; the genus *Arge* Schrank, from ἀργής (argēs), white: to indicate affinity with the genus in which 1620 *Melanargia galathea* was formerly placed. Macleod incorrectly derived the name from Argus, the hundred-eyed guardian of Io (see 1571 and 1597).

1614 **aegeria** (Linnaeus, 1758) – Egeria (Aegeria), one of the Camenae, prophetic nymphs resident in ancient Italy. Egeria instructed Numa Pompilius, the second king of Rome.

subsp. **tircis** (Godart, 1821) – derivation not traced.

subsp. **oblita** Harrison, 1949 – *oblitus*, smeared: probably from the purple suffusion on the hindwing underside.

subsp. **insula** Howarth, 1971 – *insula*, an island: from occurrence on the Isles of Scilly.

Lasiommata Humphreys & Westwood, 1841 – λάσιος (lasios), hairy; ὄμματα (pl.) (ommata), eyes: 1614 *Pararge* and *Lasiommata* are the only British satyrine genera to have adults with hairy eyes.

1615 **megera** (Linnaeus, 1767) – Megaera (Megēra), one of the Furies (see 1532).

1616 **maera** (Linnaeus, 1758) – Maera, the faithful dog of Icarius. Icarius was taught how to make wine by Dionysius (Bacchus), but when he gave it to some peasants, they thought they had been poisoned and murdered him. His daughter Erigone was eventually led to his grave by Maera, who was rewarded by being placed in the stars by Zeus. There may be a pun on *maero*, to mourn, since the butterfly is sombre-coloured, Linnaeus' description reading '*alis fusco-nebulosis*', with fuscous-clouded wings.

Erebia Dalman, 1816 – Ἔρεβος, Erebus, the region of darkness situated between earth and Hades (the Underworld): from the dusky ground colour.

1617 **epiphron** (Knock, 1783) – ἐπίφρων (epiphrōn), thoughtful: from the dark, staid coloration of the butterfly.

subsp. **mnemon** (Haworth, 1812) – μνήμων (mnēmōn), mindful: a name with the same meaning as *epiphron*.

subsp. **scotica** Cooke, 1943 – the subspecies occurring in Scotland.

1618 **aethiops** (Esper, 1777) – 'Αιθίοψ, Aethiops, an Ethiopian: from the dark coloration.

1619 **ligea** (Linnaeus, 1758) – the name of one of the Nereids (sea-nymphs) or Sirens.

Melanargia Meigen, 1828 – μέλαν (neut.) (melan), black; ἀργής (argēs), white: from the black and white coloration of the next species, perhaps influenced by *Arge* Schrank; cf. 1614 *Pararge*.

1620 **galathea** (Linnaeus, 1758) – Galatea, a nymph beloved by Polyphemus who was rejected by her in favour of Acis. Handel's opera based on the legend may have been known to Linnaeus.

subsp. **serena** Verity, 1913 – *serenus*, clear, unclouded: from the clear, white markings.

Hipparchia Fabricius, 1807 – Hipparchus, a celebrated Greek astronomer of the 2nd century B.C. His catalogue of the stars was preserved by Ptolemy. Perhaps the prominent ocelli on the wings of some species suggested stars to Fabricius. Originally a family name for all the satyrines (see p. 30).

1621 **semele** (Linnaeus, 1758) – Semele, a beautiful mortal who was beloved by Zeus and became the mother of Dionysius (Bacchus) by him. Hera, the wife of Zeus and always jealous (with good reason), persuaded her to ask Zeus to appear in his full splendour. Accordingly he did so as the god of thunder and she was consumed by the lightning. Zeus saved her son and later immortalized her.

subsp. **thyone** Thompson, 1944 – Thyone was Semele's name after she had been immortalized (see above).

subsp. **atlantica** Harrison, 1946 – the subspecies occurring in the Hebrides which are washed by the Atlantic Ocean.

subsp. **scota** (Verity, 1911) – the subspecies occurring in Scotland.

subsp. **hibernica** Howarth, 1971 – the subspecies occurring in Ireland.

subsp. **clarensis** de Lattin, 1952 – the subspecies occurring in the Burren district of Co. Clare, western Ireland.

1622 **fagi** (Scopoli, 1763) – *Fagus*, the beech genus: the foodplants are grasses, but the butterfly is often found resting on the trunks of trees.

Chazara Moore, 1893 – possibly from χάζω (khazō), to deprive of, to bereave, Achilles having been deprived of Briseis (see next species), but more probably, like many of Moore's names, a meaningless neologism.

1623 **briseis** (Linnaeus, 1764) – Briseis was a beautiful Greek maiden beloved by Achilles. When Agammemnon seized her, Achilles sulked in his tent and much of the future outcome of the Trojan War was the outcome of this incident.

Arethusana de Lesse, 1951 – formed from the specific name which follows.

1624 **arethusa** ([Denis & Schiffermüller], 1775) – the name of a Nereid (sea-nymph) who was changed into a fountain by Artemis so that she could escape the unwelcome attentions of the river-god Alpheus.

Pyronia Hübner, 1819 – πυρωνία (purōnia), the purchase of wheat: the next word in the lexicon is πυρωπός (purōpos), fiery-eyed, descriptive of the prominent subapical ocellus set in an orange ground colour of 1625 *tithonus*. Hübner may have inadvertently transcribed the wrong word.

1625 **tithonus** (Linnaeus, 1771) – Tithonus, a Trojan youth beloved by Eos (Aurora), the goddess of dawn. She persuaded the gods to grant him immortality, but forgot to ask for eternal youth as well, so that he became a kind of Struldbrug (Gulliver's Travels). His plight is described by Tennyson in the poem *Tithonus* which I, and perhaps some of my readers, rendered into Latin hexameters as a schoolboy.

subsp. **britanniae** (Verity, 1914) – the subspecies occurring in Britain.

Maniola Schrank, 1801 – Maniola, dim. of Mania or Manes, the souls of the departed. I have had difficulty in understanding what Schrank wrote; he appears to regard them as the children of '*furva Proserpina*', 'dusky Proserpina', and to link them with the mania or frenzy that may be regarded as hell-born. The intention, however, is clear. He states that even the prettiest of the butterflies have a gloomy look (in spite of his having included 1585 *Apatura iris*), and so treats them as denizens of the murky nether regions, in contrast to the members of 1549 Pieris, the white butterflies, which inhabit the bright upper air. Maniola was one of the five families into which Schrank divided the butterflies and so is the oldest name for what are now known as the Satyrinae; it is now restricted to a small genus.

1626 **jurtina** (Linnaeus, 1758) – there is no such name and it is probably a typographical error for Jurturna, the name of the nymph of a fountain near Rome, the waters of which were reputed to have healing properties. Linnaeus supposed the sexes to be distinct species, this being the female; lower on the same page he lists the male as *janira*, Janira being a Nereid, a nymph of the Mediterranean Sea. He says that it resembles *jurtina* but lacks the yellow blotch on the forewing upperside, and has three fuscous spots on the hindwing underside; he gives the habitat of *jurtina* as grassland and that of *janira* as woodland. In the past *janira* was the more favoured name but *jurtina* is now considered to have priority.

subsp. **insularis** Thompson, 1969 – *insularis*, pertaining to an island: the subspecies occurring in the British Isles.

subsp. **iernes** Graves, 1930 – Ierne, a mainly poetic variant of Hibernia, Ireland: the subspecies occurring in Ireland.

subsp. **cassiteridum** Graves, 1930 – gen. pl. of Cassiterides, the Tin Islands, the name given by the Phoenicians to the Isles of Scilly which they visited for the purpose of obtaining tin: the subspecies occurring in the Isles of Scilly.

subsp. **splendida** White, 1872 – *splendidus*, brilliant, magnificent: the richly-coloured subspecies occurring in western Scotland.

1627, 1628 } see below 1629

Aphantopus Wallengren, 1853 – ἄφαντος (aphantos), made invisible; πoῦς (pous), the foot: from the degenerate foreleg, though this is a family, not generic, character.

1629 **hyperantus** (Linnaeus, 1758) – Hyperanthus, like Pamphilus (1627), one of the 50 sons of Aegyptus, whose brother, Danaus, had 50 daughters, among them Palaeno (1542), Hyale (1543) and Daplidice (1552), qq.v. Hyperanthus and his brothers sought to marry their cousins, but Danaus gave each girl a knife and they murdered all but one of their bridegrooms. The suggestion of Pickard *et al.* that Hyperanthes, a son of Darius, the Persian king, was intended is certainly wrong because Linnaeus wrote in a footnote 'Danaorum Candidorum *nomina a filiabus Danai, Aegypti* Festivorum *a filiis mutuatus sum*', I have derived the names of the white Danai from the daughters of Danaus, of the gay Danai from the sons of Aegyptus (I have transposed the comma from after to before 'Aegypti', since this appears to be a typographical error; the order of words is chiastic).

Coenonympha Hübner, 1819 – κοινός (koinos), shared in common; νύμφη (numphē), a nymph, a nymphalid butterfly (Latreille (1804) included the Satyri in Nymphalis, see p. 31): a genus containing nymphalid butterflies that are widespread, although this is not strictly the meaning of κοινός.

1627 **pamphilus** (Linnaeus, 1758) – Pamphilus, one of the 50 sons of Aegyptus; see 1629 *hyperantus*. Pamphilus, 'loved by all', was also used as a personal name by the Romans.

subsp. **rhoumensis** Harrison, 1948 – the subspecies occurring on the island of Rum (Rhum to Sassenachs).

1628 **tullia** (Müller, O. F., 1764) – Tullius, fem. Tullia, a Roman name as in Marcus Tullius Cicero, known as 'Tully' to 18th century writers. The name, therefore, has affinity with 1627 *pamphilus*, q.v.; cf. also 1583 *julia* and 2057 *caja*.

subsp. **davus** (Fabricius, 1777) – Davus, another Roman name like Tullia.

subsp. **polydama** (Haworth, 1803) – Polydamas, a Trojan friend of Hector.

subsp. **scotica** Staudinger, 1901 – the subspecies found in Scotland.

DANAINAE (1630)

Danaus Kluk, 1802 – Danaus, King of Argos (see 1629), after whom the Argives and often the Greeks as a whole were called Danai by Homer. The Danai formed one of the six tribes into which Linnaeus divided the butterflies, characterized by having *'alis integerrimis'*, wings fully rounded, without tails, prominences or excavations; he divided them into the Danai candidi, the white species now in the Pieridae, and the Danai festivi, the gaily-coloured species now in the Nymphalidae, especially the Danainae and Satyrinae (Pl. II). See also 1552 *daplidice*.

1630 **plexippus** (Linnaeus, 1758) – the name of a Greek hero who took part in the boar hunt at Calydon and was killed by Meleager because he tried to take the prize for success from Atalanta (1590).

BOMBYCOIDEA

Bombyx was the first of the seven sections into which Linnaeus divided the Phalaenae, the moths other than the hawk-moths (Pl. III); it is derived from βόμβυξ (Lat. bombyx), the silk-worm or silk-moth, which he included as *Bombyx mori* (*morus*, a mulberry-tree). The superfamily is notable for the cocoons spun by its members and the words βομβύκιον (bombukion) and βομβυλιὸς (bombulios) were used for cocoons as well as for buzzing insects such as wasps (Liddell & Scott, 1968; Davies & Kathirithamby, 1986).

LASIOCAMPIDAE (1636)

Poecilocampa Stephens, 1828 – ποίκιλος (poikilos), varied, changeful; κάμπη (kampe) a larva: from the variability of the larva's pattern, 'it varies much'. 'Larva is covered with black spots' (Macleod); his own invention, there being no hint of this in Stephen's diagnosis.

1631 **populi** (Linnaeus, 1758) – *Populi* (gen.), of poplar, which is listed as a foodplant by Linnaeus.

Trichiura Stephens, 1828 – θρίξ, τριχός (thrix, trikhos), hair; οὐρά (oura), a tail: from the strongly developed anal tuft in the adult, especially the female, of the next species.

1632 **crataegi** (Linnaeus, 1758) – *Crataegi* (gen.), of hawthorn, which is listed as a foodplant by Linnaeus.

Eriogaster Germar, 1811 – ἔριον (erion), wool; γαστήρ (gastēr), the belly: from the woolly anal tuft of the female adult.

1633 **lanestris** (Linnaeus, 1758) – *lanestris*, woollen, *'anus Phalaenae valde lanatus albidus'*, the anus of the moth strongly furnished with white wool.

Malacosoma Hübner, 1820 – μαλακός (malakos), soft; σῶμα (sōma), the body: from the flaccid body of the larva.

1634 **neustria** (Linnaeus, 1758) – a problematic name. Pickard *et al.* write 'Neustria, a name formerly applied to a portion of France, including Normandy, Brittany and Anjou'. Macleod gives the same explanation, adding, with disregard for fact, 'from type locality'. There are a number of objections. First, although Réaumur is included among the ten references Linnaeus gives, there is nothing to associate the moth with France; second, Linnaeus would have written Neustriae (gen.), had he given the name as a noun; and third, the rest of Linnaeus' very few geographical names are adjectival, e.g. 1037 *holmiana*; 1567 *boeticus*. Linnaeus' brief diagnosis reads as follows 'P. [Phalaena] Bombyx *elinguis*, *alis reversis flavescentibus: fascia grisea sesquialtera; subtus unica'*, *Bombyx* sp. without a haustellum: wings yellowish and reversed (i.e. with the hindwing projecting in front of the forewing in repose); with one and a half grey fasciae; beneath with only one. His specimen was evidently of the form shown by Skinner (1984), Pl. 4, fig. 14, in which the inner fascia is incomplete and does not reach the dorsum. This 'one and a half fasciae' is the most likely source for a name. A fascia is a band traversing the whole wing; a stria is a short dash: the inner marking is something between the two. Suppose that in a longer description from which the one in *Systema Naturae* is a condensation Linnaeus had written *'neu fascia neu stria'*, neither a fascia nor a stria, and then telescoped the last two words to serve as the name, and you have a possible solution. Although this is fantasy, it is less unlikely than the conventional derivation.

1635 **castrensis** (Linnaeus, 1758) – *castrensis*, pertaining to a camp: from the tent-like spinnings made by the larva which is gregarious when young; *'migratque saepius novo tentorio'*, and it changes rather often to a new tent.

Lasiocampa Schrank, 1802 – λάσιος (lasios), hairy; κάμπη (kampe), a larva. This was originally a name for the whole family, with roughly the same coverage as Lasiocampidae today.

1636 **trifolii** ([Denis & Schiffermüller], 1775) – *Trifolii* (gen.), of the clover and trefoil genus: the larva is polyphagous on low-growing herbs and frequently chooses *Trifolium* spp.

subsp. **flava** Chalmers-Hunt, 1962 – *flavus*, yellow: from the ground colour of the race occurring at Dungeness, Kent.

1637 **quercus** (Linnaeus, 1758) – *Quercūs* (gen.), of the oak genus: oak is one of the foodplants of this polyphagous species.

subsp. **callunae** Palmer, 1847 – *Callunae* (gen.), of the heather genus: the northern subspecies feeds on heathers and bilberry.

Macrothylacia Rambur, 1866 – μακρός (macros), large; θύλακος (thulakos), a bag, a pouch: from the large cocoon.

1638 **rubi** (Linnaeus, 1758) – *Rubi* (gen.), of the genus *Rubus*, bramble, given as foodplant by Linnaeus.

Dendrolimus Germar, 1811 – δένδρον (dendron), a tree; λιμός (limos), hunger, famine: from the capacity of the gregarious larvae to defoliate.

1639 **pini** (Linnaeus, 1758) – *Pini* (gen.), of pine, the foodplant.

Euthrix Meigen, 1830 – εὖ (eu), well, intensive prefix; θρίξ (thrix), hair: from the hairy adult.

= **Philudoria** Kirby, 1892 – φίλος (philos), loving, liking; ὕδωρ (hudōr), water: from the larval liking for drinking drops of rain or dew on its foodplant.

1640 **potatoria** (Linnaeus, 1758) – *potatorius*, pertaining to drinking: from the habit described above.

Phyllodesma Hübner, 1820 – φύλλον (phullon), a leaf; δεσμός (desmos), a bond, a fetter: from the resemblance of the resting adult to a bunch of leaves.

1641 **ilicifolia** (Linnaeus, 1758) – *Ilex*, the holm-oak genus; *folium*, a leaf: see 1642 *Gastropacha quercifolia*.

Gastropacha Ochsenheimer, 1810 – γαστήρ (gastēr), the belly; παχύς (pakhus), thick, stout: from the moth's large abdomen. Macleod's derivation from φακός (phakos), a lentil, is ridiculous.

1642 **quercifolia** (Linnaeus, 1758) – *quercus*, an oak-tree; *folium*, a leaf: the foodplants Linnaeus lists for this and the last species do not include oak or holm-oak; in each case he was referring to the distinctive attitude of the adult at rest which mimics a bunch of leaves.

SATURNIIDAE (1643a)

Pavonia Hübner, 1819 – see below.

1643 **pavonia** (Linnaeus, 1758) – *pavonius*, pertaining to the peacock (*pavo*): from the prominent ocelli on the wings.

Saturnia Schrank, 1802 – Saturnus, Saturn, was the oldest of the Roman gods and the names Saturnius, Saturnia were given as titles to his sons and daughters. Both Juno and Vesta were styled Saturnia, and here the former is intended, since the peacock (1643 *pavonia*, formerly in this genus) was sacred to her.

1643a **pyri** ([Denis & Schiffermüller], 1775) – *Pyri* (gen.), of the pear genus: from the larval foodplant.

ENDROMIDAE (1644)

Endromis Ochsenheimer, 1810 – ἐν (en), in; δρόμος (dromos), running; ἐνδρομίς (endromis), a term used both for a running shoe and a tracksuit worn to keep the runner warm after the race. Here the latter is intended and the allusion is to the dense hair-scales clothing the moth's abdomen.

1644 **versicolora** (Linnaeus, 1758) – *versus*, turned, reversed; *color*, colour: from the ante- and postmedian fasciae which reverse their colour because they are black, edged inwardly and outwardly respectively with white, '*strigis nigro-albis*'; Linnaeus quoted a similar description from an earlier author. Pickard *et al.* and Spuler miss the point when they explain as 'of various colours' and Macleod is clearly wrong when he writes 'larva turns from black to green' since Linnaeus did not know the life history.

GEOMETROIDEA (1666)

DREPANIDAE (1646)

Falcaria Haworth, 1809 – *falx*, a sickle, a reaping-hook; *falcarius* strictly means a sickle-maker, but here Haworth is using it adjectivally to mean sickle-shaped, from the falcate forewing.

1645 **lacertinaria** (Linnaeus, 1758) – *lacerta*, a lizard; the termination '*-aria*' because Linnaeus described it in the Geometrae pectinicornes (see GEOMETRAE, Pl. III; p. 21): from the larva; Linnaeus describes its structure and must have have fancifully supposed it to resemble a lizard.

Drepana Schrank, 1802 – δρέπανον (drepanon), a reaping-hook: from the falcate forewing.

1646 **binaria** (Hufnagel, 1767) – *bini*, two, a pair; *-aria*, geometrid termination: from the two conspicuous black discal spots on the forewing.

1647 **cultraria** (Fabricius, 1775) – *culter*, a knife or ploughshare; *-aria*, geometrid termination: from the falcate forewing which is compared to a curved ploughshare.

1648 **falcataria** (Linnaeus, 1758) – *falcatus*, falcate; *-aria*, geometrid termination: from the falcate forewing.

subsp. **scotica** Bytinski-Salz, 1939 – the subspecies occurring in Scotland.

1649 **curvatula** (Borkhausen, 1790) – dim. of *curvatura*, a bending: from the falcate forewing.

Sabra Bode, 1907 – perhaps a meaningless neologism. Bode gives no explanation.

1650 **harpagula** (Esper, 1786) – dim. of *harpago*, a grappling-hook: from the falcate forewing. Denis & Schiffermüller (1775) introduced the termination '*-ula*' for the 'hook-tips', even altering the Linnaean names 1645 *lacertinaria* and 1648 *falcataria* to *lacertula* and *falcula* respectively.

Cilix Leach, 1815 – Κίλιξ (Cilix), an inhabitant of Cilicia in the south-east of Asia Minor or, more probably, Cilix, a son of Agenor, king of Phoenicia, who was the mythical ancestor of the Cilicians. There is no entomological application. Spuler derives from κίλλιξ (killix), which he spells incorrectly, an ox with crooked horns, supposing that there is such a mark on the forewing (the 'Chinese character').

1651 **glaucata** (Scopoli, 1763) – *glaucus*, bluish grey; *-ata*, the geometrid termination: from the steely blue metallic scales on the tawny dorsal blotch.

THYATIRIDAE (1652)

Thyatira Ochsenheimer, 1816 – a city of Asia Minor, best known from biblical references in Acts of the Apostles and Revelation.

1652 **batis** (Linnaeus, 1758) – βάτος (batos), bramble: the foodplant, correctly given by Linnaeus.

Habrosyne Hübner, 1821 – ἁβροσύνη (habrosunē), splendour: from the moth's beauty.

1653 **pyritoides** (Hufnagel, 1766) – πυρίτης (puritēs), of fire, the mineral copper pyrites which strikes fire; εἶδος, ὠδ- (eidos, ōd-), form: from the brassy yellow colour present in the markings of the forewing.

Tethea Ochsenheimer, 1816 – *tetheus*, an adjective coined from Τηθύς, Tēthys, the wife of Oceanus and the mother of the river gods and the Oceanides (nymphs of the open sea): a name suggested by the stream-like forewing fasciae ('rivulets' or, changing the metaphor, 'lute-strings').

1654 **ocularis** (Linnaeus, 1767) – *ocularis*, pertaining to the eyes: from the eye-like stigmata on the forewing.

subsp. **octogesimea** (Hübner, 1786) – *octogesimus*, eightieth: from the stigmata on the forewing which together resemble the number '80'.

1655 **or** ([Denis & Schiffermüller], 1775) – from the stigmata on the forewing which together resemble the letters 'O', 'R'.

subsp. **scotica** (Tutt, 1888) – the subspecies occurring in Scotland.

subsp. **hibernica** (Turner, 1927) – the subspecies occurring in Ireland.

Tetheella Werny, 1966 – dim. of 1654 *Tethea*, q.v.

1656 **fluctuosa** (Hübner, 1803) – *fluctuosus*, wavy: from the 'rivulets' or 'lute-strings' on the forewing.

Ochropacha Wallengren, 1871 – ὠχρός (ōkhros), pale yellow; παχύς (pakhus), thick, broad: probably from the broad, yellowish brown median fascia of the next species, although Wallengren's diagnosis deals only with the wing venation and the slender abdomen.

1657 **duplaris** (Linnaeus, 1761) – *duplaris*, containing double: from the two black dots present in the reniform stigma of the forewing.

Cymatophorima Spuler, 1908 – a replacement name for the genus *Cymatophora* Treitschke, 1825, preoccupied by *Cymatophora* Hübner, 1812, from κῦμα (kuma), a wave; φορέω (phoreō), to carry: from the wavy lines, 'rivulets' or, to change the metaphor, 'lutestrings' on the forewings of the members of what was formerly a more extensive genus.

1658 **diluta** ([Denis & Schiffermüller], 1775) – *dilutus*, washed off: from the pale ('dilute') coloration of the forewing.

subsp. **hartwiegi** (Reisser, 1927) – i.h.o. Dr Fritz Hartwieg (1877–1962), a German entomologist and friend of Reisser; he lived at Brunswick.

Achlya Billberg, 1820 – an adjectival form from ἀχλύς (akhlus), mist, darkness: the other species placed in the genus by Billberg were 2311 *Ipimorpha retusa*, 2312 *I. subtusa* and 2315 *Dicycla oo*, but the name does not seem applicable to the pattern of any of them. It probably refers to the time of flight and is a loose Greek rendering of 2107 *Noctua*; cf. 2380 *Charanyca* and 2466 *Lygephila* of the same author. 'From clouded appearance of markings on forewings of some specimens' (Macleod).

1659 **flavicornis** (Linnaeus, 1758) – *flavus*, yellow; *cornu*, a horn, the antenna of an insect: '*antennis luteis*', with yellow antennae, the 'yellow-horned'.

subsp. **scotica** (Tutt, 1888) – the subspecies occurring in Scotland.

subsp. **galbanus** (Tutt, 1891) – *galbanus*, greenish yellow: from the ground colour of the English subspecies which is distinctly greener than that of the nominate subspecies.

Polyploca Hübner, 1821 – πολύς (polus), many; πλοκή (plokē), a twisting: from the numerous wavy 'lutestrings'.

1660 **ridens** (Fabricius, 1787) – *ridens*, laughing: application obscure. Macleod's explanation 'from bright appearance of forewings, head and thorax' is unconvincing since these are very drab in most British specimens, though they may have been brighter in the original type material. A possible explanation is that the reference is to the 'lutestrings' which resemble the wrinkles of a laughing face, but these are relatively weakly developed in this species. Fabricius himself gives no reason for the name.

GEOMETRIDAE (1666)

Linnaeus divided the Geometridae (Geometrae) into two main divisions, *Geometrae pectinicornes* with pectinated antennae, and *Geometrae seticornes* with bristle-like (simple) antennae (Pl. III). He gave the termination *-aria* to the specific names of the former and *-ata* to those of the latter, a practice followed by the majority of his successors. However, many species were given the 'wrong' termination and a large proportion have synonyms consisting of the same or a different name with the other ending. Many of the adjectives formed with these terminations are non-classical; this will not be repeated below, only the source word being given.

ARCHIEARINAE (1661)

Archiearis Hübner, 1823 – ἀρχή, ἀρχι- (arkhē, arkhi-), beginning; ἔαρ (eär), the spring: from the emergence of the moths very early in the year.

1661 **parthenias** (Linnaeus, 1761) – παρθενίας (parthenias), the son of a concubine (παρθένος (parthenos), a maiden), according to the lexicon, a word not involving disgrace. This is one of the series of names of moths with red, yellow or orange hindwings taken from brides, fiancées or mistresses; cf. 446, 1662, 2107, 2109, 2452–2455a, etc. For a light-hearted comment, see 2452.

1662 **notha** (Hübner, 1803) – νόθος, fem. νόθη (nothos, nothē), a bastard: a name chosen to indicate affinity with the last species. Macleod considers both the names 'pointless'.

OENOCHROMINAE

From the Indo-Australian genus *Oenochroma* Guenée, 1858 – οἶνος (oinos), wine; χρῶμα (krōma), colour: this coloration is not to be seen in the British representative.

Alsophila Hübner, 1825 – ἄλσος (alsos), a grove; φίλος (philos), loving: from the habitat.

1663 **aescularia** ([Denis & Schiffermüller], 1775) – *aesculus*, strictly a species of oak but in post-classical times applied to the horse-chestnut (*Aesculus hippocastanum*): oak is a foodplant, horse-chestnut probably not.

GEOMETRINAE (1666)

Aplasta Hübner, 1823 – ἄπλαστος (aplastos), not moulded, simple, natural: probably from the weak markings of the moths.

1664 **ononaria** (Fuessly, 1783) – ὄνωνις (*Onōnis*), rest-harrow: the foodplant.

Pseudoterpna Hübner, 1823 – ψεῦδος (pseudos), falsehood; τερπνός (terpnos), delightful: from the deceptively attractive green coloration which soon fades.

1665 **pruinata** (Hufnagel, 1767) – *pruina*, hoar-frost: from the whitish irroration on the green wings.
subsp. **atropunctaria** (Walker, 1863) – *ater*, black; *punctum*, a spot: from the dark speckling in the ante- and postmedian fasciae.

Geometra Linnaeus, 1758 – γεομετρέω (geometreō), to measure land: from the 'looping' gait of the larva.

1666 **papilionaria** Linnaeus, 1758 – *Papilio*, a butterfly: from superficial resemblance. This may not really be the species intended by Linnaeus because in a reference to Wilkes (1747–49) he states that Wilkes' English moth is twice the size of the Swedish one.

Comibaena Hübner, 1823 – κώμυς (kōmus), a bundle; βαίνω (bainō), to go: from the larval habit of attaching fragments of oak-bracts, etc., to their bodies. Macleod attempted derivation from κῶμος (kōmos), a revel.

1667 **bajularia** ([Denis & Schiffermüller], 1775) – *bajulus*, a porter, a carrier: from the larval behaviour (see genus).
= **pustulata** (Hufnagel, 1767) *nec* (Müller, 1764) – *pustula*, a blister: the 'blotched' emerald.

Thetidia Boisduval, 1840 – Θέτις, Θέτιδος, Thetis, a sea-goddess, sister of the Nereids and mother of Achilles. The species which follows is not coastal in France, so the selection by Boisduval of a 'sea'-goddess has no significance.

1668 **smaragdaria** (Fabricius, 1787) – *smaragdus*, an emerald: from the coloration of the adult.
subsp. **maritima** (Prout, 1935) – *maritimus*, pertaining to the sea: from the coastal habitat of the English subspecies.

Hemithea Duponchel, 1829 – ἡμίθεος (hēmitheos), a demigod; in appreciation.

1669 **aestivaria** (Hübner, 1799) – *aestivus*, pertaining to summer: from the season of the adult's appearance; perhaps also from the green ground colour characteristic of the summer.

Chlorissa Stephens, 1831 – χλωρός (khlōros), pale green, Chloris, a flower-goddess: from the colour of the species, perhaps influenced by the girl's name used by 18th century lyricists.

1670 **viridata** (Linnaeus, 1758) – *viridis*, green: from the wing coloration of this, according to Linnaeus, oak-feeding species.

Chlorochlamys Hulst, 1896 – χλωρός (khlōros), pale green; χλαμύς (khlamus), a cloak or mantle: from the coloration of the species.

1671 **chloroleucaria** (Guenée, 1857) – χλωρός (khlōros), pale green; λευκός (leukos), white: from the pale green coloration.

Thalera Hübner, 1823 – θαλερός (thaleros), blooming, fresh, youthful: from the fresh, spring-like green coloration of the species.

1672 **fimbrialis** (Scopoli, 1763) – *fimbriae* (pl.), a fringe, the cilia on an insect's wing: from the red-checkered terminal cilia.

Hemistola Warren, 1893 – ἡμι- (hēmi-), half; στολή (stolē), a garment: from the thin wing-scaling, 'scaling loose and thin, so that the wings appear semi-diaphanous'. 'From paleness of green wings' (Macleod), an explanation at variance with Warren's diagnosis.

1673 **chrysoprasaria** (Esper, 1794) – *chrysoprasus*, chrysoprase or green chalcedony, χρυσός (khrusos), gold; *prasius*, *prasinus*, leek-green; from the coloration.

Jodis Hübner, 1823 – ἰώδης (iōdēs), rust-like; from the degeneration of the fresh green wing coloration with the passage of time, 'Lay not up for yourselves treasures upon earth, where moth and rust doth corrupt' (Matthew 6:19). 'From verdigris colour of wings when very fresh' (Macleod); very unlikely, because rust is not verdigris and neither of the species which Hübner included in the genus has wings of that colour.

1674 **lactearia** (Linnaeus, 1758) – *lacteus*, milky: from the very pale green wings which quickly fade to white and were, in fact, described as white by Linnaeus.

STERRHINAE (1696)

Cyclophora Hübner, 1822 – κύκλος (kuklos), a ring; φορέω (phoreō), to carry: from the annular discal spot on the forewing of some of the species.

1675 **pendularia** (Clerck, 1759) – *pendulus*, hanging, but probably from *Betula pendularia* (silver birch), the foodplant not of this species but of 1677 *C. albipunctata* which was formerly known as *C. pendularia*. After careful scrutiny of Clerck's figure (in the Rare Books Room at the University Library, Cambridge, where I could not take specimens for comparison), I inclined to the opinion that the birch-feeding species was the one depicted. No text accompanies Clerck's figure to give guidance. There is a tradition (Pickard *et al.*; Macleod) that the name refers to the suspension of the pupa from a leaf; Clerck figures the pupa, but it is not shown as suspended.

= **orbicularia** (Hübner, 1799) – *orbiculus*, a small ring: from the annular discal markings on both forewing and hindwing. The name used for this species when *pendularia* was used for 1677 *C. albipunctata*.

1676 **annulata** (Schulze, 1775) – *annulus*, a small ring: from the annular discal markings on both forewing and hindwing.

1677 **albipunctata** (Hufnagel, 1767) – *albus*, white; *punctum*, a spot: from the white pupils of the annular discal spots on each wing.

1678 **puppillaria** (Hübner, 1799) – *pupilla*, a ward, the pupil of the eye: from the annular discal spots.

1679 **porata** (Linnaeus, 1767) – *porus*, tufa, travertine, a porous limestone deposit pale in colour and sometimes with darker freckles: from the speckled wings.

1680 **punctaria** (Linnaeus, 1758) – *punctum*, a dot: *'ordine transverso punctorum atrorum'*, with a transverse series of black dots, these being situated in the postdiscal area.

1681 **linearia** (Hübner, 1799) – *linea*, a line: from the dark median fascia, possibly also the obsolescent subbasal and postdiscal fasciae which give the English name 'clay triple-lines'.

Timandra Duponchel, 1829 – the daughter of Tyndareus and Leda and the sister of Clytemnestra.

1682 **griseata** Petersen, 1902 – *griseus* (late Lat.), grey: from the ground colour.

= **amata** sensu auctt. – *amatus*, loved: a misidentification; the species which Linnaeus 'loved' was in fact 1680 *Cyclophora punctaria*; see p. 39.

Scopula Schrank, 1802 – *scopula*, a small broom: from the expansible tuft of scales on the posterior tibia in the male of certain species. See p. 30.

1683 **immorata** (Linnaeus, 1758) – meaning uncertain: *immoror* means to stay or linger near. Macleod's explanation 'from tendency to stay in one locality' applies well enough to the

'Lewes wave' in Sussex but there is no evidence that it has similar habits on the Continent or, if it does, that Linnaeus knew this in 1758. *'In'* in compounds also means 'not' as in 1692 *immutata*, the species which precedes this one in *Systema Naturae*, and Linnaeus writes in his diagnosis *'fasciis obsoletis albidis'* (with obsolescent white fasciae); obsolescent markings are fugitive and that, possibly, is the meaning of the name.

1684 **nigropunctata** (Hufnagel, 1767) – *niger*, black; *punctum*, a spot: from the sprinkling of black dots on the forewing.

1685 **virgulata** ([Denis & Schiffermüller], 1775) – *virgulatus*, striped: descriptive.

1686 **decorata** ([Denis & Schiffermüller], 1775) – *decoratus*, adorned: descriptive.

1687 **ornata** (Scopoli, 1763) – *ornatus*, adorned: descriptive.

1688 **rubiginata** (Hufnagel, 1767) – *robigino* (*rubigino*), to rust: from the rusty ground colour – the 'tawny wave'.

1689 **marginepunctata** (Goeze, 1781) – *margo, marginis*, an edge; *punctum*, a spot: from the terminal series of black dots.

1690 **imitaria** (Hübner, 1799) – *imitor*, to imitate, to counterfeit: probably from similarity to another species, e.g. 1682 *Timandra griseata* (Hübner would have known it as *amata* Linnaeus, 1758). Macleod's suggestion 'from larva's habit of feigning death' is ingenious and amusing, but Hübner was an artist depicting adults and it is unlikely that he had made observations of the larval behaviour.

1691 **emutaria** (Hübner, 1809) – *emuto*, to change: probably from the change in colour when the delicate pink of newly emerged specimens begins to fade. Five species in this genus have names derived from *muto* or its compounds.

1692 **immutata** (Linnaeus, 1758) – *immutatus*, unchanged: *'alis albidis concoloribus'*, wings with both the upper and under surfaces white (see *Systema Naturae*, p. 464, footnote, for Linnaeus' definition of *'concolores alae'*). 'Unchanged, i.e. with few variations' (Macleod); such observations were not made in the infancy of entomology, when variations were regarded as distinct species.

1693 **floslactata** (Haworth, 1809) – *flos*, flower; *lactis* (gen.), of milk, i.e. cream: from the ground colour; *lactata*, named by Haworth at the same time, is now reduced to synonymy with this species. cf. 75 *floslactella*.

subsp. **scotica** Cockayne, 1951 – the subspecies occurring in Scotland.

1694 **ternata** (Schrank, 1802) – *terni*, three each: from the three fasciae on all four wings. 'From the three pale lines along the larva's back' (Macleod); why, then, three *each*? In any case, the life history may not have been known in 1802.

1695 **limboundata** (Haworth, 1809) – *limbus*, a border; *undatus*, waved: descriptive of the forewing; pronounce *limboündata*.

Idaea Treitschke, 1825 – Ἰδαῖος (Idaios), pertaining to Mt. Ida, the grandstand from which the gods and goddesses watched the battles of the Trojan War.

= **Sterrha** Hübner, 1825 – στερρός (sterrhos), hard, stiff: the name has puzzled authors and Pickard *et al.* and Macleod offer no explanation. Probably it refers to the expansible tuft of 'stiff' setae on the male hindtibia in certain species.

1696 **ochrata** (Scopoli, 1763) – *ochra*, a kind of yellow earth: from the ground colour.

subsp. **cantiata** (Prout, 1913) – *Cantium*, Kent: the English subspecies which occurs in Kent.

1697 **serpentata** (Hufnagel, 1767) – *serpens*, a snake: from the wriggly fasciae.

1698 **muricata** (Hufnagel, 1767) – *murex*, a gasteropod mollusc which yielded the famous Tyrian purple dye: from the purple markings.

1699 **vulpinaria** (Herrich-Schäffer, 1851) – *vulpinus*, pertaining to a fox: from the reddish fascia on the forewing.

subsp. **atrosignaria** Lempke, 1967 – *ater*, black; *signum*, a mark: the British subspecies in which the fascia is black, almost completely lacking the reddish 'foxy' tinge of the nominate subspecies.

1700 **laevigata** (Scopoli, 1763) – *laevigo*, to pulverize, or to polish with grinding powder: so either speckled or glossy, or both; the first meaning seems likely for this adventive species. cf. 227 *laevigella* and 401 *laevigatella*.

1701 **sylvestraria** (Hübner, 1799) – *sylvestris*, of woodland: the habitat in Britain is heathland, not woodland, so it is just possible that Hübner bestowed this name i.h.o. Israel Sylvestre, a French entomologist and friend of Fabricius.

1702 **biselata** (Hufnagel, 1767) – *bis*, twice, *seta*, a bristle: from the large tuft on the tibia of each hindleg in the male. The name is a typographical error for *bisetata* and most of the earlier authors emended it accordingly, but this is not permitted under the rules of the I.C.Z.N.

1703 **inquinata** (Scopoli, 1763) – *inquinatus*, defiled, unclean: from the cloudy markings.

1704 **dilutaria** (Hübner, 1799) – *dilutus*, washed out, faint: from the pale ground colour and the obsolescent markings.

1705 **fuscovenosa** (Goeze, 1781) – *fuscus*, dusky; *venosus*, veined: from the dark streak between vein 11 (R_1) and the costa; the whole wing is not streaked as supposed by Macleod.

1706 **humiliata** (Hufnagel, 1767) – *humilis*, near the ground, small, base, insignificant: any or more than one of these meanings may be intended; it is a small, low-flying species without the rich coloration of, say, 1698 *I. muricata*, named in the same work. Macleod's explanation 'slight, its wings being slighter than those of [1705] *S. fuscovenosa*' is an anachronism.

1707 **seriata** (Schrank, 1802) – *series*, a row, a chain: from the wing markings which consist of fasciae broken into a series of dots.

1708 **dimidiata** (Hufnagel, 1767) – *dimidio*, to divide into half: from the subterminal markings which are strongly developed only in the dorsal half of the forewing. Haworth implied a different interpretation when he stated that it was only half the size of 1712 *I. emarginata*.

1709 **subsericeata** (Haworth, 1809) – *sub-*, somewhat; *sericus*, silken: from the relatively weak gloss on the wings.

1710 **contiguaria** (Hübner, 1799) – *contiguus*, touching, bordering upon: perhaps because the most strongly expressed markings are touching the costa, or because it is close in appearance to another species, e.g. 1689 *Scopula marginepunctata*.

subsp. **britanniae** (Müller, L., 1936) – the subspecies occurring in Britain.

1711 **trigeminata** (Haworth, 1809) – *trigeminus*, threefold, triple, a triplet: from the postdiscal fascia, which is broken into three spots.

1712 **emarginata** (Linnaeus, 1758) – *emarginatus*, deprived of the edge, having parts of the margin missing: from the emarginate wings.

1713 **aversata** (Linnaeus, 1758) – *aversus*, belonging to the hinder or, as here, the under part: from the stronger expression of the discal spot on the lower surface. '*In aversa charta*' (Martial, 1st century A.D.) signified 'on the back of the paper'; Linnaeus wrote '*punctum in pagina inferiore magis saturatum*', the spot more deeply coloured on the underside of the 'page', 'page' here meaning the surface of the wing. 'Turning away, perhaps from shape of band across wings' (Macleod); the diagnostic significance of the subcostal angulation of the fascia was not realized until 1715 *I. straminata* was named thirty-six years later and Linnaeus makes no mention of this character.

1714 **degeneraria** (Hübner, 1799) – *degener*, unworthy of one's race, degenerate: application obscure; Hübner may have noted some aberrant character, or he may be implying that it is one of the less noteworthy members of the family.

1715 **straminata** (Borkhausen, 1794) – *stramen*, straw: from the ground colour.

Rhodometra Meyrick, 1892 – ῥόδον (rhodon), a rose; μέτρον (metron), a measure, a length: from the rose-purple fascia on the forewing of the next species.

1716 **sacraria** (Linnaeus, 1767) – *sacer*, holy; *-aria*, the conventional termination for the Geometrae pectinicornes (see p. 21); or, *sacraria*, a female keeper of a temple, a priestess, a vestal virgin: perhaps because priestesses wore, or Linnaeus thought they wore, saffron robes, or just because the simple yet beautiful pattern of the moth suggests chastity. English collectors must have assumed that the second derivation was the correct one when they gave the moth the vernacular name 'the vestal'.

LARENTIINAE (1745)

Lythria Hübner, 1823 – λύθρον (luthron), gore: from the purple colour of the next species, the only one included by Hübner in the genus.

1717 **purpuraria** (Linnaeus, 1758) – *purpureus*, purple-coloured: from the two purple fasciae on the forewing.

Phibalapterix Stephens, 1829 – Stephens explains as follows, 'φίβαλος (phibalos), *gracilis* (slender); πτερόν (pteron), *ala* (wing)'; in his diagnosis he describes the thorax and abdomen, but not the wings, as slender, so πτερόν probably means simply a winged moth. The word 'φίβαλος' is not in my lexicon; φιβάλεως (phibaleōs) meant a kind of early fig and was used metaphorically for a lean, dried-up person. Stephens was a very good entomologist but not a classical scholar; see, for example, 1002, genus.

= **Mesotype** sensu auctt. – μέσος (mesos), middle; τύπος (typos), an image, a pattern: *Mesotype* Hübner, 1825 is a junior synonym of 1801 *Perizoma* Hübner, 1825, and refers to the median fascia of some of the species in that genus.

1718 **virgata** (Hufnagel, 1767) – *virga*, a rod, a staff, a coloured stripe on a garment: the 'oblique striped'.

Orthonama Hübner, 1825 – ὀρθός (orthos), straight; νᾶμα (nāma), a stream: from the straight lines across the forewing, which do not meander like the 'rivulets' of, for example, 1721 *Xanthorhoe*.

1719 **vittata** (Borkhausen, 1794) – *vitta*, a band, a chaplet: in entomological descriptions '*vitta*' usually signifies a stripe parallel to the dorsum, but here it applies to the diagonal stripe extending from the apex.

1720 **obstipata** (Fabricius, 1794) – *obstipus*, bent to one side: from the diagonal apical dash.

Xanthorhoe Hübner, 1825 – ξανθός (xanthos), yellow; ῥωή (rhōē), a stream: from the yellowish wavy lines ('rivulets') on the forewings of some species; cf. 1719 *Orthonama*.

1721 **biriviata** (Borkhausen, 1794) – *bi-*, two; *rivus*, a stream: from the presence of two 'rivulets' on the forewing, presumably the two subbasal wavy lines.

1722 **designata** (Hufnagel, 1767) – *designo*, to mark out, to define: probably because the median fascia is more sharply defined than in related species.

1723 **munitata** (Hübner, 1809) – *munio*, to fortify, *munitus*, fortified: from the central fascia which in one form is reduced to ante- and postmedian lines which 'wall in' the area between. Hübner borrowed this concept from Scopoli (1729 *moeniata*) and used it again in 1730 *peribolata*.

subsp. **hethlandica** Prout, 1901 – the subspecies occurring on 'heathland' in 'Shetland', a pun apparently being intended. Rebel borrowed this name for 1809 *Perizoma didymata hethlandica*.

1724 **spadicearia** ([Denis & Schiffermüller], 1775) – *spadix*, nut-brown: from the colour of the median fascia. This name would have been more appropriate for the next species.

1725 **ferrugata** (Clerck, 1759) – *ferrugo*, iron rust, its colour: from the median fascia. This name would have been more appropriate for the last species.

1726 **quadrifasiata** (Clerck, 1759) – a typographical error for *quadrifasciata*, *quadri-*, four; *fascia*, a band: from the fasciae on the forewing which may be supposed to number four.

1727 **montanata** ([Denis & Schiffermüller], 1775) – *montanus*, mountain-: the type locality is in the Vienna district where the moth may well have first been found on mountains. In Britain it is a ubiquitous species, occurring on highland and lowland alike.

subsp. **shetlandica** (Weir, 1880) – the subspecies occurring in Shetland.

1728 **fluctuata** (Linnaeus, 1758) – *fluctus*, a wave: from the wavy transverse lines or 'rivulets' on the forewing.

subsp. **thules** (Prout, 1896) – the subspecies occurring in '*Ultima Thulē*' (Shetland); see 14 *Hepialus humuli thulensis*.

Scotopteryx Hübner, 1825 – σκότος (skotos), darkness; πτέρυξ (pterux), a wing: from the shaded markings on the forewing, especially the median fascia, e.g. in 1732, the 'shaded broad-bar'.

1729 **moeniata** (Scopoli, 1763) – *moenia*, city walls, defences: see 1723 *munitata*. Scopoli includes the word *propugnaculum*, rampart, in his diagnosis.

1730 **peribolata** (Hübner, 1817) – περιβολή (peribolē), a throwing round, an encirclement, the walls encircling a town: a name indicating affinity with the previous species.

1731 **bipunctaria** ([Denis & Schiffermüller], 1775) – *bi-*, two; *punctum*, a spot: from the two black discal spots, placed one above the other.

subsp. **cretata** (Prout, 1937) – *creta*, Cretan earth, chalk: 'the chalk carpet'.

1732 **chenopodiata** (Linnaeus, 1758) – *Chenopodium*, the goosefoot genus: given, apparently incorrectly, as the genus of the foodplants by Linnaeus.

1733 **mucronata** (Scopoli, 1763) – *mucronatus*, pointed: from the acute apex of the forewing.

subsp. **umbrifera** (Heydemann, 1925) – *umbrifer*, casting a shade, shady: the subspecies with paler ground colour, thereby making the pattern appear darker.

subsp. **scotica** (Cockayne, 1940) – the subspecies occurring in Scotland.

1734 **luridata** (Hufnagel, 1767) – *luridus*, wan, pale yellow, ghastly; *luridatus* (from the same root), besmeared, defiled: neither meaning gives an apt description of the form found in Britain.

subsp. **plumbaria** (Fabricius, 1775) – *plumbus*, lead: from the leaden ground colour. It is 1733 *S. mucronata* that bears the English name 'lead belle' in translation of Fabricius' Latin one, both having long been regarded as conspecific.

Catarhoe Herbulot, 1951 – κατά (kata), a prefix often, as here, without precise meaning; ῥωή (rhōē), a stream: a genus with affinity with 1721 *Xanthorhoe*.

1735 **rubidata** ([Denis & Schiffermüller], 1775) – *rubidus*, reddish: the 'ruddy carpet'.

1736 **cuculata** (Hufnagel, 1767) – a typographical error for *cucullata, cucullus*, a cowl or hood: from the appearance of the dark thorax and wing-base when the moth is at rest, the 'royal mantle'. *Cuculus* means a cuckoo.

Epirrhoe Hübner, 1825 – ἐπιρρωή (epirrhōē), a flood, a river: from the wavy 'rivulets' on the forewing.

1737 **tristata** (Linnaeus, 1758) – *tristis*, sad: from the black and white coloration which suggests mourning.

1738 **alternata** (Müller, O. F., 1764) – *alternare*, to alternate: from the alternation of black and white fasciae on the forewing.

subsp. **obscurata** (South, 1888) – the subspecies occurring in the Hebrides in which the dark markings are largely absorbed in the pale ground colour 'the whole of the central third of the forewing is whitish' (South, 1908). Macleod writes '*obscurus*, dark, from forewings', which is the exact opposite of what South said.

1739 **rivata** (Hübner, 1813) – *rivus*, a river: from the wavy 'rivulets'.

1740 **galiata** ([Denis & Schiffermüller], 1775) – *Galium*, the bedstraw genus, on which the larva feeds.

Costaconvexa Agenjo, 1949 – *costa*, a rib, the anterior margin of a wing; *convexus*, arched, convex: from the costa of the forewing, although this is hardly arched more than in related genera.

1741 **polygrammata** (Borkhausen, 1794) – πολύς (polus), many; γράμμα, pl. γράμματα (gramma, grammata), line, lines: the 'many-lined'.

Camptogramma Stephens, 1831 – καμπτός (kamptos), bent; γράμμα (gramma), a line: from the 'flexuous' cross-lines, especially the outer margin of the median fascia.

1742 **bilineata** (Linnaeus, 1758) – *bi-*, two; *linea*, a line: from the 'double' lines bounding the median fascia which are fuscous edged with white, '*fascia repanda margine fusco alboque*', with a curved fascia, having its margin fuscous and white. 'From two bands across each forewing' (Macleod): this may be so, but it is not in Linnaeus' diagnosis.

subsp. **atlantica** (Staudinger, 1892) – the subspecies occurring in the Hebrides and Shetland.

subsp. **hibernica** Tutt, 1892 – the subspecies occurring in Ireland.

subsp. **isolata** Kane, 1896 – *isolatus*, an invented Latin word for isolated, which itself came to us via the Italian from the Latin *insula*, an island: the *isolated* subspecies occurring on the Blasket *Islands*, off the south-western coast of Ireland.

Entephria Hübner, 1825 – ἔντεφρος (entephros), ash-coloured: from the ground colour of 1744 *E. caesiata*.

1743 **flavicinctata** (Hübner, 1813) – *flavus*, yellow; *cinctatus*, encircled: from the irroration of yellow-brown superimposed on the dark markings.

subsp. **ruficinctata** (Guenée, 1857) – *rufus*, red: the subspecies occurring in Scotland, in which the irroration has a redder tinge.

1744 **caesiata** ([Denis & Schiffermüller], 1775) – *caesius*, bluish grey, often used of the eyes: from the ground colour.

Larentia Treitschke, 1825 – Acca Laurentia (Larentia), the wife of a shepherd and the foster-mother of Romulus and Remus after they had been taken from the she-wolf. Although without entomological application, *Larentia* has given its name to the subfamily Larentiinae.

1745 **clavaria** (Haworth, 1809) – Haworth states that he took this name from the 'Marsh MS' but does not explain it; perhaps from *clavus*, a stripe on Roman togas, broad for senators and narrow for the equestrian order, and referring to the median fascia, which is broad like the *latus clavus*, but lacks its purple colour. Macleod's derivation from *clava*, a club, also with reference to the median fascia, is likewise possible.

Anticlea Stephens, 1831 – the name of the mother of Ulysses who died in grief because of his long absence at the Trojan War.

1746 **badiata** ([Denis & Schiffermüller], 1775) – *badius*, brown, bay-coloured: from the ground colour.

1747 **derivata** ([Denis & Schiffermüller], 1775) – *derivo*, to divert a stream (*rivus*): from the postmedian subcostal strigula (a 'rivulet') which is diverted towards the termen, giving the species its distinctive character.

Mesoleuca Hübner, 1825 – μέσος (mesos), middle; λευκός (leukos), white: the next species differs from most others that are closely related in having the median area clear white without a fascia.

1748 **albicillata** (Linnaeus, 1758) – '*albus*, white; *cilla*, naturalists' Latin for tail, cf. *Motacilla*, wagtail; from white abdomen, contrasted with dark front part of body' (Macleod). Although Linnaeus makes no mention of the abdomen in his diagnosis, Macleod's explanation is the most probable.

Pelurga Hübner, 1825 – πηλουργός (pēlourgos), a worker in clay: from the supposedly clay-coloured ground colour of the next species.

1749 **comitata** (Linnaeus, 1758) – *comitor*, to attend, accompany, be a companion to: probably from resemblance to 1732 *Scotopteryx chenopodiata*, the species immediately preceding this one in *Systema Naturae*, the two descriptions being worded almost identically. Macleod thought it was the next species, *dotata*, now = 1756 *Eulithis populata*, but here there is virtually no similarity in phrasing.

Lampropteryx Stephens, 1831 – λαμπρός (lampros), bright; πτέρυξ (pterux), a wing, from the strong gloss on the forewings of the next species.

1750 **suffumata** ([Denis & Schiffermüller], 1775) – *sub-*, somewhat; *fumatus*, smoked, smoky: from the coloration of the wings; the authors may have had ab. *piceata* Stephens before them. Stephens states that ab. *piceata* (*piceus*, pitch-black) also has very glossy forewings (see genus).

1751 **otregiata** (Metcalfe, 1917) – *Otregia*, the Latin name for Ottery St Mary, Devon, the type locality.

Cosmorhoe Hübner, 1825 – κόσμος (kosmos), good order, adornment; ῥωή (rhōē), a stream: a genus with affinity with 1721 *Xanthorhoe*, but more beautiful.

1752 **ocellata** (Linnaeus, 1758) – *ocellatus*, eyed: from the two (sometimes only one) subapical black spots on the forewing, '*maculaque apicis ocellari didyma*', and with a double apical eye-spot; see MBGBI 7(2), Pl. C, fig. 5.

Nebula Bruand, 1846 – *nebula*, fog, a cloud: from the umbrageous wing coloration.
= **Coenotephria** Prout, 1914 – κοινός (koinos), in common; τέφρα (tephra), ashes: a genus sharing with 1743 *Entephria* the possession of ash-coloured moths.

1753 **salicata** (Hübner, 1799) – *Salix*, the willow genus, but not the correct foodplant.
subsp. **latentaria** (Curtis, 1830) – *latens*, lying concealed, lurking: although Curtis and Dale took the moth abundantly on rocks at a situation near Ambleside, Cumbria, the cryptic wing pattern rendered it difficult to find. 'From larva's habit of resting by day among leaves' (Macleod): a bad guess and bad entomology.

Eulithis Hübner, 1821 – εὔλιθος (eulithos), of goodly stone: the moths are attractive, several with a yellowish ground colour suggestive of sandstone. Hübner may have been influenced by the meaning of 1755 *testata*, q.v.

1754 **prunata** (Linnaeus, 1758) – *Prunus*, the plum genus, but, at any rate in Britain, the larva feeds on *Ribes* spp.

1755 **testata** (Linnaeus, 1761) – *testa*, a piece of burnt clay, a brick, a tile: the colour of tiles varies from district to district; this moth has a yellowish brown ground colour.

1756 **populata** (Linnaeus, 1758) – *Populus*, the poplar genus: in Britain the larva is polyphagous; although not recorded on poplar, it will feed on other Salicaceae.

1757 **mellinata** (Fabricius, 1787) – a word coined from the root of *mel*, honey: from the yellow, honey-coloured forewing.

1758 **pyraliata** ([Denis & Schiffermüller], 1775) – Pyralis, the spelling used in 1775 for the Pyralidae: signifying resemblance, but the termination *-ata* indicates that the authors placed it correctly in the Geometridae.

Ecliptopera Warren, 1894 – ἐκλείπω (ekleipō), to fail, be wanting; ὄψ, ὄπος (ops, opos), the face: in his generic diagnosis Warren wrote 'face not rounded but obliquely flat, the lower part produced into a short point'. '*Eclipes*, deficient, *peras*, tip; from blunt apex of forewings' (Macleod): this contradicts Warren's own words 'apex bluntly produced, almost subfalcate' and in any case πέρας (peras) means an end in time or an aim, objective, never an end in place, a tip. The next species was not included by Warren in his genus.

1759 **silaceata** ([Denis & Schiffermüller], 1775) – *sil*, a kind of yellowish earth, yellow ochre; *silaceus*, having the colour of *sil*: from the irroration of yellowish scales in the subbasal and postmedian areas of the forewing.

Chloroclysta Hübner, 1825 – χλωρός (khlōros), greenish yellow; κλύζω (kluzo), to wash off or away: from the fugitive nature of the greenish ground colour of the next two species. 'From greyish-green colour and wavy marking of forewings' (Macleod). Cf. 1858 *Chloroclystis*.

1760 **siterata** (Hufnagel, 1767) – σιτηρός (sitēros), pertaining to corn: as growing corn turns from green to golden as it ripens, so the forewing of this species fades from green to yellow after death. 'Gr. *sitos*, corn, from colour of forewings' (Macleod): this misses the point, the natural colour of the wings not being that of corn.

1761 **miata** (Linnaeus, 1758) – *mio*, to make water, urinate: from the pea-green colour which turns yellowish. Macleod derives from the Greek capital mu ('M'), claiming that the forewing has such a marking.

1762 **citrata** (Linnaeus, 1761) – *citrus*, the citron tree, originally an African tree with fragrant wood, possibly a species of *Thuya*, but later, as here, applied to the orange family: from the orange colour often present in the markings of the forewing.

subsp. **pythonissata** Millière, 1872 – Πύθων, Pythōn, a serpent slain by Apollo; νίσσομαι (nissomai), to go: from the sinuous cross-lines going across the forewing of the variable subspecies occurring in Shetland.

1763 **concinnata** (Stephens, 1831) – *concinnus*, well-adjusted, elegant: this species (or race) is often more brightly coloured than *C. truncata*.

1764 **truncata** (Hufnagel, 1767) – *trunco*, to cut short: from the median fascia which often ends abruptly in the disc.

Cidaria Treitschke, 1825 – a title of Ceres, the goddess who protected agriculture, an explanation given by Treitschke himself; he may have had 1760 *Chloroclysta siterata*, q.v., in mind, since he included it together with other yellowish species in his genus. Macleod's explanation '*kidaris*, tiara, from markings on forewing' is at variance with Treitschke's diagnosis.

1765 **fulvata** (Forster, 1771) – *fulvus*, tawny yellow: from the ground colour.

Plemyria Hübner, 1825 – πλημμυρίς, πλημυρίς (plēmmuris, plēmuris), the flood tide: a variant on the 'rivulet' theme.

1766 **rubiginata** ([Denis & Schiffermüller], 1775) – *robigino*, *rubigino*, to rust: from the sprinkling of reddish scales in the basal patch and median fascia of the forewing.

subsp. **plumbata** (Curtis, 1837) – *plumbum*, lead, the colour of lead: the Scottish subspecies which has a more extensive grey terminal area on the forewing.

Thera Stephens, 1831 – an island in the Aegean Sea: no entomological application.

1767 **firmata** (Hübner, 1822) – *firmo*, to make stable, to confirm: the species in this genus are variable and hard to determine, 1769 *T. britannica*, for example, having been correctly identified only in the last decade; here Hübner may be confirming that this species is distinct from the next which he had named thirty-five years previously. Alternatively, he may be implying that this species is less variable than [1769] *T. variata*, having misunderstood the meaning of that name.

1768 **obeliscata** (Hübner, 1787) – ὀβελίσκος (obeliskos), a spit, any pointed instrument, a spire: from the median fascia which tapers to a point on the dorsum. A reference to the printer's obelus or dagger sign (†), as suggested by Macleod, is unlikely, although he claims to see such a marking on the forewing.

1769 **britannica** (Turner, H. J., 1925) – the species of *Thera* occurring in Britain.
= **variata** sensu auctt. *nec* ([Denis & Schiffermüller], 1775) – *variatus*, variegated: from the variegated wing pattern. Macleod's explanation 'varying from *T. obeliscata*' is an anachronism.

1770 **cognata** (Thunberg, 1792) – *cognatus*, related by birth: from affinity with 1768 *T. obeliscata* or 1769 *T. variata*, both already named.

1771 **juniperata** (Linnaeus, 1758) – *Juniperus communis*, juniper: the foodplant.
subsp. **scotica** (White, 1871) – the subspecies occurring in Scotland.
subsp. **orcadensis** Cockayne, 1950 – the subspecies occurring in the Orkneys (Orcadia).

1771a **cupressata** (Geyer, [1831]) – *Cupressus sempervirens*, the common cypress: the larval foodplant.

Eustroma Hübner, 1825 – εὖ (eu), intensive prefix; στρῶμα (strōma), something spread, a mattress, the bed-frame that supports it: here the last of these is meant, the name referring to the reticulate pattern of the next species.

1772 **reticulata** ([Denis & Schiffermüller], 1775) – *reticulatus*, netted: the 'netted carpet'.

Electrophaes Prout, 1923 – ἤλεκτρον (ēlektron), a shining substance, amber or an alloy of gold and silver; φάος (phaos), light, brightness: from the yellowish brown irroration on the forewing of the next species.

1773 **corylata** (Thunberg, 1792) – *Corylus avellana*, hazel: one of the many foodplants (*pace* Macleod, who said it was not so).

Colostygia Hübner, 1826 – Hübner's own emendation for his *Calostigia* of 1825, κόλος (kolos), docked, stunted; Στύγιος (Stugios), pertaining to the River Styx, the black River of Hate in the Underworld: from the black 'rivulets' which are reduced to spots on the costa. 'Substitute for *Calostigia*, 'beautiful dotted'; *calos*, beautiful, *stizo*, prick; with special reference to *C. multistrigaria*' (Macleod), a species not included by Hübner in his genus. The use of the junior *Colostygia* instead of *Calostigia* can be justified only by regarding it not as an emendation but as a different name.

1774 **olivata** ([Denis & Schiffermüller], 1775) – *oliva*, an olive: from the olive-green ground colour.

1775 **multistrigaria** (Haworth, 1809) – *multus*, much, *multi* (pl.), many; *striga*, a furrow, a line: from the numerous transverse fasciae on the forewing.

1776 **pectinataria** (Knoch, 1781) – *pectinatus*, toothed like a comb: from the male's strongly pectinated antenna.

Hydriomena Hübner, 1825 – perhaps ὑδρία (hudria), a water-pot; μένω (meno), to remain: an echo of the suggested interpretation of 1778 *impluviata*, q.v.

1777 **furcata** (Thunberg, 1784) – *furca*, a two-pronged fork: in this variable species the dark markings of the forewing may coalesce in more than one area to form such a marking.

1778 **impluviata** ([Denis & Schiffermüller], 1775) – *impluvium*, the square basin in the central court of a Roman house into which rainwater was directed. Pickard *et al.* suggest 'like an *impluvium*, i.e. having a square border', but this carries no conviction. The larva feeds in a spinning on alder and as every microlepidopterist knows spinnings may become waterlogged after rain (*pluvia*). Denis and Schiffermüller may have reared specimens from larvae found in such spinnings and the meaning of Hübner's generic name, 'abiding in a water-pot', may echo their imagery.

1779 **ruberata** (Freyer, 1831) – *ruber*, red: the 'ruddy highflier'.

Coenocalpe Hübner, 1825 – κοινός (koinos), in common; κάλπις (kalpis), a vessel for drawing water: perhaps indicating affinity with 1777 *Hydriomena*, which it follows, though not immediately, in Hübner's work. Macleod's suggestion 'perhaps from Lat. *coenum*, mud, Gr. *calpe*, rock of Gibraltar, the moth being found on marshes and hills' shows how enigmatic some of Hübner's names can be.

1780 **lapidata** (Hübner, 1809) – *lapis, lapidis*, a stone: either from the ground colour, suggestive of a reddish sandstone, or from the mountain habitat.

Horisme Hübner, 1825 – ὅρισμα (horisma), a boundary: from the well-defined black terminal line on both fore- and hindwing. Macleod describes the line as 'whitish'.

1781 **vitalbata** ([Denis & Schiffermüller], 1775) – *Clematis vitalba*, traveller's-joy, is the foodplant.

1782 **tersata** ([Denis & Schiffermüller], 1775) – *tersus*, wiped clean, neat, wiped off; from the forewing pattern which may be supposed to have been made paler by wiping, but nevertheless is neat and attractive.

1783 **aquata** (Hübner, 1813) – *aqua*, water: perhaps because the pattern looks washed-out, Hübner having the name of the last species in mind.

Melanthia Duponchel, 1829 – μέλας (melas), black; ἄνθος (anthos), blossom, a flower; μελανθής (melanthēs), having black blossoms: a genus containing a black-marked species of great beauty.

1784 **procellata** ([Denis & Schiffermüller], 1775) – *procella*, a storm of rain: from the sprinkling of dark scales on the white ground colour – 'weisser brandschwarzstreisigter sp.' 'From the stormy flight of the male' (Spuler); 'thrown down, from collectors' custom of beating it out of hedges' (Macleod); however Pickard *et al.* got it right.

Pareulype Herbulot, 1951 – παρά (para), beside, signifying affinity; the genus *Eulype* Hübner, 1825, = 1787 *Rheumaptera* – εὖ (eu), well, good; λύπη (lupē), pain, sorrow: from the smart, mourning garb of 1787 *R. hastata*.

1785 **berberata** ([Denis & Schiffermüller], 1775) – *Berberis vulgaris*, barberry, is the foodplant.

Spargania Guenée, 1857 – *Sparganium*, the bur-reed genus but, at any rate in Britain, the next species does not feed on any member.

1786 **luctuata** ([Denis & Schiffermüller], 1775) – *luctus*, mourning: from the black and white pattern of the forewing.

Rheumaptera Hübner, 1822 – ῥεῦμα (rheuma), a stream; πτερόν (pteron), a wing: another variant of Hübner's favourite 'rivulet' theme.

1787 **hastata** (Linnaeus, 1758) – *hasta*, a spear: Linnaeus describes the wings as black with white fasciae which are '*hastato dentatis*', toothed spearwise. Today we would reverse the colours and say that the black terminal fascia on both fore- and hindwing is broken by a spear-shaped patch of the white ground colour.

subsp. **nigrescens** (Prout, 1914) – *nigrescens*, tending to blackness: the smaller, but not necessarily darker, subspecies occurring in Scotland.

1788 **cervinalis** (Scopoli, 1763) – *cervinus*, pertaining to a stag: either from the fawn colour or because the median fascia often forks like an antler towards the costa. The termination *-alis* shows that Scopoli placed it in Pyralis, presumably from the disposition of the wings in repose (see p. 21).

1789 **undulata** (Linnaeus, 1758) – *undulatus*, marked as with waves: from the numerous wavy transverse fasciae. This is the archetypal name derived from the 'rivulet' theme.

Triphosa Stephens, 1829 – τρι- (tri-), three, triply; φάος, φῶς (phaos, phōs), light: from the strong gloss on the wings.

1790 **dubitata** (Linnaeus, 1758) – *dubitatus*, doubtful: '*statura* Geometrae, *magnitudo* Noctuae, *facies* Tineae', the build of a geometrid, the size of a noctuid and the look of a tineid. 'Doubtful, i.e. whether distinct species, being very similar to *Rheumaptera cervinalis*' (Macleod); a double blunder, being also an anachronism, and one which he repeats on p. 10 of his Introduction. 'From erratic flight of moth' (Spuler).

Philereme Hübner, 1825 – φιλέρημος (philerēmos), fond of solitude: perhaps from the larval habit of secluding itself between spun leaves.

1791 **vetulata** ([Denis & Schiffermüller], 1775) – *vetulus*, oldish: because the rather drab, greyish brown wings lack the bloom of youth, or because the transverse fasciae are reminiscent of wrinkles.

1792 **transversata** (Hufnagel, 1767) – *transversus*, going across: from the fasciae, in this species particularly in evidence on the forewing.
subsp. **britannica** Lempke, 1968 – the subspecies occurring in Britain.

Euphyia Hübner, 1825 – εὐφυΐα (euphuia), goodness of shape: laudatory.

1793 **biangulata** (Haworth, 1809) – *biangulatus*, two-angled: from the double dentate process on the distal margin of the median fascia.

1794 **unangulata** (Haworth, 1809) – *unangulatus*, single-angled: from the single dentate process on the distal margin of the median fascia.

Epirrita Hübner, 1822 – ἐπιρρέω (epirrheo), to flow, to stream onwards: another name based on Hübner's favourite concept of 'rivulets'.

1795 **dilutata** ([Denis & Schiffermüller], 1775) – *dilutus*, washed, washed out, pale: from the pale grey, rather colourless forewing of some forms.

1796 **christyi** (Allen, 1906) – i.h.o. W. M. Christy (1863–1939), the British entomologist who took the type material at Rannoch in Scotland.

1797 **autumnata** (Borkhausen, 1794) – from the time of appearance of the adult.

1798 **filigrammaria** (Herrich-Schäffer, 1846) – *filum*, a thread; *gramma* (γραμμή), a line: from the relatively narrow fasciae.

Operophtera (Hübner, 1825) – a typographical error for *Oporophthera*, ὀπώρα (opōra), fruit; φθείρω (phtheirō), to destroy: from the damage done to fruit-trees by the larva of the winter moth.

1799 **brumata** (Linnaeus, 1758) – *bruma*, a contraction of *brevissima*, shortest, i.e. the shortest day and so winter in general: from the time of appearance of the adult.

1800 **fagata** (Scharfenberg, 1805) – *Fagus*, the genus of the beech: the larva, however, feeds mainly on birch and rosaceous trees.

Perizoma Hübner, 1825 – περίζωμα (perizōma), a girdle (περί (peri), round; σῶμα (sōma), the body): from the resemblance of the fascia or fasciae on each forewing to a belt or belts when the wings are folded. It is odd that Hübner's favourite 'rivulet' theme is absent from his name for the genus that includes the species with our vernacular name 'rivulet'.
= **Mesotype** Hübner, 1825 – μέσος (mesos), middle; τύπος (tupos), the mark of a blow, a marking: from the median fascia of some species. See 1718, genus.

1801 **taeniata** (Stephens, 1831) – *taenia*, a ribbon, a fillet: from the fascia on the forewing.

1802 **affinitata** (Stephens, 1831) – *affinitatus*, related: from affinity to the next species from which it had not previously been separated.

1803 **alchemillata** (Linnaeus, 1758) – '*habitat in* Alchemilla', the lady's-mantle genus: not the correct foodplant.

1804 **bifaciata** (Haworth, 1809) – a typographical error for *bifasciata*, *bi-*, two; *fascia*, a band: from the two pale bands bounding the dark median fascia in one form; Haworth named another form of this species *unifasciata*.

1805 **minorata** (Treitschke, 1828) – *minor*, smaller: the smallest British member of the genus.
subsp. **ericetata** (Stephens, 1831) – *Erica*, the heath genus, *ericetum*, an area where it grows, a heath: from the habitat.

1806 **blandiata** ([Denis & Schiffermüller], 1775) – *blandus*, pleasant: the 'pretty pinion'.
subsp. **perfasciata** (Prout, 1914) – *per*, through, very; *fasciatus*, marked with a band: the subspecies found in the Hebrides which has the median fascia complete, as opposed to being reduced to a costal blotch.

1807 **albulata** ([Denis & Schiffermüller], 1775) – *albulus*, whitish: from the ground colour.
subsp. **subfasciaria** (Boheman, 1852) – *sub-*, under-; *fascia*, a band: the almost unmarked, 'underfasciated' subspecies occurring in Shetland.

1808 **flavofasciata** (Thunberg, 1792) – *flavus*, yellow; *fascia*, a band: from the yellowish fasciae on the forewing.

1809 **didymata** (Linnaeus, 1758) – *didymus* (δίδυμος), twin, double: the 'twin-spot' carpet; Linnaeus describes the spot as single but bilobed.
subsp. **hethlandica** (Rebel, 1910) – the heathland subspecies occurring in Shetland; cf. 1723 *Xanthorhoe munitata hethlandica*, which probably influenced this name.

1810 **sagittata** (Fabricius, 1787) – *sagitta*, an arrow: the median fascia is strongly dentate on its distal margin, giving the appearance of an arrowhead.

Eupithecia Curtis, 1825 – εὖ (eu), well, good, goodly; πίθηξ, πίθηκος (pithēx, pithēkos), a dwarf: from the attractive appearance and small size of the moths. 'These pretty moths form a most natural genus and when alive are characterized (as Mr Haworth has observed) by the elegant attitude in which they repose, with their wings beautifully expanded, lying close to the surface on which they rest, as moths are displayed for our cabinets by the London collectors'. Authors have derived the name from πίθηκος, πιθήκου (pithēkos, pithēkou), an ape, but Curtis gives no hint that this was his intention and I have quoted him *in extenso* because I believe that such an idea was far from his thoughts. For those who would nevertheless prefer this derivation, it can be explained as follows. Haworth (1809: 358) stated that the moths were called pugs because the hindwing is shorter than the forewing, just as the lower lip of a pug-dog is shorter than its upper lip. The word 'pug', originally meaning a goblin, can be applied to a monkey as well as a dog, and Curtis deliberately changed animals. I consider this very unlikely. In his section entitled *Common Names* Macleod wrote 'Pug, here used in the original sense of "monkey" and therefore merely translation of scientific name *Eupithecia*'. 'Pug' is in fact the senior name by at least 16 years and 'monkey' is not its original sense.

1811 **tenuiata** (Hübner, 1813) – *tenuis*, slender: the 'slender pug'.

1812 **inturbata** (Hübner, 1817) – *inturbatus*, not confused: from the relatively well-defined wing-pattern. 'Undisturbed, but actually the moth gets very agitated if molested' (Macleod).

1813 **haworthiata** Doubleday, 1856 – i.h.o. A. H. Haworth (1767–1833), the British entomologist and author of *Lepidoptera Britannica*.

1814 **plumbeolata** (Haworth, 1809) – *plumbeolus*, dim. of *plumbeus*, leaden: the 'lead-coloured pug'.

1815 **abietaria** (Goeze, 1781) – *Picea abies*, Norway spruce: the larva feeds in the cones.

1816 **linariata** ([Denis & Schiffermüller], 1775) – *Linaria*, the toadflax genus: the 'toadflax pug'.

1817 **pulchellata** Stephens, 1831 – *pulchellus*, pretty little: the moth is one of the more colourful members of the genus.

subsp. **hebudium** Sheldon, 1899 – the subspecies first found in the Hebrides (Lat. *Ebudes*, the Inner Hebrides).

1818 **irriguata** (Hübner, 1813) – *irriguatus*, well-watered: from the 'rivulets' (Hübner's favourite imagery) forming the dorsal half of the median fascia.

1819 **exiguata** (Hübner, 1813) – *exiguus*, very small: not small among pugs, but small among the members of the genus in which Hübner placed it (*Geometra*).

subsp. **muricolor** Prout, 1938 – *mus, muris*, a mouse; *color*, colour: from the grey colour of the subspecies occurring in Aberdeenshire.

1820 **insigniata** (Hübner, 1790) – *insignis*, marked, and thereby distinguished: from the sharply defined, striking pattern of the forewing which renders the moth noteworthy.

1821 **valerianata** (Hübner, 1813) – *Valeriana officinalis*, common valerian: the larval foodplant.

1822 **pygmaeata** (Hübner, 1799) – *pygmaeus*, a pygmy: one of smallest members of the genus in which Hübner placed it (*Geometra*).

1823 **venosata** (Fabricius, 1787) – *venosus*, marked with veins: from the network of black markings which are vein-like but do not correspond with the wing venation.

subsp. **hebridensis** Parkinson Curtis, 1944 – the subspecies occurring in the Hebrides.

subsp. **fumosae** Gregson, 1887 – *fumosus*, smoky: the dark race occurring in Orkney and Shetland.

subsp. **ochracae** Gregson, 1886 – *ochraceus*, ochreous: a so-coloured form also found in Orkney.

subsp. **plumbea** Huggins, 1962 – *plumbeus*, leaden: a lead-grey race occurring in the west of Ireland.

1824 **egenaria** Herrich-Schäffer, 1848 – *egenus*, poor, needy: a poorly marked species (the 'pauper pug').

1825 **centaureata** ([Denis & Schiffermüller], 1775) – *Centaurea*, the knapweed genus, which includes some of the many foodplants.

1826 **trisignaria** Herrich-Schäffer, 1848 – *tri-*, three; *signum*, a mark: from the black discal spot and two dark clouds on the costa, vestiges of the usual transverse fasciae (the 'triple-spotted pug').

1827 **intricata** (Zetterstedt, 1839) – *intricatus*, tangled, intricate: from the above-average number of fasciae on the forewing. 'Entangled, from the various races included in the species' (Macleod); unsound entomology. The recognition that a species has races does not precede but follows its naming, and in any case the races are generally at first supposed to be specifically distinct.

subsp. **arceuthata** (Freyer, 1842) – ἄρκευθος (arkeuthos), a juniper bush: the larval foodplant. Freyer regarded it as a distinct species.

subsp. **millieraria** Wnukowsky, 1929 – i.h.o. P. Millière (1811–87), a French entomologist. This Scottish subspecies had been misidentified as *helveticaria* Boisduval, 1840. Millière detected the mistake and bestowed the name *anglicata* on our moth in 1872. When it was realized that this name was preoccupied by *E. pusillata anglicata* Herrich-Schäffer, 1863, Wnukowsky renamed the subspecies in honour of the man who had first realized that it was not *helveticaria*.

subsp. **hibernica** Mere, 1964 – the subspecies found in the Burren, western Ireland.

1828 **satyrata** (Hübner, 1813) – *satyrus*, a satyr, one of the woodland deities associated with the worship of Bacchus: from one of the habitats in which it is found. *Satyrus* can also mean a small ape, possibly the chimpanzee, and so Macleod has 'a kind of ape, for the same reason as generic name'; this name was bestowed 12 years before the generic name, so there obviously is no connection. See also generic name.

subsp. **callunaria** Doubleday, 1850 – the subspecies found on northern moorland where heather (*Calluna vulgaris*) grows.

subsp. **curzoni** Gregson, 1884 – the subspecies found in Orkney and Shetland and named i.h.o. E. R. Curzon, who had taken it on Hoy.

1829 **cauchiata** (Duponchel, 1831) – καύχη (kaukhē), a boast, a thing to boast about: a handsome species Duponchel was proud to name.

1830 **absinthiata** (Clerck, 1759) – *Artemisia absinthium*, wormwood: a larval foodplant.

1831 f. **goossensiata** Mabille, 1869 – i.h.o. T. Goossens (1827–89), French naturalist. Lhoste (1987) spells his name 'Goosens'.

1832 **assimilata** Doubleday, 1856 – *assimilis*, similar: i.e. to the last species.

1833 **expallidata** Doubleday, 1856 – *expallidus*, exceedingly pale: the 'bleached pug'.

1834 **vulgata** (Haworth, 1809) – *vulgus*, common: the 'common pug'.

subsp. **scotica** Cockayne, 1951 – the subspecies occurring in Scotland.

subsp. **clarensis** Huggins, 1962 – the subspecies occurring in Co. Clare, western Ireland.

1835 **tripunctaria** Herrich-Schäffer, 1852 – *tri-*, three; *punctum*, a spot: from the subterminal white line which is often broken into three spots.

1836 **denotata** (Hübner, 1813) – *denotatus*, distinct, marked out: deemed distinct from related species rather than distinctly marked.

subsp. **jasioneata** Crewe, 1881 – *Jasione montana*, sheep's-bit: the foodplant of this mainly coastal subspecies.

1837 **subfuscata** (Haworth, 1809) – *subfuscus*, somewhat dusky: from the ground colour (the 'grey pug').

1838 **icterata** (Villers, 1789) – ἴκτερος (ikteros), *icterus*, a yellow bird, the sight of which was said to cure jaundice, whereupon the bird died; jaundice itself: from the tawny ground colour of the forewing.

subsp. **subfulvata** (Haworth, 1809) – *sub*, somewhat; *fulvus*, tawny: from the ground colour of the forewing; the 'tawny-speckled pug'.

subsp. **cognata** Stephens, 1831 – *cognatus*, related, i.e. to *E. sobrinata* (Hübner, 1817), now 1854 *E. pusillata*. Macleod, who had not consulted the source, thought relationship to another subspecies was intended. For the meaning of *sobrinata* see 2116 *sobrina*.

1839 **succenturiata** (Linnaeus, 1758) – *succenturiatus*, a recruit received into the century (a company of 100 men in the Roman army) in place of another man, a substitute in general: from the tegulae which appeared to Linnaeus as miniature wings, '*thorax albus ad latus tectus duabus valvulis, alarum rudimentis, basi alarum anteriorum incumbentibus*', thorax white, covered laterally by two tegulae (*valvulae*), the rudiments of wings, lying over the base of the forewings. These rudimentary 'wings' were the young recruits joining the ranks of the veteran wings proper. The tegulae are less strongly developed than those of 1424 *Endotricha* (q.v.) *flammealis*, but are notable in that they terminate in a pecten of long, setose hair-scales. 'From resemblance to *E. centaureata*' (Macleod): completely wrong, an imputation of bad spelling to Linnaeus and an anachronism by seventeen years.

1840 **subumbrata** ([Denis & Schiffermüller], 1775) – *sub*, somewhat; *umbratus*, shaded: the 'shaded pug'.

1841 **millefoliata** Rössler, 1866 – *Achillea millefolium*, yarrow: the larval foodplant.

1842 **simpliciata** (Haworth, 1809) – *simplex, simplicis*, simple: from the lightly marked pattern (the 'plain pug').

1843 **distinctaria** Herrich-Schäffer, 1848 – *distinctus*, distinct: i.e. from 1845 *E. pimpinellata*, 'Es ist kaum möglich, dass das hier gelieferte Thier zu *pimpinellaria* [sic] gehort'. 'Divided, from division of forewings by three cross lines' (Macleod); Spuler interprets as distinguished, adorned. Neither had consulted Herrich-Schäffer.

subsp. **constrictata** Guenée, 1857 – *constrictus*, narrowed: from the slenderness of the forewing fasciae.

1844 **indigata** (Hübner, 1813) – *indiges*, poor, needy: from the forewing which is bereft of almost all markings.

1845 **pimpinellata** (Hübner, 1813) – *Pimpinella saxifraga*, burnet saxifrage, is the larval foodplant.

1846 **nanata** (Hübner, 1813) – *nanus*, a dwarf: from the small size.

subsp. **angusta** Prout, 1938 – *angustus*, narrow: the 'narrow-winged pug'.

1847 **extensaria** (Freyer, 1844) – *extensus*, stretched out: from the elongate forewing.

subsp. **occidua** Prout, 1914 – *occiduus*, western: the subspecies occurring in western Europe.

1848 **innotata** (Hufnagel, 1767) – *innotatus*, unmarked: this species does not occur in Britain.

1849 **fraxinata** Crewe, 1863 – *Fraxinus excelsior*, the ash, is a foodplant.

1850 **tamarisciata** (Freyer, 1836) – *Tamarix anglica*, tamarisk, is the foodplant of this doubtfully British species.

1851 **virgaureata** Doubleday, 1861 – *Solidago virgaurea*, goldenrod, is the foodplant.

1852 **abbreviata** Stephens, 1831 – *abbreviatus*, shortened: from the hindwing which is short in comparison to the forewing; Haworth (1809) had given the title '*Abbreviatae*' to the section of the Geometridae in which he placed the pugs and Stephens also gives this as one of the main characters of the genus. See also Haworth's explanation of the name 'pug' given under 1811 *Eupithecia*. 'From interrupted bars across forewings' (Macleod); another incorrect guess.

1853 **dodoneata** Guenée, 1857 – Dodona, a town of Epirus, in north-western Greece, where there was an ancient oracle. The oracular responses were given by the wind blowing through a sacred grove of oak-trees, amplified by brazen vessels suspended from the branches. Hence *dodoneus* came to mean belonging to the oak, which is one of the larval foodplants of this species. Cf. 124 *dodonea* and 2014 *dodonaea*.

1854 **pusillata** ([Denis & Schiffermüller], 1775) – *pusillus*, very small, petty: from small size compared with neighbouring species, all, in 1775, placed in Geometra.

subsp. **anglicata** Herrich-Schäffer, 1863 – the subspecies formerly occurring in England, where it was confined to the Dover area.

1855 **phoeniceata** (Rambur, 1834) – *Juniperus phoenicea* is a foodplant on the Continent.

1855a **ultimaria** Boisduval, 1840 – *ultimus*, last, furthest: perhaps with reference to the terminal cilia of the forewing which are checkered light and dark brownish grey, more conspicuously on the underside; cf. 671 *ultimella*, a name which certainly refers to the terminal area.

1856 **lariciata** (Freyer, 1842) – *Larix decidua*, the European larch, is the foodplant.

1857 **tantillaria** Boiduval, 1840 – *tantillus*, so little: a smallish species, even for a pug.

Chloroclystis Hübner, 1825 – χλωρός (khlōros), pale green; κλυστ-, from the root of κλύζω (kluzō), to wash, wash away: from the fugitive green coloration of the species in the genus. Cf. 1760 *Chloroclysta*.

1858 **v-ata** (Haworth, 1809) – from the prominent V-shaped black subcostal mark on the forewing; cf. 409 *ivella* on the same author.

1859 **chloerata** (Mabille, 1870) – χλόη (khloē), the pale green of spring vegetation: from the colour of the forewing.

1860 **rectangulata** (Linnaeus, 1758) – *rectangulus* (late Lat.). rectangular: *'alae subtus fascia rectangula e punctis nigricantibus'*, the wings having on the underside a rectangular fascia formed of blackish dots; not on the upperside as implied by Macleod. Linnaeus describes the upperside as 'subfasciated'.

1861 **debiliata** (Hübner, 1817) – *debilis*, weakened: from the pale green of the forewing which quickly fades to greyish white.

Gymnoscelis Mabille, 1868 – γυμνός (gumnos), naked; σκέλος (skelos), the leg: referring here to the posterior tibia which is devoid of middle spurs.

1862 **rufifasciata** (Haworth, 1809) – *rufus*, red; *fascia*, a band: from the reddish fasciae on the forewing.

Anticollix Prout, 1938 – ἀντι- (anti-), instead of; the genus *Collix* Guenée, 1857, to which the next species had been wrongly assigned. *Collix* (κολλίξ) means a roll or loaf of coarse bread: 'M. Guenée confesses this name to be "sans étymologie"' (Pickard *et al.*, addenda).

1863 **sparsata** (Treitschke, 1828) – *sparsus*, bespattered: from the black and white irroration, especially along the veins of the forewing. 'Spread out, from ample forewings' (Macleod); possible, but a rare meaning of *sparsus*.

Chesias Treitschke, 1825 – Chesias, stated by Treitschke to be a title of Diana. It is derived from the promontory of Chesium on the island of Samos, where the goddess had a temple.

1864 **legatella** ([Denis & Schiffermüller], 1775) – *legatus*, an ambassador, a provincial governor. An ambassador or governor of senatorial rank would wear the *latus clavus* (broad purple stripe) on his toga and the most notable feature of this species is the broad stripe (the 'streak') extending towards the apex, as described by Denis and Schiffermüller. They give this species the tineid termination '*-ella*' and place it among the addenda at the end. Their uncertainty arose from the moth's habit of resting with unextended overlapping wings, wholly untypical of a geometrid; see p. 21 and MBGBI 7 (2): Pl. C, fig. 12.

1865 **rufata** (Fabricius, 1775) – *rufus*, red; from the reddish postdiscal suffusion.
subsp. **scotica** Richardson, 1952 – the subspecies occurring in Scotland.

Carsia Hübner, 1825 – κάρσιος (karsios), a word occurring mainly in compounds, across, athwart: from the prominent transverse fasciae of the next species.

1866 **sororiata** (Hübner, 1813) – *soror*, a sister: sistered, from resemblance to the next species which is placed by Hübner together with this one in *Carsia*.
subsp. **anglica** Prout, 1937 – the subspecies occurring in England.

Aplocera Stephens, 1827 – ἅπλως (haplōs), simple; κέρας (keras), a horn: having the antenna simple (to the naked eye). As in 653 *Aplota*, Stephens has missed the rough breathing (the accent indicating an aspirate).

1867 **plagiata** (Linnaeus, 1758) – *plaga*, a stripe: from the conspicuous fasciae on the forewing.
subsp. **scotica** Richardson, 1952 – the subspecies occurring in Scotland.

1868 **efformata** (Guenée, 1857) – *ex*, out of, implying divergence; *formatus*, formed: differently formed from the last species in respect to the abdomen.

1869 **praeformata** (Hübner, 1826) – *prae-*, in a high degree; *formatus*, well fashioned: from the sharply defined wing pattern; this name may have influenced the last one.

Odezia Boisduval, 1840 – ὁδός (hodos), a road; ἕζομαι (hexomai), to sit; from the next species' habit of settling on damp, bare ground. The dropping of the 'aitch' is normal with French authors. 'Perhaps from the type locality Odessa' (Macleod): genera have type species, not type localities.

1870 **atrata** (Linnaeus, 1758) – *ater*, black: from the ground colour. The type locality is Europe (see genus).

Lithostege Hübner, 1825 – λίθος (lithos), a stone; στέγη (stegē), a roof, a covering: from the next species which has an irroration of grey scales, giving the aspect of grey stone.

1871 **griseata** ([Denis & Schiffermüller], 1775) – *griseus* (late Lat.), grey: from the ground colour.

Discoloxia Warren, 1895 – δίσκος (diskos), a round plate, a quoit, the disc of an insect's wing;

λοξός (loxos), slanting: from the prominent black margin of the median fascia which is outwards oblique in the costal half of the forewing of the next species.

1872 **blomeri** (Curtis, 1832) – i.h.o. Captain C. Blomer (d. 1835), the British collector who reared the first specimen from an unstated locality in 1830. The specimens mentioned by Macleod as having been taken in Durham in 1832 were not part of Blomer's type material. He lived at Cheltenham.

Venusia Curtis, 1839 – Venus, goddess of love and beauty: Curtis describes the next species, which is the type species, as a 'pretty little moth'.

1873 **cambrica** Curtis, 1839 – *Cambria*, Latinized from Welsh Cymru, Wales: the type locality is Hafod in Cardiganshire.

Euchoeca Hübner, 1823 – εὖ (eu), well, good; χοϊκός (khoikhos), of earth or clay: from the ground colour of the next species which is of a 'goodly' clay colour.

1874 **nebulata** (Scopoli, 1763) – *nebula*, a cloud: from the posterior fuscous suffusion on the forewing.

Asthena Hübner, 1825 – ἀσθενής (asthenēs), weak: from the indistinct pattern or the delicate structure of the next species.

1875 **albulata** (Hufnagel, 1767) – *albulus*, whitish: from the ground colour.

Hydrelia Hübner, 1825 – ὑδρηλός (hudrēlos), watery: probably from the 'rivulets' on the forewing of 1877 *H. sylvata*, the 'waved carpet', the only species included by Hübner in his genus, rather than the habitat (Macleod), which is unapt.

1876 **flammeolaria** (Hufnagel, 1767) – *flammeolus*, flame-coloured: from the ochreous orange striae on the wings.

1877 **sylvata** ([Denis & Schiffermüller], 1775) – *silva* (*sylva*), a wood: from the habitat.

Minoa Treitschke, 1825 – *Minous*, Cretan: no entomological application.

1878 **murinata** (Scopoli, 1763) – *murinus*, mouse-coloured: from the ground colour.

Lobophora Curtis, 1825 – λοβός (lobos), a lobe, especially that of the ear; φορέω (phoreō), to bear: from the lobe on the dorsum of the male hindwing.

1879 **halterata** (Hufnagel, 1767) – *halter*, a dumb-bell used by gymnasts: from the lobe on the male hindwing. The modified hindwings of Diptera are termed halteres.

Trichopteryx Hübner, 1825 – τρίχους (trikhous), holding three; πτέρυξ (pterux), a wing: from the lobe on the male hindwing, which resembles a third wing. Macleod incorrectly derives from θρίξ (thrix), hair, saying that the hairs are the fasciae on the forewing.

1880 **polycommata** ([Denis & Schiffermüller], 1775) – πολύς (polus), many; κόμμα (komma), the stamp or impression of a coin, a mark, a comma: from the markings of the forewing, which include numerous black streaks on the veins.

1881 **carpinata** (Borkhausen, 1794) – *Carpinus*, the hornbeam genus, though this is not the usual foodplant.

Pterapherapteryx Curtis, 1825 – πτερόν (pteron), a feather, a wing; φέρω (pherō), to carry; πτέρυξ (pterux), a wing: 'wing-bearing-wing', from the lobe, resembling an ancillary wing, borne by the male hindwing. Curtis probably pronounced the name 'Ptéra-phéra-ptéryx'.

1882 **sexalata** (Retzius, 1783) – *sex*, six; *ala*, a wing: from the lobe on the male hindwing, which appears to be an ancillary wing.

Acasis Duponchel, 1845 – ἀ-, alpha privative; κάσις (kasis), a brother: brotherless, because the genus was monotypic when first erected, but this is no longer the case.

1883 **viretata** (Hübner, 1799) – *viretum*, a green spot, greenness in general: from the ground colour. Macleod's explanation 'grassy place, from larva's habit, when young, of feeding on flowers' is wrong because the flowers it eats are those of shrubs, not of herbaceous plants.

ENNOMINAE (1911)

Abraxas Leach, 1815 – a Coptic word, said to have been coined by the Egyptian Gnostic Basilides (2nd century A.D.) to express 365 (the number of days in the year) by the addition of the numerical values of the Greek letters. Gems to be used as charms were inscribed with the word together with mystical figures of combined animal and human forms. The conjurors' word 'abracadabra' is from the same source. There is no entomological significance.

1884 **grossulariata** (Linnaeus, 1758) – *Ribes grossularia* (now R. *uva-crispa*), the gooseberry: one of the foodplants.

1885 **sylvata** (Scopoli, 1763) – *silva, sylva*, a wood: from the habitat.

1886 **pantaria** (Linnaeus, 1767) – *panthera*, a panther: from the spotted pattern.

Lomaspilis Hübner, 1825 – λῶμα (lōma), a hem, a fringe; σπίλος (spilos), a spot, a stain: from the 'clouded border' of the next species.

1887 **marginata** (Linnaeus, 1758) – *margo, marginis*, a margin, a border: from the dark termen of both fore- and hindwing – 'the clouded border'.

Ligdia Guenée, 1857 – perhaps without meaning. Macleod's tentative derivation from λίζω (lizō), to scrape, from the appearance of the underside of the wings of the next species, is unconvincing.

1888 **adustata** ([Denis & Schiffermüller], 1775) – *adustus*, scorched: 'the scorched carpet'; the 'scorching' is more evident on the underside which is suffused with reddish brown.

Stegania Guenée, 1845 – στεγανός (steganos), closely covered, sheathed: perhaps because the larva feeds concealed in a spinning.

1888a **trimaculata** (Villers, 1789) – *tri-*, three; *macula*, a spot: from the pattern of the forewing.

Semiothisa Hübner, 1818 – σημεῖον (sēmeion), a mark, a character; ὠθίζω (ōthizō), to thrust, struggle: Hübner's diagnosis stresses the alternation of bands of contrasting colours and the idea may be that they vie with each other for dominance; cf. 1890 *alternaria*.

1889 **notata** (Linnaeus, 1758) – *nota*, a mark, letter or character: the 'mark' is the postdiscal blackish spot, characteristically cut by the veins into several components; Linnaeus says there are four and that they resemble fly frass.

1890 **alternaria** (Hübner, 1809) – *alternus*, alternate: from the alternation in colour between the darker fasciae and the paler ground colour. Macleod's explanation 'from larva's change of colour before pupation' is pure guesswork: Hübner does not mention and probably did not know the life history. See genus above.

1891 **signaria** (Hübner, 1809) – *signum*, a mark: from the dark subquadrate spot on the postdiscal fascia. The name is probably influenced by 1889 *notata*, q.v.

1892 **praeatomata** (Haworth, 1809) – *prae-*, in a high degree; *atomatus*, an adj. coined from *atomus* (ἄτομος), an atom, to mean speckled: from the numerous fuscous dots on the forewing.

1893 **liturata** (Clerck, 1759) – *litura*, a blot or smear: from the suffused subterminal fascia.

1894 **clathrata** (Linnaeus, 1758) – *clathratus*, furnished with a grate or lattice, set with bars: the 'latticed heath'.

subsp. **hugginsi** (Baynes, 1959) – the Irish subspecies named i.h.o. H. G. Huggins (1891–1977), an English collector who also studied Irish Lepidoptera. Huggins had already named a subspecies i.h.o. E. S. A. Baynes (1532 *Erynnis tages baynesi*).

1895 **carbonaria** (Clerck, 1759) – *carbonarius*, pertaining to charcoal: from the blackish pattern, accentuated by the white ground colour.

1896 **brunneata** (Thunberg, 1784) – *brunneus* (late Lat.), brown: from the ground colour.

1897 **wauaria** (Linnaeus, 1758) – *vau*, the word for the letter 'v' (vee); 'wau' may be understood as the double 'v' or double-you: *'alis . . . anticis fasciis quatuor nigris abbreviatis inaequalibus'*, forewings with four truncate black fasciae of differing length. Only one is V-shaped but the point seems to be that the four costal marks are like the tops of the two letters.

Hypagyrtis Hübner, 1818 – probably malformed from ὑπαργυρίζω (huppargurizō), to be silver-grey, referring to *Erastria pustularia* Hübner (non-British), originally the only species in the genus.

1898 **unipunctata** (Haworth, 1809) – *unus*, one; *punctum*, a spot: from a white subapical spot.

Isturgia Hübner, 1823 – ἱστουργία (histourgia), weaving: the application is not clear; perhaps from the speckling of black on the wings of the next species, the 'frosted orange'. 'From marking on wings of [1895] *I. carbonaria*' (Macleod): a bad guess, since the next species was the only one included by Hübner in his genus.

1899 **limbaria** (Fabricius, 1775) – *limbatus*, bordered: from the fuscous costa and termen which contrast with the orange ground colour.

Nematocampa Guenée, 1857 – νῆμα, νήματος (nēma, nēmatos), a thread; κάμπη (kampē), a larva; καμπή (kampē), a bending: probably a pun, signifying a slender looper caterpillar.

1900 **limbata** (Haworth, 1809) – *limbatus*, bordered: the 'bordered chequer'.

Cepphis Hübner, 1823 – κέπφος (kepphos), a small sea-bird (? a stormy petrel), a simpleton, a 'gull': probably without entomological meaning.

1901 **advenaria** (Hübner, 1790) – *advena*, a stranger, a visitor: perhaps referring to the origin of the specimen figured by Hübner; a wide circle of friends and correspondents sent him material from diverse localities.

Petrophora Hübner, 1811 – πέτρος (petros), a rock; φορέω (phoreō), to bear: from the stone-coloured fasciae on the forewing of the next species, to which he had given the name *petraria* in 1799 (a junior synonym). Later, in 1821, he renamed the genus *Ortholitha* (another junior synonym) from ὀρθός (orthos), straight, and λίθος (lithos), a stone, using the same imagery and referring to the straight forewing fascia.

1902 **chlorosata** (Scopoli, 1763) – χλωρός (khlōros), pale green or simply pale: from the ground colour. Scopoli describes the forewing as *'albida, seu colore eodem fere ut vultus in* Cachexia virginea', white, or almost the colour of the face in *Cachexia virginea*, a malignant wasting illness often accompanied by a yellowish pallor of the skin. Scopoli trained and practised as a doctor before becoming a university professor and specializing in natural history and geology. He appears to have been the first entomologist to apply the term 'diagnosis' to the description of a species.

Plagodis Hübner, 1823 – πλάγιος (plagios), slanting; εἶδος (eidos), shape: from the termen of the forewing of 1904 *dolabraria*, the only species included by Hübner, which is angled inwards dorsad of a central projection. There is no reference to 'creased wings' as supposed by Macleod; he may have been muddling *P. dolabraria* with 1910 *Apeira syringaria*.

1903 **pulveraria** (Linnaeus, 1758) – *pulvis, pulveris*, dust: from the powdering of reddish fuscous scales over the wings.

1904 **dolabraria** (Linnaeus, 1767) – *dolabra*, a pickaxe: a name that has puzzled entomologists. 'From shape of markings on wings' (Pickard *et al.*); from fanciful resemblance of tornal blotch on each wing to axe-head' (Macleod); 'P. [Phalaena] ailes en doloire', geometrid with wings in the form of a pickaxe (Latreille, 1804). Latreille's interpretation is correct. Linnaeus did not set the specimens in his collection and so saw them in their natural resting position, and it is to the shape of the wings in repose and not their pattern that the name refers. If one looks at the photograph by M. W. F. Tweedie (MBGBI 7(2), Pl. C, fig. 8) and in imagination extends the abdomen to a ridiculous length to represent the haft, the axe-head is apparent, the cutting edges being formed by the squared upper half of the termen of each wing.

Pachycnemia Stephens, 1829 – παχύς (pakhus), thick; κνήμη (knēmē), the shin, the tibia of an insect's leg: from the male hindtibia, which is dilated.

1905 **hippocastanaria** (Hübner, 1799) – *Aesculus hippocastanum*, the horse chestnut: the choice of name is obscure because the foodplant was then unknown and *Aesculus* is seldom utilized by Lepidoptera; Hübner could, however, have been misinformed. The grey forewings sometimes have a faint purplish flush, but are never chestnut-coloured. Later, Duponchel was to suppose wrongly that the moth was associated with sweet chestnut (*Castanea sativa*).

Opisthograptis Hübner, 1823 – ὄπισθεν (opisthen), behind, at the back; γραπτός (graptos), painted, marked with letters: Hübner seems to have understood ὄπισθεν to mean 'on', not 'at', the back, a feature of the next species being that the red-brown subapical markings of the forewing are as strongly expressed on the underside as on the upperside.

1906 **luteolata** (Linnaeus, 1758) – *luteolus*, yellowish: from the ground colour of both fore- and hindwing.

Epione Duponchel, 1829 – the name of the wife of Aesculapius, the 'blameless physician' of Homer, later regarded as the god of medicine; Aesculapius and Epione were the parents of Machaon (1539) and Podalirius (1540).

1907 **repandaria** (Hufnagel, 1767) – *repandus*, bent back: from the sharply angled antemedian fascia.

1908 **paralellaria** ([Denis & Schiffermüller], 1775) – a typographical error for *parallelaria*; *parallelus* (παράλληλος), parallel: from the subterminal line which is parallel to the termen; Denis and Schiffermüller had named the previous species *apiciaria* because in that species the subterminal line is directed towards the apex, but their name of 1775 is a junior synonym of Hufnagel's of 1767 and may not be used.

Pseudopanthera Hübner, 1823 – ψεῦδος (pseudos), false; πάνθηρ, a panther, a leopard: from the black spots on a yellow ground colour.

1909 **macularia** (Linnaeus, 1758) – *macula*, a spot: the 'speckled yellow'.

Apeira Gistl, 1848 – ἄπειρος (apeiros), without end: from the median and postmedian fasciae which do not extend to the costa but end abruptly on reaching the subapical purple patch.

1910 **syringaria** (Linnaeus, 1758) – *Syringa vulgaris*, lilac, cited by Linnaeus as a foodplant.

Ennomos Treitschke, 1825 – ἔννομμος (ennomos), lawful, within the law, translated by Treitschke 'rechtmässig', legal: perhaps he regarded this genus as containing the archetypal geometrids (it has given its name to the subfamily), the other genera being 'the lesser breeds without the law'.

1911 **autumnaria** (Werneburg, 1859) – *autumnus*, autumn: the flight period of the adult extends into early autumn.

1912 **quercinaria** (Hufnagel, 1767) – *Quercus*, the oak genus: a larval foodplant.

1913 **alniaria** (Linnaeus, 1758) – *Alnus*, the alder genus: correctly given as a foodplant by Linnaeus.

1914 **fuscantaria** (Haworth, 1809) – *fuscans*, becoming dusky: from the purplish fuscous suffusion in the subterminal area of the forewing.

1915 **erosaria** ([Denis & Schiffermüller], 1775) – *erosus*, gnawed away, eroded: from the dentate termen, with areas 'bitten' out.

1916 **quercaria** (Hübner, 1813) – *Quercus*, the oak genus: a larval foodplant.

Selenia Hübner, 1823 – σελήνη (selēnē), the moon: from the lunate discal spots.

1917 **dentaria** (Fabricius, 1775) – *dens, dentis*, a tooth: from the dentate termen.

1918 **lunularia** (Hübner, 1788) – *lunula*, a little moon: from the lunate discal spots.

1919 **tetralunaria** (Hufnagel. 1767) – *tetra-*, four; *luna*, the moon: from the four lunate discal spots, one on each wing.

Odontopera Stephens, 1831 – ὀδούς, ὀδόντος (odous, odontos), a tooth; πέρας (peras), the end, strictly in time but here loosely in position, the termen: from the dentate wing margins.

1920 **bidentata** (Clerck, 1759) – *bidens*, having two teeth: from the dentate processes on the termen, two of which may be regarded as larger than the rest.

Crocallis Treitschke, 1825 – κροκός, the crocus, its yellow colour; κάλλος (kallos), beauty; *crocallis*, an unknown precious stone mentioned by Pliny may also have been in Treitschke's mind: from the yellow ground colour of the next species.

1921 **elinguaria** (Linnaeus, 1758) – *elinguis*, speechless, without a tongue: the haustellum is degenerate.

Ourapteryx Leach, 1814 – οὐρά (oura), a tail; πτέρυξ (pterux), a wing: from the tail-like projection on the hindwing of the next species.

1922 **sambucaria** (Linnaeus, 1758) – *Sambucus*, the elder genus: elder is a foodplant, as stated by Linnaeus.

Colotois Hübner, 1823 – κόλος (kolos), docked, stunted; ὠτώεις (ōtōeis), with ears or handles: probably from the greatly reduced labial palpus, now regarded as an organ of scent, but perhaps as one of hearing by Hübner. Macleod suggests a typographical error for *Calotois*, 'from *calos*, beautiful, *ous* (gen. *otos*), ear; from moth's ear-like tufts'. He does not say what these 'tufts' are. Errors in spelling were generally corrected by Hübner or Geyer in later publications, but not in this case.

1923 **pennaria** (Linnaeus, 1761) – *penna*, a feather: from the strongly bipectinate, and so feather-like, antenna.

Angerona Duponchel, 1829 – the name of the goddess of silence.

1924 **prunaria** (Linnaeus, 1758) – *Prunus spinosa*, blackthorn: given by Linnaeus as the foodplant.

Apocheima Hübner, 1825 – ἀπό (apo), from, after; χεῖμα (kheima), winter: a genus for moths appearing in February.

1925 **hispidaria** ([Denis & Schiffermüller], 1775) – *hispidus*, shaggy; from the hair-scales on the abdomen.

1926 **pilosaria** ([Denis & Schiffermüller], 1775) – *pilosus*, shaggy; from the hair-scales on the abdomen.

Lycia Hübner, 1825 – probably λύκειος (lukeios), pertaining to a wolf, shaggy like a wolf: from the hairy abdomen; possibly from Lycia, the country in Asia Minor. The former interpretation is more in keeping with Hübner's policy in coining generic names.

1927 **hirtaria** (Clerck, 1759) – *hirtus*, shaggy: from the hairy thorax and abdomen.

1928 **zonaria** ([Denis & Schiffermüller], 1775) – *zonarius*, belted: from the reddish ochreous margins to the abdominal segments.

subsp. **britannica** (Harrison, 1912) – the subspecies occurring in Britain.

subsp. **atlantica** (Harrison, 1938) – the subspecies occurring in the Hebrides.

1929 **lapponaria** (Boisduval. 1840) – *Lapponicus*, occurring in Lapland.

subsp. **scotica** (Harrison, 1916) – the subspecies occurring in Scotland.

Biston Leach, 1815 – a son of Mars and ancestor of the Bistones, a Thracian tribe which worshipped Bacchus. Leach's generic name 1651 *Cilix* is of similar derivation.

1930 **strataria** (Hufnagel, 1767) – *stratum*, something spread, a quilt: from the embroidered appearance of the forewing; cf. the vernacular name 'carpet' used extensively in the family.

1931 **betularia** (Linnaeus, 1758) – *Betula*, the birch genus: one of the foodplants cited by Linnaeus.

Agriopis Hübner, 1825 – ἄγριος (agrios), wild; ὤψ, ὠπός (ὄψ, ὄπος) (ōps, ōpos (ops, opos)), the face: from the rough scales clothing the frons.

1932 **leucophaearia** ([Denis & Schiffermüller], 1775) – λευκόφαιος (leukophaios), whitish grey: from the coloration of some specimens.

1933 **aurantiaria** (Hübner, 1799) – *aurum*, gold; *antiae*, the hair growing on the forehead, the forelock: from the head and frons which, like the wings, are golden orange. But see also 1233 *Pammene aurantiana*.

1934 **marginaria** (Fabricius, 1777) – *margo, marginis*, a border: the 'dotted border'.

Erannis Hübner, 1825 – ἐραννός (erannos), lovely: the verdict of an artist.

1935 **defoliaria** (Clerck, 1759) – *de-*, privative prefix; *folium*, a leaf: from the periodic abundance of the larvae which then defoliate their foodplants.

Menophra Moore, 1887 – μήνη (mēnē), the moon; ὀφρύς (ophrus), an eyebrow, Lat. *cilia* and hence the cilia or fringe of an insect's wing: from the strongly dentate termen, especially of the hindwing, causing the cilia to form a series of crescents. 'From shape of lines across wings' (Macleod).

1936 **abruptaria** (Thunberg, 1792) – *abruptus*, broken off: from the reddish brown shade extending from the apex into the disc, where it ends abruptly.

Peribatodes Wehrli, 1943 – περί (peri), round; βατώδης (batōdēs), overgrown with thorns: perhaps from a habitat.

1937 **rhomboidaria** ([Denis & Schiffermüller], 1775) – *rhomboides* (ῥομβοειδής), a rhomboid, a four-sided figure whose opposite sides and angles are equal: from the dark spot on the dorsum of the forewing, formed by the conjunction of the median and postmedian fasciae.

1937a **secundaria** (Esper, 1794) – *secundus*, following: the next species to be described. Possibly i.h.o. Caius Plinius Secundus (23–79 A.D.), author of *Naturalis Historia*.

Selidosema Hübner, 1823 – σελίς, σελίδος (selis, selidos), a gangway between rowing benches, the benches themselves, a row of seats in a theatre, a page marked with lines, a column on a page; σῆμα (sēma), a mark: the general sense is of a broad row or band, here referring to the wide dark terminal area of the forewing of the next species. Another species (non-British) included by Hübner in the genus is *S. taeniolaria* (Hübner, 1813) – *taeniola*, a small ribbon.

1938 **brunnearia** (Villers, 1789) – *brunus* (late Lat.), brown: from the ground colour.
subsp. **scandinaviaria** Staudinger, 1901 – the subspecies found in Scandinavia.
subsp. **tyronensis** Cockayne, 1948 – the subspecies formerly found on a bog in Co. Tyrone, Northern Ireland.

Cleora Curtis, 1825 – the wife of Agesilaus, king of Sparta in the 4th century B.C.

1939 **cinctaria** ([Denis & Schiffermüller], 1775) – *cinctus*, girdled: from the discal spot of the forewing, which is white ringed with black (the 'ringed carpet').
subsp. **bowesi** Richardson, 1952 – the subspecies occurring in Scotland, named i.h.o. A. J. L. Bowes, who discovered it in Perthshire and was later killed in the second World War.

Deileptenia Hübner, 1825 – δείλη (deilē), the afternoon, strictly the early part, but occasionally used for the late part or early evening; πτηνός (ptēnos), winged: dusky-winged, the dark pattern being the character stressed by Hübner in his diagnosis. 'Dusk-flying' (Macleod).

1940 **ribeata** (Clerck, 1759) – *Ribes*, the currant genus: from a misconception regarding the foodplant.

Alcis Curtis, 1826 – a daughter of Aegyptus; she had 50 brothers (see 1629 *hyperantus*).

1941 **repandata** (Linnaeus, 1758) – *repandus*, bent, sinuous: from the dentate black marginal line of the hindwing (*'posticis margine repando atro'*).
subsp. **muraria** Curtis, 1826 – *murus*, a wall: from the situation where Curtis found the moth in the Isle of Arran, Scotland.
subsp. **sodorensium** (Weir, 1881) – gen. pl. of *Sodorenses*, the Hebrides, a Norse name surviving in the bishopric of 'Sodor and Man': the subspecies occurring in the Hebrides.

1942 **jubata** (Thunberg, 1788) – *jubatus*, having a mane or crest: possibly from the rather weak, bifid metathoracic crest, but then the junior synonym *glabraria* (Hübner, 1799), derived from *glaber*, bald, would appear to be an antonym. The name might refer to the foodplant, lichen, which hangs like a mane from its substrate. The name appears in a Doctorial Dissertation by one of Thunberg's pupils and I have not been able to locate a copy to check whether he knew the life history.

Boarmia Treitschke, 1825 – βοῦς, βοός (bous, boös), an ox; ἁρμός (harmos), a fitting together, as in 'harmony'; thence βοαρμία (Boarmia), the Ox-yoker, an epithet of Athena.

1943 **roboraria** ([Denis & Schiffermüller], 1775) – *Quercus robur, roboris*, the oak: a larval foodplant.

Serraca Moore, 1887 – *serrago*, sawdust: from the irroration on the wings of the next species.

1944 **punctinalis** (Scopoli, 1763) – *punctum*, a spot; *-alis*, the pyralid termination (cf. 1788 *cervinalis*): from the dense irroration of fuscous scales on the wings. Scopoli also describes the fascia as composed of dots.

Cleorodes Warren, 1894 – the genus 1939 *Cleora*, q.v.; εἶδος, ὠδ- (eidos, od-), a likeness: a genus closely resembling *Cleora*.

1945 **lichenaria** (Hufnagel, 1767) – λιχήν, lichēn: the larval food.

Fagivorina Wehrli, 1943 – *Fagus*, the beech genus; *voro*, to devour: the larva of the next species eats lichens growing on beech-trunks, not the beech itself. The name may be a syncopated form of *Fagivoratrina*, from *voratrina*, an eating house.

1946 **arenaria** (Hufnagel, 1767) – *arena*, sand: from the yellow irroration on the forewing. The species is now extinct in Britain.

Ectropis Hübner, 1825 – ἔκτροπος (ektropos), turning out of the way: from the sinuous course of the fasciae.

1947 **bistortata** (Goeze, 1781) – *bi-, bis-*, twice; *tortus*, twisted: from the double sinuation in the postmedian fascia.

1948 **crepuscularia** ([Denis & Schiffermüller], 1775) – *crepusculum*, twilight: from the time of flight.

Paradarisa Warren, 1894 – παρά (para), alongside; the oriental genus *Darisa* Moore; a closely related genus, but differing in the structure of the antenna.

1949 **consonaria** (Hübner, 1799) – *consonus*, fitting, harmonious: perhaps in approbation or perhaps from the moths' habit of resting on tree-trunks, where they harmonize with their substrate. From resemblance to related species (Spuler); 'name seems pointless' (Macleod).

1950 **extersaria** (Hübner, 1799) – *extersus*, wiped off: from the subterminal white spot on the forewing which resembles a patch denuded of scales.

Aethalura McDunnough, 1920 – αἴθαλος (aithalos), smoke; οὖρον (ouron), a boundary: from the fuscous subterminal suffusion of the forewing.

1951 **punctulata** ([Denis & Schiffermüller], 1775) – *punctulum*, a small prick, a small spot: from the irroration of fuscous scales.

Ematurga Lederer, 1853 – ἦμαρ, ἤματος (ēmar, ēmatos), a poetic variant of ἡμέρα (hēmera), the day; ἔργον (ergon), work: a 'day-worker' because the next species flies by day.

1952 **atomaria** (Linnaeus, 1758) – *atomus*, an atom: '*alis . . . fasciis atomisque fuscis*', the wings with fuscous fasciae and specks.

subsp. **minuta** Heydemann, 1925 – *minutus*, very small: the dwarf race found on northern heathland.

Tephronia Hübner, 1825 – τεφρός (tephros), ash-coloured: from the ground colour of the next species.

1953 **cremiaria** (Freyer, 1838) – *cremo*, to burn, *cremia*, dry firewood: from the ashen colour. '*Cremor*, cream, from colour' (Macleod).

Bupalus Leach, 1815 – name of a Greek sculptor. Cf. 1651 *Cilix* and 1930 *Biston*.

1954 **piniaria** (Linnaeus, 1758) – *Pinus*, the pine-tree genus: the foodplant, as stated by Linnaeus.

Cabera Treitschke, 1825 – name of the daughter of Proteus, the prophetic old man of the sea who kept changing his shape to avoid being caught and having to make prophesies.

1955 **pusaria** (Linnaeus, 1758) – *pusa*, a little girl (*pusus*, dim. of *puer*, a boy); from the delicate complexion. Cf. 721 *pusiella*.

1956 **exanthemata** (Scopoli, 1763) – ἐξάνθημα (exanthēma), an efflorescence, an eruption on the skin: from the irroration of dots on the wings. Scopoli says they are reddish, Macleod grey. A name drawn from Scopoli's experience as a physician; cf. 1902 *chlorosata*.

Lomographa Hübner, 1825 – λῶμα (lōma), a border; γραφή (graphē), a drawing: from the terminal markings of the forewings.

1957 **bimaculata** (Fabricius, 1775) – *bi-*, two; *macula*, a spot: from the two black spots on the costa of the forewing.

1958 **temerata** ([Denis & Schiffermüller], 1775) – *temero*, to pollute, stain: from the subterminal fuscous suffusion which 'stains' the pure white ground colour.

Aleucis Guenée, 1845 – ἀ-, alpha privative; λευκός (leukos), white: a genus related to the last one but differing in containing a species in which the wings are not white.

1959 **distinctata** (Herrich-Schäffer, 1839) – *distinctus*, distinct: a species distinct from *pictaria* Thunberg, 1788 (synonym of 1945 *Cleorodes lichenaria*), a name wrongly applied to this species by Curtis.

Theria Hübner, 1825 – θήρειος (thēreios), of wild beasts: from the presence in the genus of *rupicapraria*, named from the wild goat. '*Thereios* [θέρειος], summer-; but the moth is most in evidence in January and February' (Macleod).

1960 **primaria** (Haworth, 1809) – *primarius*, of the first, here the first part of the year, early spring: from the time of the adult's appearance.

= **rupicapraria** sensu auctt. *nec* ([Denis & Schiffermüller], 1775) – *rupicapra*, the chamois, the 'rock-goat': from similarity in colour. *T. rupicapraria* does not occur in Britain.

Campaea Lamarck, 1816 – καμπή (kampē), a bending, a change of direction; κάμπη (kampē), a caterpillar: Lamarck is probably punning on both senses, the species having a looper caterpillar that bends. Macleod's explanation 'from tendency of wings to change colour' is unlikely, since καμπή does not mean that kind of change.

1961 **margaritata** (Linnaeus, 1767) – *margarita*, a pearl: from the pale green colour.

Hylaea Hübner, 1822 – ὑλαῖος (hulaios), belonging to the wood: from the habitat.

1962 **fasciaria** (Linnaeus, 1758) – *fascia*, a band: from the broad reddish fascia on the forewing ('*fascia lata ferruginea*').

Gnophos Treitschke, 1825 – γνόφος (gnophos), darkness: from the dusky coloration.

1963 **obfuscata** ([Denis & Schiffermüller], 1775) – *obfuscus*, dusky: from the ground colour.

1964 **obscurata** ([Denis & Schiffermüller], 1775) – *obscurus*, dark: from the dark coloration of some forms.

Psodos Treitschke, 1827 – σποδός (spodos), wood-ash, according to Treitschke himself an intentional anagrammatic variation; nevertheless, probably influenced by ψόδος (psodos), soot, which would suit the next species well. Perhaps he found the name in Ochsenheimer's papers, was puzzled by it and misinterpreted the meaning.

1965 **coracina** (Esper, *post* 1796) – *corax*, a raven, *coracinus*, raven-black: the 'black mountain moth'.

Siona Duponchel, 1829 – from Mt. Sion, on account of 'its barrenness of markings (Stephens), but more probably from the town and canton of the same name in Switzerland.

1966 **lineata** (Scopoli, 1763) – *linea*, a line: from the black veins on the underside (the 'black-veined moth').

Aspitates Treitschke, 1825, corrected to *Aspilates*, Treitschke, 1827; however, according to the I.C.Z.N. rules the original spelling must stand – ἀσπιλάτης (aspilatēs), an Arabian precious stone mentioned by Pliny.

1967 **gilvaria** ([Denis & Schiffermüller], 1775) – *gilvus*, yellow: from the ground colour.
subsp. **burrenensis** Cockayne, 1951 – the subspecies occurring in the Burren, Co. Clare.

1968 **ochrearia** (Rossi, 1794) – *ochra*, ochre, a kind of yellow earth: from the ground colour.

Dyscia Hübner, 1825 – δύο, δυ- (duo, du-), two; σκιά (skia), a shadow: from the two subterminal dark blotches on the forewing of the next species.

1969 **fagaria** (Thunberg, 1784) – *Fagus*, the beech genus: the larva, however, feeds on heather.

Perconia Hübner, 1823 – περι- (peri-), in comp. all round; κόνιος (konios), dusty: from the dusty irroration on the wings of the nine species included by Hübner in his genus.

1970 **strigillaria** (Hübner, 1787) – *strigilla*, a coined dim. of *striga*, a furrow, a line: from the fine fasciae on the forewing.

SPHINGOIDEA (1976)

SPHINGIDAE (1976)

SPHINGINAE (1976)

Agrius Hübner, 1819 – the name of one of the giants who waged war against the gods and was killed by the Fates (cf. 1979 *Mimas*): the moths are gigantic.

1971 **cingulata** (Fabricius, 1775) – *cingulatus*, girdled: from the bright pink bands round the abdomen.

1972 **convolvuli** (Linnaeus, 1758) – *Convolvulus arvensis*, field bindweed: the principal foodplant.

Acherontia Laspeyres, 1809 – Acheron, the River of Pain in the Underworld: an appropriate home for the next species.

1973 **atropos** (Linnaeus, 1758) – Atropos, one of the three Fates, from ἀ, alpha privative; τρόπος (tropos), a turn: she cut the thread of life and death could not be averted. In the third volume of his *Amoenitates Academicae* (1756) Linnaeus published a disquisition by his pupil Forsskahl in which this species is called *caput mortuum*, the death's head, but in the binomial nomenclature of *Systema Naturae* (1758) the specific name had to be made to consist of a single word; Linnaeus retained, however, the theme of death.

Manduca Hübner, 1807 – *manduco*, a glutton: from the voracious appetite of the huge larvae. It is unusual for Hübner to derive a generic name from the Latin. He first proposed this name in his *Tentamen* [1806] with [1973 *Acherontia*] *atropos* as the example.

1974 **quinquemaculatus** (Haworth, 1803) – *quinque*, five; *maculatus*, spotted: from the five spots placed laterally on each side of the abdomen.

1975 **sexta** (Johansson, 1763) – *sextus*, sixth: the sixth species in the series described.

1975a **rustica** (Fabricius, 1775) – *rusticus*, rustic, a countryman: from the habitat.

Sphinx Linnaeus, 1758 – a monster, usually represented with the bust of a woman and the body of a lioness. She lived at the Egyptian Thebes, posing riddles and murdering those who failed to solve them. When Oedipus succeeded, she committed suicide. It has been suggested that the myth symbolizes the once mysterious annual overflow of the R. Nile, bringing fertility but at the same time the threat of death to those in its path. No entomological analogy is necessary, but Linnaeus may have seen a similarity between the enigmatic face of the statue close to the Great Pyramids and a larva with its head raised in repose. 'Many of the Linnaean generic names are, however, destitute of any direct application to the insects which such genera contain, this author having employed the old natural history names, which he met with in the early authors, in the most senseless fashion' (Westwood, 1838). All the names taken from mythology are 'senseless'. Any serious attack on Linnaeus is quite out of harmony with the general tenor of Westwood's theme. This name started the fashion for generic names to be enigmatic and those like 166 *Zygaena* Fabricius follow in the Linnaean tradition.

1976 **ligustri** Linnaeus, 1758 – *Ligustrum*, the privet genus, includes larval foodplants.

1977 **drupiferarum** Smith *in* Abbot, 1797 – *drupa*, an olive, a fleshy fruit with a stone; *fero*, to bear: presumably from the foodplants. A North American species only once recorded in Britain.

Hyloicus Hübner, 1819 – ὕλη (hulē), a wood; οἶκος (oikos), a house, a dwelling place: from the habitat. However, the other generic names bestowed by Hübner on the Sphingidae in 1819 are taken from mythology, and this may be one that I have failed to trace. The same may be true of 1974 *Manduca*.

1978 **pinastri** (Linnaeus, 1758) – *Pinus pinaster*, maritime pine: a possible foodplant, but Linnaeus himself gives only pine.

Mimas Hübner, 1819 – one of the Giants who waged war on the gods; he was overcome by Hephaestus who buried him under Vesuvius.

1979 **tiliae** (Linnaeus, 1758) – *Tilia*, the lime-tree genus: from the principal larval foodplant.

Smerinthus Latreille, 1802 – σμήρινθος, = μήρινθος (smērinthos, mērinthos), a thread; perhaps the thread of life woven by the Fates and cut by Atropos, so echoing the imagery of 1973 *Acherontia atropos*. Pickard *et al.* and Macleod, however, think the thread refers to the oblique lateral stripes on the larva but since the name was erected for a family (p. 32), it is unlikely to be derived from the larval markings of certain species. Spuler much more plausibly suggests that it refers to the degenerate, thread-like haustellum.

1980 **ocellata** (Linnaeus, 1758) – *ocellatus*, eyed: the 'eyed hawk-moth'.

Laothoe Fabricius, 1807 – the name of one of the mistresses of Priam, King of Troy. Originally a family name with the same coverage as 1980 *Smerinthus*.

1981 **populi** (Linnaeus, 1758) – *Populus*, the poplar genus: from the foodplants.

MACROGLOSSINAE (1984)

Hemaris Dalman, 1817 – ἡμέρα (hēmera), the day: from the diurnal behaviour of the adult.

1982 **tityus** (Linnaeus, 1758) – a giant who made a pass at Diana and as a punishment was made to lie outstretched on the ground in the Underworld while vultures or snakes devoured his self-renewing liver; according to Ovid (*Metamorphoses*, bk. 4), his bulk extended over nine acres. In *Systema Naturae* his name is preceded by Tantalus and followed by Ixion, both of whom likewise underwent terrible punishments in Hades.

1983 **fuciformis** (Linnaeus, 1758) – *fucus*, a drone; *forma*, the shape: from a fancied resemblance to a hymenopteron; see 370 *Sesia*.

Macroglossum Scopoli, 1777 – μακρός (makros), large, long; γλώσση (glōssē), the tongue. Linnaeus (1758) recorded in a footnote that hawk-moths take nectar whilst hovering in front of flowers, but he apparently failed to observe that in some species (e.g. 1981 *Laothoe populi*) the haustellum is degenerate. Scopoli did so, but as he applied this name to '*stellatarum, etc.*', it is not possible to tell how many of the long-tongued species he wished to be included. The subfamily takes its name from this genus.

1984 **stellatarum** (Linnaeus, 1758) – Stellatae, a synonym of Rubiaceae, the family containing bedstraw (*Galium* spp.) and wild madder (*Rubia peregrina*), cited as foodplants by Linnaeus.

Proserpinus Hübner, 1819 – see next species.

1984a **proserpina** (Pallas, 1772) – Proserpina or Persephone, the daughter of Jupiter and Ceres (Zeus and Demeter). She was carried off by Pluto (Dis), who made her Queen of the Underworld, but through the intervention of Jupiter she was allowed to return to earth for part of each year. The myth symbolizes the return of spring and the sprouting of the grain buried in the soil.

Daphnis Hübner, 1819 – the name of a son of Hermes who became a Sicilian shepherd skilled in music and poetry.

1985 **nerii** (Linnaeus, 1758) – *Nerium*, the oleander genus, that of the foodplants.

Hyles Hübner, 1819 – the name of one of the Centaurs, a savage race, half man and half horse, who were supposed to live in Thessaly.

1986 **euphorbiae** (Linnaeus, 1758) – *Euphorbia*, the spurge genus, that of the larval foodplants.

1987 **gallii** (Rottemburg, 1775) – a typographical error for *galii*, from *Galium*, the bedstraw genus, that of the larval foodplants.

1988 **nicaea** (de Prunner, 1798) – Nicaea, a city in Asia Minor. An ecumenical council which met there in 325 A.D. drew up the Nicene Creed.

1989 **hippophaes** (Esper, 1793) – *Hippophae rhamnoides*, sea-buckthorn, the larval foodplant.

1990 **lineata** (Fabricius, 1775) – *lineatus*, lined: from the white veins on the forewing, the 'striped hawk-moth'.

subsp. **livornica** (Esper, 1780) – Livorno, Leghorn: this is odd, since the type locality is given as Germany; perhaps there was a historical connection, an early specimen having been taken there, for instance by Dr David Krieg (see 1584 Limenitinae and *camilla*).

Deilephila Laspeyres, 1809 – δείλη (deilē), the evening; φιλέω (phileo), to love: from the crepuscular habits of many of the moths in what was once a much larger genus.

1991 **elpenor** (Linnaeus, 1758) – Elpenor, one of the companions of Odysseus (Ulysses) who were metamorphosed by Circe into swine; the small, retractile head and narrow thoracic segments of the larva give the impression of a pig's snout. The choice of name may have been influenced by the next one.

1992 **porcellus** (Linnaeus, 1758) – *porcellus*, a piglet: for the same reason as the last species. Linnaeus did not originate the name himself, but took it from Thomas Mouffet's *Insectorum sive minimorum animalium Theatrum*. This was written about 1585 but not published until 1634, soon becoming a popular text-book for British naturalists. Linnaeus knew it well and gives many references to it in his *Systema Naturae*. Mouffet gave the name to the larva which he figured with its characteristic 'snout', but one cannot be entirely certain that it does not represent *D. elpenor*; it is unlikely that he also knew the adult. Mouffet did not normally bestow names, but he did so here and to 1995 *vinula* which immediately precedes it.

Hippotion Hübner, 1819 – the name of a prince who fought on the Greek side at Troy.

1993 **celerio** (Linnaeus, 1758) – Pickard *et al.* and Macleod derive the name from *celer*, swift, the latter adding 'from flight'. Since Linnaeus described the hawkmoths as rather slow and cumbersome ('*tardiores, ponderosiores*'), this explanation is unlikely. Linnaeus spells the name with a capital letter, indicating a substantival rather than an adjectival origin; it may, therefore, be the title of a Homeric hero or a god such as Hermes.

NOCTUOIDEA (2107)

NOTODONTIDAE (2000)

Phalera Hübner, 1819 – φαληρός, φαλαρός (phalēros, phalaros), having a white patch: from the pale apical patch on the forewing of the next species.

1994 **bucephala** (Linnaeus, 1758) – βουκέφαλος (boukephalos), bull-headed, an epithet applied to horses, probably because they were branded with a bull's head; Bucephalus was the name of the horse of Alexander the Great. I cannot say whether Bucephalus had a white blaze, or whether Linnaeus thought so; if he did, the name could have the same meaning as the generic name. Alternatively, the name could refer to the rather large and conspicuous bull head of the adult, as suggested by Macleod. There is, however, no need for an entomological explanation; 1616 *maera* is another example of the use of the name of an animal mentioned in the classical literature.

Cerura Schrank, 1802 – κέρας (keras), a horn; οὐρά (oura), a tail: from the larva's modified anal claspers which take the form of long appendages; 'horn' either because they are bifurcate or chitinous, or both. Schrank considered that the puss-moth group were sufficiently distinctive to merit family status.

1995 **vinula** (Linnaeus, 1758) – dim. of *vinum*, wine: from the colour of the larva's dorsal saddle. Linnaeus borrowed this name from Mouffet, whom he includes in his references; cf. 1992 *porcellus*. Mouffet wrote '*Vinula sequitur, elegans mehercule Eruca et supra fidem speciosa: cauda bifurcata ex vinaceo nigricans. Corpus totum veluti crassiore rubicundioreque vino intinctum*' (*Vinula* follows, by Jove an elegant caterpillar and handsome beyond belief: it has a bifurcate tail springing from a blackish grape-stone. The whole body is dyed as it were with a full-bodied ruby wine). P. B. M. Allan (1947), who claimed to have consulted Mouffet, wrote 'Mouffet gives no hint why he calls the larva "*vinula*"' – an extraordinary misstatement from such a scholarly writer! There is no evidence that Mouffet knew what sort of moth his *vinula* turned into, or even that it became a moth at all. He figures the larva.

Pickard *et al.* write 'Linné describes the larva as '*e rima sub capite humorem acrem exspuens*' (spitting a bitter fluid from an orifice beneath the head), implying that this fluid is the 'wine'. Spuler was mystified by the name. Macleod accepted the explanation given by Pickard *et al.* and wrote in his introduction 'Does not the name convey more to us when we learn, for instance, that *Cerura vinula* means "The horn-tailed wine-dispenser"?' It doesn't.

Furcula Lamarck, 1816 – *furcula*, a small fork: from the larva's two anal appendages.

1996 **bicuspis** (Borkhausen, 1790) – *bis, bi-*, two; *cuspis*, a sharp point: from the larva's two anal appendages.

1997 **furcula** (Clerck, 1759) – not necessarily for the same reason as the generic name. Linnaeus (1767) incorporated most of Clerck's species including this one, but makes no mention of the larva as he almost certainly would have done had he known it. The median and subterminal fasciae, when the latter is strongly expressed, converge on the dorsum, so forming a fork-shaped pattern. It is also possible that Clerck knew the larva but did not tell Linnaeus or that they assumed that the larva would resemble those of its relatives.

1998 **bifida** (Brahm, 1787) – *bifidus*, cleft in two: from the larva's anal appendages.

Stauropus Germar, 1811 – σταῦρος (stauros), a cross; ποῦς (pous), the foot: possibly from the aspect of the anal segments of the larva when raised in repose, the modified anal claspers and those on abdominal segment 6 forming a cross through divergence (see South, 1961, **1**: pl. 27). Pickard *et al.*, followed by Macleod, say that the name refers to the shape of the forelegs, but this seems less likely.

1999 **fagi** (Linnaeus, 1758) – *Fagus sylvatica*, the beech, is the foodplant.

Notodonta Ochsenheimer, 1810 – νῶτος (nōtos), the back; ὀδούς, ὀδόντος (odous, odontos), a tooth: from the dorsal scale-tufts on the forewings, which project upwards when the wings are folded. It is likely that Ochsenheimer also had in mind the dorsal humps of the larvae of some species.

2000 **dromedarius** (Linnaeus, 1767) – *dromedarius*, a dromedary, a camel: from the scale-tufts on the dorsum of the forewing which project upwards like humps when the moth is at rest. 'From larva's humps' (Macleod); Linnaeus did not know the life history.

2001 **torva** (Hübner, 1803) – *torvus*, stern, gloomy: from the moth's sombre coloration.

Tritophia Kiriakoff, 1967 – from the name of the next species.

2002 **tritophus** ([Denis & Schiffermüller], 1775) – a typographical error for *trilophus*, τρίλοφος (trilophos), with three crests: the 'three-humped'.

Eligmodonta Kiriakoff, 1967 – ἕλιγμος (eligmos), rolling, winding, writhing; ὀδούς, ὀδόντος (odous, odontos), a tooth: from the flexed posture of the larva at rest and its dorsal humps.

2003 **ziczac** (Linnaeus, 1758) – Latinized from the German 'zickzack', zigzag: from the zigzag profile of the humped larva, enhanced by its posture.

Harpyia Ochsenheimer, 1810 – one of the Ἄρπυιαι (Harpuiai), Harpies or Snatchers. They were monsters with the head of a girl and the wings and body of a bird of prey. Sometimes they carried off persons bodily, sometimes their food, or they would befoul it with their excrement. The Indian kite has a lot in common with them as I learnt to my cost in the war.

2004 **milhauseri** (Fabricius, 1775) – i.h.o. J. A. Milhauser (*fl.* 18th century), who figured the species. Fabricius gives the reference as 'Milhauser *Monogr.* Dresd., 1763, tab. 1'.

Peridea Stephens, 1828 – περιδεής (perideiēs), very timid: 'the larva and imago, when touched, tremble as if in fear' (Stephens), the name being descriptive of the next species.

2005 **anceps** (Goeze, 1781) – *anceps*, doubtful: from the uncertain, tremulous movement of the larva and imago. 'Doubtful, i.e. whether separate species' (Macleod).

Pheosia Hübner, 1819 – φέως (pheōs), a prickly plant, a spine: from the wedge-shaped dorso-tornal dash, its shape diagnostic between the next two species, the only ones placed by Hübner in the genus. I have rejected Macleod's explanation 'phaios [φαιός], dusky; from colour of forewings' because Hübner, the artist, would never have described them thus; the ground colour is whitish. '*Tremula*, as a moth, I adore. He is, to me, aesthetically the perfect insect' (Allan, 1947).

2006 **gnoma** (Fabricius, 1777) – γνῶμα (gnōma), a mark: from the white dorso-tornal dash mentioned under the generic name.

2007 **tremula** (Clerck, 1759) – *tremulus*, shaking, trembling: probably from the tremulous movement of the larva and imago, perhaps also with a play on words with *Populus tremula*, the aspen, cited as foodplant by Linnaeus (1767). Had the name been taken solely from the foodplant, Clerck should have written '*tremulae*'.

Ptilodon Hübner, 1822 – πτιλόν (ptilon), a feather, a wing; ὀδούς, ὀδόντος) a tooth: from the dorsal tuft on the forewing.

2008 **capucina** (Linnaeus, 1758) – late Lat. *cappa*, a cap, whence came the French capuce, a pointed hood or cowl; this gave the name to the Capuchin monks, a branch of the Franciscan order, and to capuchin monkeys: from the moth's prominent thoracic crest.

Ptilodontella Kiriakoff, 1967 – dim. of 2008 *Ptilodon*, q.v.

2009 **cucullina** ([Denis & Schiffermüller], 1775) – *cucullus*, a hood, a cowl: from the thoracic crest.

Odontosia Hübner, 1819 – ὀδούς, οδόντος (odous, odontos), a tooth; ὦσις (ōsis) a thrusting: from the prominent dorsal scale-tooth on the forewing of the next species which thrusts itself upwards when the wings are folded.

2010 **carmelita** (Esper, 1799) – Latinized from Carmelite, one of an order of monks who, like a Capuchin, wore a cowl: from the thoracic crest.

Pterostoma Germar, 1811 – πτέρον (pteron), a feather; στῶμα (stōma), the mouth from the very long, densely scaled, labial palpus, formerly regarded as belonging to the mouth-parts. 'From feathery antennae' (Macleod).

2011 **palpina** (Clerck, 1759) – *palpus*, the labial palp which is 'very long, erect and clothed with long dense rough scales' (Meyrick, 1928).

Leucodonta Staudinger, 1892 – λευκός (leukos), white; ὀδούς, ὀδόντος (odous, odontos), a tooth: from the white forewing with a black-spotted scale-tooth of the next species.

2012 **bicoloria** ([Denis & Schiffermüller], 1775) – *bi-*, two; *color*, colour: from the forewing, which has black and orange markings on a white ground colour.

Ptilophora Stephens, 1828 – πτίλον (ptilon), a feather; φορέω (phoreō), to carry: a Greek rendering of the following specific name.

2013 **plumigera** ([Denis & Schiffermüller], 1775) – *pluma*, a feather; *gero*, to carry: from the strongly pectinate male antenna.

Drymonia Hübner, 1819 – δρυμός (drumos), an oak-coppice: both the British members of the genus occur in oak-woods.

2014 **dodonaea** ([Denis & Schiffermüller], 1775) – Dodona, a town of Epirus; see 1853 *Eupithecia dodoneata*.

2015 **ruficornis** (Hufnagel, 1767) – *rufus*, red; *cornu*, a horn: from the pale reddish antenna. This species has four synonyms derived from oak: see genus.

Gluphisia Boisduval, 1828 – γλυφίς (gluphis), the notch in an arrow: a Greek rendering of the following specific name. One would have expected *Glyphisia*.

2016 **crenata** (Esper, 1785) – *crena*, a notch: from the subcrenate termen of the forewing.
subsp. **vertunea** Bray, 1929 – Virton in Belgium is the type locality.

Clostera Samouelle, 1819 – κλωστήρ (klōstēr), a spindle: from the shape of the abdomen, which is slender and tapers in the anal segments.

2017 **pigra** (Hufnagel, 1766) – *piger*, reluctant, sluggish: from the larval habit of living in spun leaves and consequent sendentary behaviour; the synonym *reclusa* ([Denis & Schiffermüller], 1775) is derived from the same habit. 'Reluctant, from moth's rare appearance' (Macleod): he had muddled this species with the next, and supposed the name had been given to describe the British distribution.

2018 **anachoreta** ([Denis & Schiffermüller], 1775) – *anachoreta* (ἀναχωρητής, anakhōrētēs), a hermit: from the larval habit of living in a 'cell' of spun leaves.

2019 **curtula** (Linnaeus, 1758) – κυρτός (kurtos), arched, humped: from the larva, which Linnaeus describes as '*turrita*', with towers or humps. Pickard *et al.* and Macleod derive the name from *curtus*, cut short, descriptive of the chocolate tip of the forewing.

Diloba Boisduval, 1840 – δι- (di-), two; λοβός (lobos), a lobe, especially of the ear: from the 'figure of eight' on the forewing of the next species.

2020 **caeruleocephala** (Linnaeus, 1758) – *caeruleus*, sky-blue, sea-blue; κεφαλή (kephalē), the head: from the larva, described by Linnaeus. Note the hybrid derivation from Latin and Greek.

THAUMETOPOEIDAE (2021)

Thaumetopoea Hübner, 1820 – a typographical error for *Thaumatopoea*, from θαυματοποιός (thaumatopoios), wonder-working: from the remarkable processionary behaviour of the larvae.

2021 **pityocampa** ([Denis & Schiffermüller], 1775) – πιτυών (pituōn), a pine-forest; καμπή (kampē), a caterpillar; πιτυοκάμπη (pituokampē), *pituocampa*, the pine caterpillar mentioned by Galen, Dioscorides and Pliny (1st and 2nd centuries A.D.). It has urticating hairs and was regarded as poisonous (Beavis, 1988). This name, like *Cossus*, has an entomological history of nearly two millennia. Réaumur (1734–43) used the name, but since he was writing before Linnaeus he does not qualify for authorship.

2022 **processionea** (Linnaeus, 1758) – *processio*, a marching onwards, a procession: from the larval behaviour.

Trichiocercus Stephens, 1835 – a typographical error for *Trichocercus*, from θρίξ, τρίχος (thrix, trikhos), hair; κερκός (kerkos), a tail: from the pale yellow caudal tuft on the abdomen of the next species.

2023 **sparshalli** (Curtis, 1830) – i.h.o. Joseph Sparshall, who reputedly took the type specimen of this Australian species at Horning, Norfolk, on 29 August 1829.

LYMANTRIIDAE (2033)

Laelia Stephens, 1828 – the name of a Vestal Virgin: Stephens gives no explanation, but the pale, immaculate wings of the next species may have suggested chastity.

2024 **coenosa** (Hübner, 1808) – *coenosus*, marshy: from the habitat.

Orgyia Ochsenheimer, 1810 – ὄργυια (orguia), strictly the reach of the outstretched arms, a fathom, but here the outstretched arms themselves: from the posture of the moths at rest, with their forelegs extended forwards.

2025 **recens** (Hübner, 1819) – *recens*, young, recent: to indicate close relationship with, but distinction from, 2026 *O. antiqua*, q.v.

2026 **antiqua** (Linnaeus, 1758) – *antiquus*, ancient, existing long ago, of the old stamp, simple, honest and innocent. This name appears on the same page as 2028 *pudibunda*, both conveying the sense of chastity. Linnaeus describes the female as wingless and perhaps thought of her as a wife of the golden age when women stayed at home to work and did not gad about with the men. It may be relevant that the male hindwings are coloured and such species suggested the fairer sex to Linnaeus; see 2452 *nupta*, etc. 'Old, as distinguished from *O. recens*' (Macleod): an anachronism by 61 years!

Dicallomera Butler, 1881 – δι- (di-), two, twice; κάλλος (kallos), beauty; μηρός (mēros), the thigh: from the ornamental scaling on the femora, especially of the forelegs, which are either both, or doubly, beautiful.

= **Dasychira** sensu auctt. – δασύς (dasus), shaggy; χείρ (kheir), the hand: from the hairy foreleg.

2027 **fascelina** (Linnaeus, 1758) – *fasciculus*, a small bundle: from the 'tussocks' of setae on the dorsum of the larva, described in detail by Linnaeus.

Calliteara Butler, 1881 – derivation obscure, perhaps κάλλος (kallos), beauty; t, phonetic insertion; ἔαρ (eär), the spring: the genus of a beautiful species occurring relatively early in the year (May).

2028 **pudibunda** (Linnaeus, 1758) – *pudibundus*, modest: Linnaeus gives no hint to explain his choice of name; it may be from the anal tuft of the male which discreetly hides its genitalia, or from that of the female which hides the egg-batch. Macleod's 'from escaping notice by imitating feather or mass of cottony seeds' is improbable; when Linnaeus used the names 1641 *ilicifolia* and 1642 *quercifolia* to convey a somewhat similar idea he was more explicit. *Pudibundus* can also mean immodest, disgraceful, descriptive of a thing of which one should be ashamed; no well-bred lady would flaunt her legs as this moth does.

Euproctis Hübner, 1819 – εὖ (eu), good; πρωκτός (prōktos), the anus, the backside: from the well-developed anal tuft.

= **Sphrageides** Maes, 1984 – σφραγίς (sphragis), a seal, a signet; εἶδος (eidos), shape, form: possibly with reference to the subtornal spot on the forewing of 2030 *Euproctis similis*, the wings of the members of the genus being otherwise unmarked.

2029 **chrysorrhoea** (Linnaeus, 1758) – χρυσός (khrusos), gold; ῥέω (rheō), to flow: from the flowing golden tail. Haworth (1803) pointed out that this description belongs more properly to 2030 *E. similis*, and renamed this species *phaeorrhoeus* – φαιός (phaios), dusky.

2030 **similis** (Fuessly, 1775) – *similis*, like: because it is similar to the last species, and particularly to Linnaeus' description of it.

Leucoma Hübner, 1822 – λευκός (leukos), white; κόμη (komē), the hair: from the white scales which cover the whole moth.

2031 **salicis** (Linnaeus, 1758) – *Salix*, the willow/sallow genus, on whose species the larva sometimes feeds; it prefers poplar which is also mentioned as a foodplant by Linnaeus.

Arctornis Germar. 1811 – ἄρκτος (arktos), a bear, also the constellation *Ursa Major*, the Great Bear, and so the north; ὄρνις (ornis), a bird: possibly from the polar bear, which is white, and a bird, because it is winged.

2032 **l-nigrum** (Müller, O. F., 1764) – l, the letter of the alphabet; *niger*, black; from the black mark at the end of the cell on the forewing; this is 'V'-shaped rather than 'L'-shaped, though Müller no doubt had the capital Greek letter (Λ) in mind. Since authors did not realize this, amendments such as *V-nigrum* (Fabricius, 1775) and *vau-nigra* (Stephens, 1828) have been proposed.

Lymantria Hübner, 1919 – λυμαντήρ (lumantēr), a spoiler, destroyer: from the pest species, such as 2034 *L. dispar*, contained in the genus.

2033 **monacha** (Linnaeus, 1758) – *monacha*, a nun: from the black and white coloration, suggestive of a nun's habit.

2034 **dispar** (Linnaeus, 1758) – *dispar*, unalike: from the sexual dimorphism, the male having the ground colour brown, the female white. There is also a disparity in size but Linnaeus mentions only colour.

ARCTIIDAE (2057)

LITHOSIINAE (2051)

Thumatha Walker, 1866 – probably a meaningless neologism; Walker gives no explanation.

2035 **senex** (Hübner, 1808) – *senex*, old, advanced in years: perhaps from the thin wing scaling, suggestive of an elderly man getting 'rather thin on top'. Hübner described it in *Noctua* but it was later placed in 2038 *Nudaria*, q.v. 'From the wrinkled appearance of the wings' (Pickard *et al.*; Macleod), a character hard to detect; Spuler suggests that the name refers to the grey colour of the wings, which is a more likely explanation.

Setina Schrank, 1802 – σής (sēs), a moth; Schrank translates the name 'Motteule', moth-owl (cf. 2166 *Hadena*). The explanation given by Pickard *et al.* and Macleod, 'a town in Latium', is therefore certainly incorrect: geographical generic names came later.

2036 **irrorella** (Linnaeus, 1758) – *irroro*, to sprinkle with dew: Linnaeus gave no explanation, but the figure by Clerck (1759) shows a liberal sprinkling of black spots; cf. 428 *rorrella* and 429 *irrorella*. A literal explanation is also possible, since the moth may be found resting on grass-stems at dawn, its wings spattered with dew; the vernacular name 'dew moth' may either be derived from this habit or be a translation of the scientific name. 'From moth's transparent appearance when hanging from leaf' (Macleod); where does the sense of 'sprinkling' come in?

Miltochrista Hübner, 1819 – μίλτος (miltos), red earth; χριστός (khristos), anointed: from the reddish pink colour of the next species, the only one included by Hübner in the genus.

2037 **miniata** (Forster, 1771) – *minium*, red lead, vermilion: from the coloration; cf. 2183 *miniosa*.

Nudaria Haworth, 1809 – *nudus*, naked: from the thin wing scaling; '*alis omnibus denudatis*'.

2038 **mundana** (Linnaeus, 1761) – *mundus*, elegant: from the delicate appearance of the moth. Pickard *et al.* derived from *mundanus*, earthly, overlooking the fact that Linnaeus had described the moth as a tortricid and *-ana* was the family termination.

Atolmis Hübner, 1819 – ἀτολμία (atolmia), lack of courage: 2039 *A. rubricollis* is the type species and Macleod's explanation, 'from larva's habit of hiding by day in bark of trees' is probably correct, if Hübner knew the life history.

2039 **rubricollis** (Linnaeus, 1758) – *ruber*, red; *collum*, the neck: from the red patagia of the adult.

Cybosia Hübner, 1819 – κύβος (kubos), a cube, a gaming die: from the position of the costal and dorsal black spots on the forewings of the next species when seen extended, the 'four-dotted footman'. It was the only species in Hübner's genus.

2040 **mesomella** (Linnaeus, 1758) – μέσος (mesos), middle; μέλας (melas), black: Linnaeus' diagnosis reads 'Tinea *alis supra albis, subtus luteis, interne nigris*, a tineid with the wings white on the upperside and yellow with a black centre on the underside, the name, therefore, being a syncopated form of '*mesomelella*' and descriptive of this black centre. Pickard *et al.* and Macleod derive from *mesomelas*, a white stone with a black stripe mentioned by Pliny; the objection is that the moth has no black stripe.

Pelosia Hübner, 1819 – πηλός (pēlos), clay: from the brownish grey colour of the five species Hübner included in his genus.

2041 **muscerda** (Hufnagel, 1766) – *muscerda*, mouse-dung: from the scattered dots on the forewing.

2042 **obtusa** (Herrich-Schäffer, 1852) – *obtusus*, blunt: from the rounded termen of the forewing.

Eilema Hübner, 1819 – εἴλημα (eilēma), a veil: from the involute posture with the wings wrapped round the body adopted by some of the species when at rest (see 2047 *complana*). 2052 *Spiris* Hübner, 1819 has the same meaning. 'From tuft of scales under base of forewings of most species' (Macleod); a tuft is not a veil and in any case it is on the upperside.

2043 **sororcula** (Hufnagel, 1766) – *sororcula*, a little sister: 'from affinity to another species' (Macleod); Hufnagel gives no hint of this, nor are any other 'footmen' described on the same page. The name seems just to be fanciful.

2044 **griseola** (Hübner, 1803) – *griseus* (late Lat.), grey: from the colour of the forewing. The termination '*-ola*' was adopted for the footman by several other authors following the lead set here by Hübner.

2045 **caniola** (Hübner, 1808) – *canus*, grey (of hair): the 'hoary' footman.

2046 **pygmaeola** (Doubleday, 1847) – *pygmaeus*, pygmy: the 'pygmy' footman.

subsp. **pallifrons** (Zeller, 1847) – *pallidus*, pale; *frons*, the forehead: the subspecies having the frons pale grey as opposed to pale yellow.

2047 **complana** (Linnaeus, 1758) – *complano*, to make level: from the posture at rest with the wings held flat. Linnaeus included the phrase '*alis depressis*', which has the same meaning (see p. 21), in his diagnosis and placed the species in Noctua, characterized by the '*alis depressis*' method of wing-folding. It is therefore clear both from the name and diagnosis that Linnaeus had 2050 *E. lurideola* before him when he gave this name; awareness of this caused Boisduval (1834) to rename the latter species *complanula*. Had Linnaeus been describing the present species, he would have written '*alis convolutis*' and placed it in Tinea. The different method used by the two species in folding their wings is well known to collectors as a ready means of separation. 'From position of wings when resting' (Macleod); he should have written 'from the position in which the wings are *not* held when resting'. See also 683 *depressana* and 688 *applana*.

2048 **sericea** (Gregson, 1860) – *sericeus*, silky: from the texture of the wings.

2049 **deplana** (Esper, 1787) – *de-*, down, downwards; *planus*, flat, level: from the resting posture with the wings held flat over the body, having the same meaning as 2047 *complana* but differing in that the name in this case gives an accurate description. Esper described the moth in Noctua, as the method of wing-folding required. 'From moth's habit, when resting, of lowering wings close against sides' (Macleod): the antithesis of the correct meaning.

2050 **lurideola** (Zincken, 1817) – *luridus*, pale yellow: from the colour of the costal streak on the forewing, and the hindwing. See also 2047.

Lithosia Fabricius, 1798 – λίθος (lithos), a stone: 'from the colour of some species' (Macleod); more probably from the larval habit of feeding on lichens which grow on rocks. The name was set up to embrace all the footmen and is the source of the current subfamily name.

2051 **quadra** (Linnaeus, 1758) – *quadra*, a square: from the forewings of the female which each have two black spots; if these are joined in a set specimen, an almost rectangular figure is formed. Linnaeus apparently did not know the male.

ARCTIINAE (2057)

Spiris Hübner, 1819 – σπεῖρα (speira), a wrapping: with the same meaning as 2045 *Eilema*, q.v., established by Hübner in the same work.

2052 **striata** (Linnaeus, 1758) – *striatus*, streaked: from the heavy black shading along the veins of the male forewing.

Coscinia Hübner, 1819 – κόσκινον (koskinon), a sieve: from the graticulate pattern of the forewing.

2053 **cribraria** (Linnaeus, 1758) – *cribrarius*, sifted: from the dark spots sprinkled over the forewing. Described in Bombyx, so *-aria* is not a family termination.

subsp. **bivittata** (South, 1900) – *bi-*, two; *vitta*, a head-band, a streak: from the two longitudinal streaks on the forewing, one on either side of the cell, which are more strongly expressed than the others.

subsp. **arenaria** (Lempke, 1937) – *arenarius*, sandy: from the pale, sandy ground colour.

Utetheisa Hübner, 1819 – οὐτάω, in composition οὐτη- (outao, outē-), to wound; θεῖος (theios), divine: from the blood-red, yet beautiful, spots on the forewing, probably prompted by Linnaeus' use of the word *sanguineus* (bloody) in his description of the next species. '*Ou*, not, *tetheis*, fixed; from variable marking' (Macleod): he apparently misread his lexicon, the word for 'fixed' being τέθμιος (tethmios); there is no such word as τεθεις in my lexicon. Cf. 2269 *Atethmia*, where Macleod gets the word, but not the derivation, correct.

2054 **pulchella** (Linnaeus, 1758) – *pulcher*, beautiful; *-ella*, the termination for species described by Linnaeus in Tinea; '*alis albis: superioribus nigro sanguineoque punctatis* (wings white: the forewing spotted black and blood-red); this description may have suggested the generic name to Hübner. The first specimen was taken by Brander (see 1088 *branderiana*).

2055 **bella** (Linnaeus, 1758) – *bellus*, handsome. A contraction of '*bellella*', Linnaeus having described it as a tineid.

Parasemia Hübner, 1820 – παράσημον (parasēmon), a mark of distinction: from the distinctive wing pattern.

2056 **plantaginis** (Linnaeus, 1758) – *Plantago, Plantaginis*, the plantain genus: incorrectly stated by Linnaeus to be the foodplant.

subsp. **insularum** Seitz, 1910 – *insula*, gen. pl. *insularum*, an island: the subspecies described from the Orkney Islands.

Arctia Schrank, 1802 – ἄρκτος (arktos), a bear: from the hairy larva, the 'woolly bear'. Mouffet (1634) called a hairy caterpillar a 'bear-worm' and in German it is a 'barenraupe'. Originally bestowed as a family name (see p. 30).

2057 **caja** (Linnaeus, 1758) – Caia, Caja, a Roman lady's name, the feminine of Caius as in Caius Julius Caesar. See next species.

2058 **villica** (Linnaeus, 1758) – *villica*, a female housekeeper in charge of a villa. This name and the last follow Linnaeus' predilection for using feminine names for species with brightly-coloured hindwings; cf. *Catocala* spp. 2452, etc.

subsp. **britannica** Oberthür, 1911 – the subspecies occurring in Britain.

Diacrisia Hübner, 1819 – διάκρισις (diakrisis), a separating: from the sexual dimorphism of the next species, the only one in the genus.

2059 **sannio** (Linnaeus, 1758) – *sannio*, a mimic, probably of its own female. Although Linnaeus described the sexes on different pages, he wrote of *russula* (the female) '*simillima P. Bombyci Sannioni, ut facile tantum sexu differre credatur*' (so very like *P. Bombyx sannio* that it may easily be thought to differ only in sex). Linnaeus described *sannio* first so the 'mimic' preceded the 'model'. Macleod translates *sannio* as 'grimacer' and supposed that he saw a grimacing face in the wing pattern.

Spilosoma Curtis, 1825 – σπίλος (spilos), a spot; σῶμα (sōma), the body: from the bold spots on the abdomen of the adult.

2060 **lubricipeda** (Linnaeus, 1758) – *lubricipes*, swift-footed: no doubt from the speedy gait of the larva, though Linnaeus does not say so.

2061 **lutea** (Hufnagel, 1766) – *luteus*, yellow: from the ground colour of the forewing.

2062 **urticae** (Esper, 1789) – *Urtica*, the nettle genus: a possible foodplant, but not the one most frequently used in Britain.

Diaphora Stephens, 1827 – διαφορά (diaphora), distinction: from the contrast between the typically brown male and white female; cf. 2059 *Diacrisia*.

2063 **mendica** (Clerck, 1759) – *mendicus*, a beggar: from the drab, brown coloration of the male. 'Here reference is probably to Carmelite mendicant friar, whose white mantle the female moth's wings are supposed to resemble' (Macleod); Clerck figured only the male and there is no evidence that he knew the female. Had he done so, he would probably have shown both sexes.

Phragmatobia Stephens, 1828 – φράγμος, φράγματος (phragmos, phragmatos), a fence; βιόω (bioō), to live: Stephens gives this explanation without further comment; he may have found adults at rest on a fence. 'From one of the places where larva hibernates' (Macleod); the larva overwinters in heather.

2064 **fuliginosa** (Linnaeus, 1758) – *fuliginosus*, sooty: from the dusky ground colour of the forewing.

subsp. **borealis** (Staudinger, 1871) – *borealis*, northern: the subspecies occurring in the north, in the British Isles in Scotland.

Pyrrharctia Packard, 1864 – πυῤῥός (purrhos), flame-coloured; the genus 2057 *Arctia*, q.v.

2065 **isabella** (Smith, 1797) – a lady's name (cf. 2057 *caja*); the colour isabella or isabelline (yellowish grey) is out of keeping with the generic name.

Halysidota Hubner, 1819 – ἁλυσίδωτός (halusidōtos), wrought in chain fashion: perhaps from the wing pattern.

2066 **moeschleri** Rothschild, 1909 – possibly i.h.o. B. H. Möschler, but Rothschild gives no explanation.

Euplagia Hübner, 1820 – εὖ (eu), well, very; πλάγιος (plagios), oblique: from the strongly outwards-oblique fasciae on the forewing of the next species.

2067 **quadripunctaria** (Poda, 1761) – *quadri-*, four; *punctum*, a spot: from the spotting on the hindwing; if the subapical streak and the small terminal dots are overlooked, there remain two largish subcircular spots on each hindwing.

Callimorpha Latreille, 1809 – κάλλος (kallos), beauty; μορφή (morphē), shape, form: descriptive of moths in the genus, which included the next species.

2068 **dominula** (Linnaeus, 1758) – dim. of *domina*, the mistress of the household (*domus*): a name analogous with 2057 *caja* and 2058 *villica*, qq.v.

Tyria Hübner, 1819 – Τυρίος (Turios), a Graecism for Tyrian, Tyre having been famous for its purple dye.

2069 **jacobaeae** (Linnaeus, 1758) – *Senecio jacobaea*, ragwort: the foodplant; called by Linnaeus *Jacobaea senecionis*.

Ecpantheria Hübner, 1820 – ἔξ, ἐκ- (ex, ek-), outside; πάνθηρ, a panther: descriptive: cf. 1909 *Pseudopanthera*.

2069a **deflorata** (Fabricius, 1794) – *de-*, un-; *florus*, bright; *-ata*, geometrid termination.

CTENUCHIDAE

From the North American genus *Ctenucha* Kirby, 1837, raised to accommodate specimens collected in Canada during an expedition led by Sir John Franklin. The meaning is obscure; it may be derived from κτείς, κτενός (kteis, ktenos), a comb, and νύχιος (nukhios), night-, either active by night or dark as night. The 'comb' is certainly the male antenna, which Kirby describes as having very long pectinations; νύχιος probably refers to the time of flight.

SYNTOMINAE (2070)

Syntomis Ochsenheimer, 1808 – σύντομος (suntomos), cut short: from the abbreviated hindwing.

2070 **phegea** (Linnaeus, 1758) – Phegea, daughter of Phegeus, King of Psophos in Arcadia.

Dysauxes Hübner, 1819 – δυσαυξής (dusauxēs), growing slowly: from the reduced hindwing, seen as failing to keep pace in growth with the forewing.

2071 **ancilla** (Linnaeus, 1767) – *ancilla*, a maid-servant: the hindwing coloration is reminiscent of that of 2107 *Noctua pronuba* and Linnaeus habitually named species with coloured hindwings after the fairer sex (cf. 2452 *nupta*).

EUCHROMIINAE (2072)

Euchromia Hübner, 1819 – εὖ (eu), well, very; χρῶμα (khrōma), colour: a genus of brightly-coloured, mainly African, species. Cf. 1289 *Euchromius*.

2072 **lethe** (Fabricius, 1775) – Lethe, the River of Oblivion in the Underworld.

Antichloris Hübner, 1818 – ἀντί (anti), against, but here probably in the sense of the Lat. *ante*, before, in front; χλωρός (khlōros), pale green: from the green forewing of some members of the genus.

2073 **viridis** (Druce, 1884) – *viridis*, green: from the coloration of the forewing.

= **musicola** Cockerell, 1910 – *Musa sapientum*, banana; *colo*, to inhabit: the type specimen was found in a grocery store in Boulder, Colorado, and was believed to have been imported from Central America amongst bananas.

2074 **caca** (Hübner, 1818) – perhaps κακός (kakos), bad: a South American species placed on the British list for a now forgotten reason. 'Mistake for *caeca*, hidden, from larva' (Macleod); an unjustified emendation and it is most improbable that Hübner knew the life history.

2074a **eriphia** (Fabricius, 1776) – the name of an unknown plant mentioned by Pliny.

NOLIDAE (2077)

Meganola Dyar, 1898 – μέγας (megas), big; the genus 2077 *Nola*, q.v.: a closely related genus, but one containing moths that are on average larger.

2075 **strigula** ([Denis & Schiffermüller], 1775) – *strigula*, a small line: from the strigulae on the forewing.

2076 **albula** ([Denis & Schiffermüller], 1775) – *albulus*, whitish: from the ground colour.

Nola Leach, 1815 – Nola, a town in Campania (Pickard *et al.*; Macleod). Spuler suggests that the name is taken from the Latin *nolo*, I refuse, and indicates chastity, a virtue exemplified by the white wings; however, the formation of a generic name from a Latin word, other than a proper name, would be unusual. If, as is more likely, it is a geographical name, it is one of the earliest, antedating those of Ochsenheimer by a year (see p. 33), and the only one where Leach turned to this source. Leach, who placed the genus in the Tortricidae, himself gives no explanation.

2077 **cucullatella** (Linnaeus, 1758) – *cucullus*, a cowl, a hood: easily explained as referring to the dark basal area of the forewing which resembles a cloak thrown over the shoulders when the moth is seen at rest (Macleod), a character which gave rise to the English name 'the short-cloaked moth'; cf. 1736 *cuculata*. Linnaeus, however, described this species in Tinea, which means that he had before him a moth that rests with convolute wings (Pl. III); *N. cucullatella* rests with forficate wings like a pyrale and other members of the genus (2078, 2079a) were accordingly placed in Pyralis. Describing the life history, Linnaeus wrote *'habitat in* Sorbo *intra folliculum conicum confectum'*, it lives inside a small conical bag constructed on *Sorbus*. Fabricius (1794) interpreted the *'folliculum'* as the cocoon but Linnaeus' text would refer more naturally to the larval habitation. *Sorbus* is not otherwise recorded as a foodplant of this species. In his description of the adult Linnaeus wrote *'antice striga nigra recurvata'*, with a curved black streak in front, 'in front' meaning either on the forewing or that part of it nearer the base, i.e. the head when the moth is at rest; if Linnaeus was describing the subbasal fascia of *N. cucullatella*, why did he not write *'fascia'* as was his usual practice? The species which rolls its wings and makes a conical spinning on *Sorbus* (as well as *Crataegus*) is 303 *Parornix anglicella*, and it is possible that the name is descriptive of the cone made by that species.

2078 **confusalis** (Herrich-Schäffer, 1847) – *confusus*, confused: probably from uncertainty over the correct classification, the termination *-alis* showing that Herrich-Schäffer concluded that it was a pyrale. '*Confusus*, sprinkled, from dark speckling of forewings' (Macleod); unlikely, because *perfusus* means sprinkled and *confusus* was never used in that sense.

2079 **aerugula** (Hübner, 1793) – *aerugo*, the rust of copper, verdigris: there is no trace of green coloration and Hübner, as an artist, would not have supposed that there was. The intended meaning is probably 'coppery', with reference to the pale brown irroration on the forewing.

2079a **chlamytulalis** (Hübner, 1813) – χλαμύς (khlamus), a cloak; τύλη (tulē), a pad on the shoulder: meaning similar to that given first for 2077 *cucullatella*, q.v. The termination *-alis* shows that Hübner thought it was a pyrale; cf. 2078 *confusalis*.

NOCTUIDAE (2107)

NOCTUINAE (2107)

Euxoa Hübner, 1821 – εὔξοος (euxoös), well-polished: from the subhyaline hindwing.

2080 **obelisca** ([Denis & Schiffermüller], 1775) – *obeliscus*, a small spit: from the black streak which extends from beyond the stigmata inwards towards the base of the forewing, ending in a point.

subsp. **grisea** (Tutt, 1902) – *griseus* (late Lat.), grey: from the greyer coloration of the subspecies which occurs in Britain.

2081 **tritici** (Linnaeus, 1761) – *Triticum*, the wheat genus, wrongly supposed by Linnaeus to include the foodplants.

2082 **nigricans** (Linnaeus, 1761) – *nigricans*, blackish: from the ground colour of the darker forms.

2083 **cursoria** (Hufnagel, 1766) – *cursorius*, pertaining to a race-track: races took place in an arena (*arena*, sand) and there may be an oblique reference to the moth's sand-hill habitat; see 2085 *vestigialis*. '*Cursor*, runner, from roaming habit' (Macleod); most unlikely and unapt, since this is a strictly colonial and sedentary species.

Agrotis Ochsenheimer, 1816 – either ἀγρώτης (agrōtēs), of the field, or ἀγρότης (agrŏtēs), a countryman: from the grassland rather than the woodland habitat. Your pronunciation will depend on the derivation you select; although Treitschke, writing after the death of Ochsenheimer, gave the latter, I prefer the former, *Agrōtis*. The name was first proposed by Hübner in his *Tentamen* [1806] and the type species is 2087 *A. segetum*, q.v.

2084 **cinerea** ([Denis & Schiffermüller], 1775) – *cinereus*, ashy: from the ground colour.

2085 **vestigialis** (Hufnagel, 1766) – *vestigium*, a footstep, a footprint: Hufnagel gives the moth the vernacular name 'Erdlaufe', earth-course, earth-track, the name evidently having affinity of meaning with 2083 *cursoria*, bestowed in the same paper. He may have seen the moths running over sand and leaving a trail. 'From marking on forewings' (Macleod); one might just suppose the orbicular and reniform stigmata to resemble the imprint of the heel and the sole respectively, but Hufnagel seems to be indicating a trail rather a single footmark.

2086 **biconica** (Kollar, 1844) – *bi-*, two; κωνικός (kōnikos), cone-shaped: from the markings on the forewing.

= **spinifera** (Hübner, 1808) – *spina*, a spine; *fero*, to carry: from the long and conspicuous claviform stigma.

2087 **segetum** ([Denis & Schiffermüller], 1775) – *seges*, gen. pl. *segetum*, a cornfield, but loosely used of any field in which a crop is grown: from the habitat of the larva which is a pest, especially of root-crops.

2088 **clavis** (Hufnagel, 1766) – *clavis*, a key: the 'claviform' stigma, the 'club' of the vernacular name 'heart and club'.

2089 **exclamationis** (Linnaeus, 1758) – *exclamatio*, an exclamation, here an exclamation mark: from the mark on the forewing described by Linnaeus as '*lineola atra latiuscula brevi*', a short but rather broad little black line.

2090 **trux** (Hübner, 1824) – *trux*, rough: perhaps from the coarsely pectinate antenna, or (Spuler) from the hairy abdomen.

subsp. **lunigera** Stephens, 1829 – *luna*, the moon; *gero*, to carry: from the lunate mark in the reniform stigma.

2091 **ipsilon** (Hufnagel, 1766) – the Greek letter ypsilon, written as a capital (Y): from the black outer margin of the reniform stigma and the dart-shaped mark projecting from it.

2092 **puta** (Hübner, 1803) – an intriguingly puzzling name: 'Puta, a goddess who presided over the pruning of trees' (Pickard *et al.*); 'pure, but application obscure' (Macleod); it could also mean *puta*, the imperative of *puto*, to think, used adverbially as our 'e.g.'; a little girl (a spelling variant of *pusa*, cf. 721 and 1955); or it may be derived from *puteus*, a well, *puteal*, the enclosure round a well (from the forewing markings demarcated in fuscous). My own preference is from *puter*, rotten, putrid, with reference to the forewing pattern which resembles dead wood; cf. 693, 2098 and 2206. The form '*putris*' (2098) was preoccupied.

subsp. **insula** Richardson, 1958 – *insula*, an island: the subspecies occurring in the Isles of Scilly.

2093 **ripae** (Hübner, 1823) – *ripa*, gen. *ripae*, the bank of a river or, as here, the shore of the sea: from the habitat.

2094 **crassa** (Hübner, 1803) – *crassus*, stout, gross: the 'great dart'.

2094a **deprivata** Walker, 1857 – Latinized from 'deprived': Walker describes the forewing markings as only faintly indicated.

Feltia Walker, 1856 – probably a meaningless neologism.

2095 **subgothica** (Haworth, 1809) – *sub-*, somewhat resembling; 2190 *Orthosia gothica*: from fancied similiarity.

2096 **subterranea** (Fabricius, 1794) – *subterraneus*, underground: from the larval behaviour.

Actinotia Hübner, 1821 – ἀκτή (aktē), a promontory, a raised place, a prominence; νῶτον (nōton), the back: from the dorsal crests on the abdomen.

2097 **polyodon** (Clerck, 1759) – πολύς (polus), many; ὀδούς, ὀδόντος (odous, odontos), a tooth: from the series of wedge-shaped subterminal markings on the forewing – '*margine postico multidentato*' (Linnaeus, 1767). Macleod invents a tooth-like tuft of scales on the forewing.

Axylia Hübner, 1821 – ἀ, alpha copulative; ξύλον (xulon), cut wood: from the resemblance between the pattern of the next species and the grain on a plank.

2098 **putris** (Linnaeus, 1761) – *puter*, rotten, decaying: from the resemblance between the moth's forewing and touchwood.

Actebia Stephens, 1829 – ἀκτή (aktē), a headland, the coast in general; βιόω (bioō), to live: from the coastal habitat, in Britain, of the next species.

2099 **praecox** (Linnaeus, 1758) – *praecox*, appearing early in the year: from the delicate green coloration, reminiscent of fresh spring foliage; cf. 2247 *aprilina*. There is no need to suppose, as Macleod did, that the reference is to the time of the moth's appearance and that Linnaeus got it wrong.

2100 **fennica** (Tauscher, 1806) – *Fennicus*, Finnish: a Holarctic species reaching Finland as a migrant.

Ochropleura Hübner, 1821 – ὠχρός (ōkhros), pale, wan, sallow; πλευρά (pleura), a rib, the costa of a wing: from the pale costal streak of 2102 *plecta*.

2101 **flammatra** ([Denis & Schiffermüller], 1775) – *flammeus*, flame-coloured; *ater*, black: from the pinkish grey forewing and its black subbasal markings.

2102 **plecta** (Linnaeus, 1761) – πλεκτή (plektē), a twisted rope: from the pale costal streak.

2102a **leucogaster** (Freyer, 1831) – λευκός (leukos), white; γαστήρ (gastēr), the belly: from the pale colour of the abdomen.

Eugnorisma Boursin, 1946 – εὖ (eu), well; γνώρισμα (gnōrisma), a mark for recognition: a genus of well-marked moths, and a Greek rendering of 2103 *depuncta*.

2103 **depuncta** (Linnaeus, 1761) – *depunctus*, clearly designated: a species with distinct markings.

Standfussiana Boursin, 1946 – i.h.o. M. R. Standfuss (1854–1917), a German entomologist.

2104 **lucernea** (Linnaeus, 1758) – *lucerna*, a lamp: *'frequenter candelis involans'*, often flying into candles (Linnaeus, *teste* Pickard *et al.*).

Rhyacia Hübner, 1821 – ῥύαξ (rhuax), a stream: from the dentate fasciae on the forewing, reminiscent of the 'rivulets' of many geometers.

2105 **simulans** (Hufnagel, 1766) – *simulans*, pretending: possibly from the adult's habit of feigning death during its period of aestivation in August. Hufnagel may have observed the behaviour, even if he was unaware of the reason.

2106 **lucipeta** ([Denis & Schiffermüller], 1775) – *lux, lucis*, light; *peto*, to seek: named for the same reason as 2104 *lucernea*, q.v.

Noctua Linnaeus, 1758 – *noctus, noctu*, night, by night; *noctua*, the short-eared owl, sacred to Minerva: whether or not Linnaeus had the bird in mind, his successors interpreted the name in that way, authors such as Schrank incorporating 'eule' (owl) into their vernacular names for the noctuids, e.g. 2166 *Hadena*, q.v.; in Britain they have been called 'owlets'. The principal meaning, however, is the time of flight of the great majority of the species. The Noctuae constitute one of the main regiments in the Linnaean classification (Pl. III); they rest with their wings folded over the body, their antennae have bristles but not pectinations and, apart from the *Noctuae elingues* (Noctuae without tongues) separated in 1775 by Fabricius (14 *Hepialus*, q.v.), they have a coiled haustellum. Deviation from this diagnosis, e.g. the presence of pectinated antennae, has influenced the formation of certain names. It was not until after 1800 that the *Noctuae spirilingues* began to be split into smaller units.

2107 **pronuba** Linnaeus, 1758 – *pronuba*, a bridesmaid: there is no entomological application, but Linnaeus made a practice of giving species with coloured hindwings the names of women, especially in the context of marriage; see 2452 *nupta*.

2108 **orbona** (Hufnagel, 1766) – *orbus*, bereaved, an orphan; Orbona was the goddess of parents who had lost their children. There may be a hint of the marriage theme.

2109 **comes** Hübner, 1813 – *comes*, a comrade, a companion; if, as is likely, Hübner was following the marriage theme for 'underwings', *comes* is a common-law wife, like its Greek equivalent ἑταίρα (hetaira), a companion. 'From close resemblance to . . . *orbona*' (Macleod); this is also possible.

2110 **fimbriata** (Schreber, 1759) – *fimbriatus*, fringed: from the orange cilia on the hindwing, which contrast strongly with the black 'broad border'. The termination '*-ata*' suggests a member of the Geometridae and accordingly Linnaeus (1767) sought to change the name to *fimbria*.

2111 **janthina** [Denis & Schiffermüller], 1775 – ἰάνθινος (ianthinos), violet-coloured: from the clouding of that colour on the forewing.

2112 **interjecta** Hübner, 1803 – *interjectus*, placed between, added: as the last of the yellow underwings to be named to date, it had to be interposed in the series. Macleod's explanation is anachronistic and factually incorrect, 'interposed, i.e. between *Euschesis janthina* and *Lampra fimbriata*, both species having once been included in the genus *Triphaena*': *fimbriata* was named in Phalaena, *janthina* and *interjecta* in Noctua. *Triphaena* Ochsenheimer, 1816 is junior by 13 years to *interjecta*, 41 years to *janthina* and 57 years to *fimbriata*.

subsp. **caliginosa** (Schawerda, 1919) – *caliginosus*, dark, obscure: a subspecies having its ground colour darker than that of the nominate subspecies.

Spaelotis Boisduval, 1840 – meaningless or, more probably, malformed. Macleod derives from σπηλαιώτης (spēlaiōtes), a cave-dweller, from the larva's habit of living underground; the next species, however, feeds above ground. Another possibility is σπίλος (spilos), a spot; οὖς, ὠτός (ous, ōtos), the ear: from the ear-shaped reniform stigma, which has its dark outline broken into spots.

2113 **ravida** ([Denis & Schiffermüller], 1775) – *ravidus*, greyish, dark-coloured: from the forewing ground colour.

Graphiphora Ochsenheimer, 1816 – γραφίς (graphis), a stilus or style for writing on wax tablets; φορέω (phoreō), to carry: from the claviform stigma or 'dart' on the forewing. Ochsenheimer took this name from Hübner's *Tentamen* [1806].

2114 **augur** (Fabricius, 1775) – *augur*, a soothsayer, a diviner: perhaps because the broken outline of the stigmata suggested runic lettering ('*characteribus atris . . . variis*', with sundry black markings). Fabricius gives no explanation. Some of his names are puns and here he may be playing on the sense of *augeo*, to increase, from the moth's stout dimensions. 'Perhaps from belief that moth is harbinger of summer' (Macleod); unlikely, since the flight period is in mid-summer.

Eugraphe Hübner, 1821 – εὖ (eu), well; γραφή (graphē), a delineation, a design, a letter of the alphabet: the type species is *E. sigma* ([Denis & Schiffermüller]), named from the Greek letter.

2115 **subrosea** (Stephens, 1829) – *sub-*, somewhat; *roseus*, rosy: from the rosy flush on the forewing.

Paradiarsia McDunnough, [1929] – παρά (para), alongside; the genus 2120 *Diarsia*, q.v.; a collateral genus.

2116 **sobrina** (Duponchel, [1843]) – *sobrina*, a cousin german, a first cousin: from resemblance to another species, e.g. 2115 *Eugraphe subrosea*, as suggested by Spuler, 2120 *Diarsia mendica*, 2121 *D. dahlii* or 2134 *Xestia xanthographa*. Macleod's explanation 'from larva's resemblance to that of [2130 *Xestia*] *Amathes baja*' lacks foundation and presupposes that Duponchel knew the life history.

2117 **glareosa** (Esper, 1788) – *glareosus*, gravelly: probably from a habitat ('. . . gravelly soils . . .', MBGBI 9: 169; '. . . shingly beaches . . .', Skinner, 1984: 91). 'From colour of some forms' (Macleod); his gravel would have to be grey with a pink, coffee or blackish tinge.

Lycophotia Hübner, 1821 – λυκόφως (lukophōs), twilight: from the 'dusky' white pattern; there is no reference to crepuscular flight, as supposed by Macleod.

2118 **porphyrea** ([Denis & Schiffermüller], 1775) – πορφύρεος (porphureos), purple: the ground colour is purplish brown in some forms.

Peridroma Hübner, 1821 – περίδρομος (peridromos), running round, in two senses (a) surrounding and (b) gadding about: in the first sense, from the ring surrounding the reniform stigma.

2119 **saucia** (Hübner, 1808) – *saucius*, wounded: from the reddish tinge often present on the forewing, suggestive of bleeding.

Diarsia Hübner, 1821 – δίαρσις (diarsis), a raising up: application obscure; there is a tendency for the male anal tuft to stand erect in set specimens, but I suggest this without confidence. Hübner's diagnosis gives no clue.

2120 **mendica** (Fabricius, 1775) – *mendicus*, a beggar, indigent: Fabricius gives no explanation and there need not necessarily be one. Macleod's explanation 'from drab appearance' is incorrect since this is one of the prettier noctuids.

subsp. **thulei** (Staudinger, 1891) – the subspecies occurring in Shetland, often identified with the Roman *Ultima Thulē*, their 'back of beyond'.

subsp. **orkneyensis** (Bytinski-Salz, 1939) – the subspecies occurring in Orkney.

2121 **dahlii** (Hübner, 1813) – i.h.o. Georg Dahl, a Viennese collector who was the first to rear the species and supplied Hübner with a specimen to figure.

2122 **brunnea** ([Denis & Schiffermüller], 1775) – *brunneus* (late Lat.), brown: from the ground colour.

2123 **rubi** (Vieweg, 1790) – *Rubus*, the bramble genus: the species is widely polyphagous but rarely, if ever, accepts bramble.

2124 **florida** (Schmidt, 1859) – *floridus*, blooming, bright: from having a brighter ground colour than the previous species to which it is closely related.

Xestia Hübner, 1818 – ξεστός (xestos), polished, smooth: from the glossy forewings of some of the species.

2125 **alpicola** (Zetterstedt, 1839) – *colo*, to inhabit: a species inhabiting the Alps.

subsp. **alpina** (Humphreys & Westwood, 1843) – *alpinus*, alpine, mountain-: from the mountainous type locality in the Scottish Highlands; originally described as a distinct species.

2126 **c-nigrum** (Linnaeus, 1758) – *niger*, black: having a black C-shaped mark in the subcostal area of the forewing.

2127 **ditrapezium** ([Denis & Schiffermüller], 1775) – δι-, δις- (di-, dis-), two; τραπέζιον (trapezion), an irregular four-sided figure, a table: from the black subcostal stripe which is broken into two by the orbicular stigma.

2128 **triangulum** (Hufnagel, 1766) – *triangulum*, a triangle: from the shape of the pale orbicular stigma which interrupts the black subcostal stripe; the basal section of the stripe is also triangular.

2129 **ashworthii** (Doubleday, 1855) – i.h.o. Joseph H. Ashworth, who discovered the species at Llangollen, Denbighshire (Clwyd) in 1853.

2130 **baja** ([Denis & Schiffermüller], 1775) – *badius*, chestnut-coloured, whence came the English 'bay' and the French 'bai', which the authors re-latinized. The derivation put forward by Pickard *et al.* from Baiae, a town in Italy, is most unlikely; the type locality is Vienna.

2131 **rhomboidea** (Esper, 1790) – ῥομβοειδής (rhomboeidēs), rhombus-shaped, a rhombus being a parallelogram other than a square: from the shape of the section of the black subcostal streak situated between the stigmata.

2132 **castanea** (Esper, 1796) – *castaneus* (late Lat.), chestnut-coloured (*Castanea sativa*, the sweet chestnut): from the colour of the forewing, though it is often paler.

2133 **sexstrigata** (Haworth, 1809) – *sex*, six; *striga*, a furrow, a line: the 'six-striped rustic'.

2134 **xanthographa** ([Denis & Schiffermüller], 1775) – ξανθός (xanthos), yellow; γραφή (graphē), a marking: from the yellowish white stigmata.

2135 **agathina** (Duponchel, 1827) – Agatha, the Latin name for Agde, a French town close to Montpellier on the Gulf of Lyons, the type locality. Macleod's derivation from ἀγαθίς (agathis), a ball of thread, is fabrication.

subsp. **hebridicola** (Staudinger, 1901) – *colo*, to inhabit: the subspecies occurring in the Hebrides.

Naenia Stephens, 1827 – Naenia, the Roman goddess of funerals: Stephens gives no reason, presumably because there is none.

2136 **typica** (Linnaeus, 1758) – τύπος (tupos), a pattern, *typicus*, having a distinctive pattern: Linnaeus describes the pale reticulation and the reniform stigma. Macleod's 'type species' is ridiculous. How could a moth be named in 1758 as the type species of a genus established sixty-nine years later and, of course, the concept of type species and type specimens is post-Linnaean. It is far from typical and its correct systematic placing is still uncertain.

Eurois Hübner, 1821 – εὔροος (euroös), fair-flowing: from the attractive undulate fasciae ('rivulets') on the forewing.

2137 **occulta** (Linnaeus, 1758) – *occultus*, hidden, obscured: Linnaeus describes the forewings as '*fusco nebulosis*', clouded with fuscous, and Clerck (1759) figures the hindwings as very dark. 'Hidden, from moth's habit of hiding among dead leaves on the ground or crevice of bark of tree trunk' (Macleod). What evidence did he have that the moth behaved in this manner?

Anaplectoides McDunnough, 1929 – There are three stages in the development of this name. (1) *Aplecta* Guenée, 1838, a junior synonym of 2148 *Polia* – ἀ, alpha privative; πλεκτός (plektos), folded: not folded, in contrast to 2305 *Euplexia* Stephens, 1829, q.v. (2) *Aplectoides* Butler, 1878 – the genus *Aplecta*; εἶδος (eidos), form, appearance: a North American genus resembling *Aplecta*. (3) *Anaplectoides* – ἀ, alpha privative; n, phonetic insertion; the genus *Aplectoides*: not that genus but related to it. This, too, is a mainly North American genus, but the next species is Holarctic. 'From markings on forewings' (Macleod).

2138 **prasina** ([Denis & Schiffermüller], 1775) – *prasinus*, leek-green: from the colour of the forewing; cf. 2421 *prasinana*.

Cerastis Ochsenheimer, 1816 – κεράστης (kerastēs), horned: from the male antenna of the next species which is bipectinate. Since Linnaeus had defined the Noctua as having bristled but not pectinate antennae, this was an important departure. 'Perhaps from crests on thorax' (Macleod): the thorax is uncrested.

2139 **rubricosa** ([Denis & Schiffermüller], 1775) – *rubricosus*, having the colour of red ochre: from the ground colour of the forewing.

2140 **leucographa** ([Denis & Schiffermüller], 1775) – λευκός (leukos), white; γραφή (graphē), a marking: from the conspicuously pale stigmata, the 'white-marked'.

Mesogona Boisduval, 1840 – μέσος (mesos), middle; γόνυ (gonu), the knee: from the midtarsus, which has three rows of spines. 'From two converging lines one on each forewing, that, if continued, would meet near the head' (Macleod): the lines are subbasal, not median (μεσο-), and it would have been better if Macleod had consulted Boisduval.

2141 **acetosellae** ([Denis & Schiffermüller], 1775) – *Rumex acetosella*, sheep's-sorrel: the larva is polyphagous, but on trees and not herbaceous plants.

HADENINAE (2166)

Anarta Ochsenheimer, 1816 – *anarta*, a sea-cockle mentioned by Pliny. Ochsenheimer himself gives this derivation; there is no entomological application. 'Perhaps from *anartao*, hang up; but application obscure' (Macleod).

2142 **myrtilli** (Linnaeus, 1761) – *Vaccinium myrtillus*, bilberry: a reputed foodplant, but the British population feeds on *Erica* and *Calluna*.

2143 **cordigera** (Thunberg, 1788) – *cor, cordis*, the heart; *gero*, to wear, to carry from the large white heart-shaped reniform stigma (with apologies to anatomists).

2144 **melanopa** (Thunberg, 1791) – μέλας, μελαν- (melas, melan-), black; ὄψ, ὤψ (ops, ōps), the eye: from the black reniform stigma, which has a white 'pupil' and often a white outline. '*Ops*, face' (Macleod).

Discestra Hampson, 1905 – δι-, δις- (di-, dis-, two), κέστρα (kestra), a pointed instrument: from the fine, bifid, post-thoracic crest.

2145 **trifolii** (Hufnagel, 1766) – *Trifolium*, the clover genus: the larva, however, feeds on Chenopodiaceae.

Lacinipolia McDunnough, 1937 – a North American genus of fifty-seven species, so the name may not be descriptive of the two that have occurred as adventives in Britain; possibly from λακίς (lakis), a break; πολιός (polios), grey. Probably affinity with 2148 *Polia* is intended.

2146 **renigera** (Stephens, 1829) – *renes*, a kidney; *gero*, to carry: from the reniform stigma.

2146a **laudabilis** (Guenée, 1852) – *laudabilis*, praiseworthy; a laudable species.

Hada Billberg, 1820 – Ἅδης (Hadēs), the Underworld: a genus having affinity with 2166 *Hadena*, q.v.

2147 **nana** (Hufnagel, 1766) – *nanus*, a dwarf: in 1766 all noctuids were in the single genus *Noctua*, and in that assemblage it was one of the smaller species.

Polia Ochsenheimer, 1816 – πολιός (polios), grey: from the colour of some members of the genus, which was formerly more extensive. Ochsenheimer took the name from Hübner's *Tentamen* [1806].

2148 **bombycina** (Hufnagel, 1766) – *bombycinus*, silken (βόμβυξ, the silk-worm): from the gloss on the forewing – the 'pale shining brown'; *nitens* Haworth, 1800, (shining) is a synonym. There is no connection with Bombyx (1631, superfamily) as supposed by Macleod.

2149 **trimaculosa** (Esper, 1788) – *tri-*, three; *maculosus*, spotted: from the subterminal line which is clearly defined only on the costa, in mid-wing and on the dorsum.
= **hepatica** sensu auctt., *nec* (Clerck, 1759) – see 2236.

2150 **nebulosa** (Hufnagel, 1766) – *nebulosus*, cloudy: from the colour of the forewing, the species being liable to melanism.

Pachetra Guenée, 1841 – παχύς (pakhus), thick; ἦτρον (ētron), the abdomen: a genus of stout-bodied moths.

2151 **sagittigera** (Hufnagel, 1766) – *sagitta*, an arrow; *gero*, to carry: from the sagittate marks in the subterminal line.
subsp. **britannica** Turner, 1933 – the subspecies occurring in Britain.

Sideridis Hübner, 1821 – σίδηρος (sidēros), iron; εἶδος (eidos), form, appearance: from the rusty brown ('rostbraun') colour of the species in the original genus, which did not then include the next species. The name is a loose Greek translation of *ferruginea* [Denis & Schiffermüller] (a junior synonym of 2262 *Agrochola circellaris*, q.v.), one of the species Hübner did include. See p. 33.

2152 **albicolon** (Hübner, 1813) – *albus*, white; *colon* (κῶλον), originally a limb, a member, then a clause in a sentence, whence the mark of punctuation separating two such clauses: from the two white dots forming the reniform stigma.

Heliophobus Boisduval, 1828 – ἥλιος (hēlios), the sun; φοβέω (phobeō), to fear: from the nocturnal habits of the species in the genus.

2153 **reticulata** (Goeze, 1781) – *reticulatus*, net-like: from the reticulation resulting from the intersection of pale veins and fasciae on the forewing.

subsp. **marginosa** (Haworth, 1809) – *marginosus*, an adjective coined from *margo, marginis*, a border: from the subterminal shade on the hindwing.

subsp. **hibernica** Cockayne, 1944 – the subspecies found in Ireland.

Mamestra Ochsenheimer, 1816 – according to Treitschke, the capital city of Lesser Armenia; there is no entomological application. See also 2162 *Papestra*.

2154 **brassicae** (Linnaeus, 1758) – *Brassica*, the cabbage genus, which includes some of the foodplants.

Melanchra Hübner, 1820 – μέλας, μελαν- (melas, melan-), black; χρῶς (khrōs), colour, complexion: from the dark ground colour.

2155 **persicariae** (Linnaeus, 1761) – *persicaria* (late Lat.), a peach-tree, later applied to *Polygonum persicaria*, redleg, from the similarity of the leaves: a foodplant.

Lacanobia Billberg, 1820 – a typographical error for *Lachanobia*, λάχανα (lakhana, vegetables, greens; βιόω (bioō), to live: the original genus was more extensive, including vegetable-eaters such as 2154 *Mamestra brassicae*.

2156 **contigua** ([Denis & Schiffermüller], 1775) – *contiguus*, touching, adjacent: i.e. close to 2145 *Discestra trifolii* which it follows, both having been placed in the same larval group of species feeding on Chenopodiaceae spp. Pickard *et al.*, Spuler and Macleod suggest proximity to other species, the first two to 2158 *L. thalassina* and the third to 2157 *L. w-latinum*.

2157 **w-latinum** (Hufnagel, 1766) – from the two contiguous v-shaped marks in the subterminal line which together form a 'w'. The Latin alphabet has no 'w', so the letter had to be 'Latinized' to qualify as a scientific name.

2158 **thalassina** (Hufnagel, 1766) – *thalassinus* (θαλάσσινος), sea-coloured, sea-green or, possibly, sea-like: probably from the wavy subterminal line. Macleod's explanation is ingenious, 'sea-coloured, from the reddish brown forewings, the sea, according to Homer, being wine-dark'. I could produce arguments for Homer having meant 'black' rather than 'red', but what matters is what Hufnagel thought, not what I think.

2159 **suasa** ([Denis & Schiffermüller], 1775) – *suasum*, a colour produced by dyeing; '*Suasum colos appellatur, qui fit ex stillicidio fumoso in vestimento albo. Nec desunt, qui dicant, omnem colorem, qui fiat inficiendo, suasum vocari*', a colour is called '*suasum*' which is produced by a smoky spray on a white garment. Moreover, there are those who say that all colour produced by dyeing is called '*suasum*' (Sextus Pompeius Festus, a grammarian of the 4th century A.D.). The name refers, therefore, to the smoky brown ground colour of the forewing. '*Suasus*, persuaded' (Pickard *et al.*).

2160 **oleracea** (Linnaeus, 1758) – *oleraceus*, pertaining to vegetables: '*habitat in* Olerum *radicibus, quas consumit*', it lives on the roots of vegetables, which it consumes. It is a polyphagous species and sometimes a pest; it generally feeds externally (the Latin '*in*' can mean 'in' or 'on').

2161 **blenna** (Hübner, 1824) – *blennus*, a blockhead: probably without entomological application. 'Perhaps from larva's clumsiness in making cocoon' (Macleod); the pupa is subterranean.

Papestra Sukhareva, 1973 – a name formed on the analogy of 2154 *Mamestra* and as its subgenus. Spuler tentatively, and wrongly, suggested that the latter name was derived from μάμμη, μάμμα, mummy, mama, and Sukhareva, accepting his opinion, formed his name from πάππας, πάπας, papa.

2162 **biren** (Goeze, 1781) – *bi-*, two; *renes*, the kidney: from the orbicular and reniform stigmata, both of which tend to be 'reniform'. Goeze spelt the name '*bi-ren*'.

Ceramica Guenée, 1852 – κεραμικός (keramikos), pertaining to pottery, of the colour of earthenware: from the coloration of the species placed in the genus.

2163 **pisi** (Linnaeus, 1758) – *Pisum sativa*, the garden pea: given by Linnaeus as a foodplant, but one unusual even for this polyphagous species.

Hecatera Guenée, 1852 – Hecate, the goddess of the Underworld: a name having affinity with 2166 *Hadena*, q.v.

2164 **bicolorata** (Hufnagel, 1766) – *bi-*, two; *coloratus*, coloured: from the dark pattern on a white ground colour.

2165 **dysodea** ([Denis & Schiffermüller], 1775) – δυσώδης (dusōdēs), evil-smelling: probably from the experience of rearing the species in an enclosed container with the result that the foodplant (*Lactuca* spp.) turned foul.

Hadena Schrank, 1802 – Ἅδης, Hadēs, the Underworld: Schrank translates 'Trübeule', 'moping owl', and his name originally held family status; see p. 30.

2166 **rivularis** (Fabricius, 1775) – *rivulus*, a small stream: from the 'wavy' subterminal line on the forewing.

2167 **perplexa** ([Denis & Schiffermüller], 1775) – *perplexus*, intricate, confused: from complexity of pattern rather than variation between local populations since these were at first regarded as distinct species.

subsp. **capsophila** (Duponchel, 1842) – κάψα (kapsa), a chest, a case, a pod; φιλέω (phileo), to love: from the larval feeding habits, the 'pod-lover'.

2168 **irregularis** (Hufnagel, 1766) – *in*, *ir-* (before 'r'), negative prefix; *regularis*, an adj. coined from *regula*, a straight line, i.e. not straight, wavy: from the undulate subterminal line.

2169 **luteago** ([Denis & Schiffermüller], 1775) – *lūteus*, rosy yellow, *lŭteus*, clay-coloured; the termination *-ago*, see 2271: the nominate subspecies is more ochreous than the British one.

subsp. **barrettii** (Doubleday, 1864) – i.h.o. C. G. Barrett (1836–1904), the British entomologist who discovered it at Howth, Ireland, in 1861.

2170 **compta** ([Denis & Schiffermüller], 1775) – *comptus*, adorned: from the attractive pattern and colour of the adult.

2171 **confusa** (Hufnagel, 1766) – *confusus*, confused, mingled: from the elaborate wing pattern in which dark and pale markings are blended. Not from confusion with related species since this was the first of the group to be named.

2172 **albimacula** (Borkhausen, 1792) – *albus*, white; *macula*, a spot: from the white spot below the orbicular stigma, the 'white spot'.

2173 **bicruris** (Hufnagel, 1766) – *bi-*, two; *crus*, *cruris*, the leg: the 'legs' are the elongate reniform and orbicular stigmata, feet on costa.

2174 **caesia** ([Denis & Schiffermüller], 1775) – *caesius*, bluish grey: from the ground colour.

subsp. **mananii** (Gregson, 1866) – *Mananius*, an invented Latin adjective meaning 'of the Isle of Man', the type locality.

Eriopygodes Hampson, 1905 – the genus *Eriopyga* Guenée, 1852 (ἔριον (erion), wool; πυγή (pugē), the rump); εἶδος, ὠδ- (eidos, ōd-), form: a genus resembling *Eriopyga* and named from the long anal tuft of the male.

2175 **imbecilla** (Fabricius, 1794) – *imbecillus*, feeble: probably because the fat-bodied, short-winged female was formerly considered incapable of flight.

Cerapteryx Curtis, 1833 – κέρας (keras), a horn; πτέρυξ (pterux), a wing: from the branched longitudinal streak on the forewing of the next species which resembles an antler and is the source of the English name, the 'antler moth'.

2176 **graminis** (Linnaeus, 1758) – *gramen*, *graminis*, grass: correctly stated by Linnaeus to be the foodplant. This is sometimes a pest species of cultivated grassland and, earlier, Linnaeus (1746) had called it *Phalaena calamitosa*.

Tholera Hübner, 1821 – θολερός (tholeros), muddy: from the colour, but perhaps also influenced by the next name.

2177 **cespitis** ([Denis & Schiffermüller], 1775) – *caespes* (*cespes*), *caespitis*, turf: from the larval foodplants which are grasses and sedges.

2178 **decimalis** (Poda, 1761) – *decimus*, the tenth: probably simply the tenth species in a series described by Poda. The strongly pectinated male antenna excluded it from Noctua as defined by Linnaeus (see 2107, genus) and Poda seems to have got himself into a muddle. He described it in Phalaena (Geometra) which should have given the termination *-aria* (see 1661, family introduction); instead he used the termination *-alis* which belongs to Pyralis. He may have been hedging. Phalaena as a family which comprised the Linnaean Geometrae and Pyrales came later (Fabricius, 1775); see p. 29.

Panolis Hübner, 1821 – παν- (pan-), all-; ὀλοός (oloos), destructive: all-destroying because of its pest status in pine-forests. The name is a Graecism for *piniperda* Panzer, 1786 (*perdo*, to ruin), a junior synonym of the next species.

2179 **flammea** ([Denis & Schiffermüller], 1775) – *flammeus*, flame-coloured: from the reddish colour of the forewing. Fire is destructive, but an oblique reference to the pest status of this species, already recognized on the Continent, does not seem likely.

Xanthopastis Hübner, 1821 – ξανθός (xanthos), yellow; παστός (pastos), sprinkled with salt: from an irroration or peppering of yellow scales.

2179a **timais** (Cramer, 1780) – τιμαῖος (timaios), highly prized: a name for an attractive species.

Brithys Hübner, 1821 – βριθύς (brithus), heavy: seems irrelevant to the next species, but may refer to a character of one of the other three which Hübner placed in the genus.

2180 **crini** (Fabricius, 1775) – derivation not traced; possibly from a foodplant of this south European and oriental species. For the name to be derived from *crinis*, hair, would involve a typographical error quite uncharacteristic of Fabricius.

subsp. **pancratii** (Cyrillo, 1787) – *Pancratium maritimum*, sand lily: a foodplant.

Egira Duponchel, 1845 – Aegira, a Greek city situated in the Peloponnesus.

2181 **conspicillaris** (Linnaeus, 1758) – an adj. coined from *conspicilium*, a place to look out from. Linnaeus wrote *'alis . . . oculorum operculis orbiculatis'*, wings with rounded eyelids; the markings resemble eyes that are closed, but nevertheless are 'places to look out from'. Other exegetists (Pickard *et al.*; Macleod) jumped to the conclusion that Linnaeus formed an adj. from *conspicuus* (conspicuous), itself already an adj., having failed to refer to his text.

Orthosia Ochsenheimer, 1816 – ὄρθωσις (orthōsis), making straight: this genus differs from, e.g., 2156 *Lacanobia*, in having the subterminal line straight. Ὀρθωσία (Orthōsia) was also an epithet of Artemis and it is likely that Ochsenheimer had both senses in mind.

2182 **cruda** ([Denis & Schiffermüller], 1775) – *crudus*, unripe, premature: from the adult's appearance early in the year; cf. 'I come to pluck your berries harsh and crude' in *Lycidas*, Milton's lament for a young friend prematurely drowned. 'Rough, from dingy appearance' (Macleod); *crudus* can mean rough in the sense of unfeeling or merciless; it cannot mean unpolished.

2183 **miniosa** ([Denis & Schiffermüller], 1775) – *minium*, red lead, vermilion; the adjectival termination *-iosa* means much the same as the English '-ish', so, reddish: from the flush on the forewing; cf. 2037 *miniata*.

2184 **opima** (Hübner, 1809) – *opīmus*, fat, rich, splendid: not really apt in any of its meanings.

2185 **populeti** (Fabricius, 1781) – *pōpuletum*, a poplar wood: poplar is the foodplant.

2186 **gracilis** ([Denis & Schiffermüller], 1775) – *gracilis*, thin, scanty, unadorned: from the pale forewing and its indistinct markings in the typical form. 'Graceful' (Macleod): a wholly different word; the adjective derived from the Latin *grātia*, grace, is *grātiosus*. *Grăcilis* is from the same root as the Greek κολεκάνος, a long, lank, lean person.

2187 **cerasi** (Fabricius, 1775) – *Prunus cerasus*, sour cherry: a possible foodplant, the larva being polyphagous on trees.

= **stabilis** ([Denis & Schiffermüller], 1775) – *stabilis*, firm, steady, constant: less liable to variation than 2188 *O. incerta*.

2188 **incerta** (Hufnagel, 1766) – *incertus*, not fixed: from the wide range of variation.

2189 **munda** ([Denis & Schiffermüller], 1775) – *mundus*, neat, tidy: the antithesis to *gracilis* (2186).

2190 **gothica** (Linnaeus, 1758) – *Gothicus*, Gothic: from the arch on the forewing (though this looks more Norman than Gothic); Linnaeus wrote *'alae in medio arcu nigro extrorsum verso'*, with a black arch turned outwards in the middle of the wing, 'outwards' when the wings are folded.

Mythimna Ochsenheimer, 1816 – Mithimna, a town in the island of Lesbos or Mytelene. This explanation is given by Treitschke who completed *Die Schmetterlinge von Europa* after Ochsenheimer's death. Sodoffsky (1837) emended the spelling to *Mithimna*.

2191 **turca** (Linnaeus, 1761) – *turca*, Turkey red, a dye obtained from madder: from the ground colour. 'Turkish' (Macleod): the type locality is in Sweden.

2192 **conigera** ([Denis & Schiffermüller], 1775) – *conus*, a cone; *gero*, to carry: from the generally cone-shaped white spot in the dorsal half of the reniform stigma.

2193 **ferrago** (Fabricius, 1787) – *ferrugo*, the colour of red rust, a dirty red; probably deliberately altered to *ferrago* to give the termination *-ago*, see 2271 *citrago*: from the ground colour of the forewing.

2194 **albipuncta** ([Denis & Schiffermüller], 1775) – *albus*, white; *punctum*, a spot: from the conspicuous white dot in the reniform stigma.

2195 **vitellina** (Hübner, 1808) – *vitellus*, a little calf: 'from resemblance of colour of forewings to that of calfskin' (Macleod). *Vitellus* also signified the yolk of an egg and this may be the meaning, though neither colour seems quite right. The botanists use *vitellinus* to mean yellow with a reddish tinge.

2196 **pudorina** ([Denis & Schiffermüller], 1775) – an adjective formed from *pudor*, a sense of shame, influenced by *pudoricolor*, shame-coloured, blushing: from the pinkish flush on the forewing.

2197 **straminea** (Treitschke, 1825) – *stramineus*, straw-coloured: from the pale forewing.

2198 **impura** (Hübner, 1808) – *impurus*, unclean, dirtied: from the fuscous suffusion on the hindwing which contrasts with the purer white of the next species.

subsp. **scotica** (Cockayne, 1944) – the subspecies occurring in Scotland.

2199 **pallens** (Linnaeus, 1758) – *pallens*, pale: from the colour of the forewing.

2200 **favicolor** (Barrett, 1896) – *favus*, a honeycomb; *color*, colour: the ground colour tends to be rather more tawny than that of the last species.

2201 **litoralis** (Curtis, 1827) – *litoralis*, pertaining to the sea-shore: from the habitat.

2202 **l-album** (Linnaeus, 1767) – the letter 'L'; *album*, a white mark: from the white median streak on the forewing and the white dot representing the reniform stigma; the two converge to form an L-shaped mark.

2203 **unipuncta** (Haworth, 1809) – *unus*, one; *punctum*, a spot: from the single white spot in the reniform stigma.

2204 **obsoleta** (Hübner, 1803) – *obsoletus*, old, obscure, undeveloped (of markings), with the markings obscure (the 'obscure wainscot'). Lorimer (MBGBI 9: 269) draws attention to the fact that Hübner figured a form not found in Britain. 'From larva's habit of remaining in faded state throughout winter' (Macleod).

2205 **comma** (Linnaeus, 1761) – *comma*, a mark, not necessarily shaped like the modern punctuation mark (cf. 1529): from the black subdorsal streak.

2206 **putrescens** (Hübner, 1824) – *putrescens*, decaying: from the resemblance between the colour of the forewing and decaying wood; cf. 693 and 2098.

2207 **commiodes** (Guenée, 1852) – 2205 *M. comma*; εἶδος (eidos), form: from resemblance.

2208 **loreyi** (Duponchel, 1827) – i.h.o. Dr Lorey of Dijon.

Senta Stephens, 1834 – possibly from *sentio*, to feel: Stephens erected the genus because of certain characters of the labial palpus (a 'feeler'); he gives no explanation.

2209 **flammea** (Curtis, 1828) – *flammeus*, flame-like, flame-coloured: 'superior wings with a brown flame-like space along the centre (narrowed at the base)'.

Graphania Hampson, 1905 – γραφή (graphē), a design, writing; ἀνία (ania), pain, distress: perhaps because the wings are feebly marked.

2210 **dives** (Philpott, 1930) – *dives*, rich: the moth, however, appears pallid and poorly marked (South, 1961, pl. 176, fig. 11).

CUCULLIINAE (2211)

Cucullia Schrank, 1802 – *cucullus*, a cowl, a hood: from the prominent cowl-like thoracic crest; originally a family name (p. 30), it has been used for the subfamily, although the taxonomic relationship between this and the other genera is not close.

2211 **absinthii** (Linnaeus, 1761) – *Artemisia absinthium*, wormwood: the foodplant.

2212 **argentea** (Hufnagel, 1766) – *argenteus*, silvery: from the silvery white head and thorax.

2213 **artemisiae** (Hufnagel, 1766) – *Artemisia*, the wormwood genus, which includes the foodplants.

2214 **chamomillae** ([Denis & Schiffermüller], 1775) – χαμαίμηλον (khamaimēlon), *chamomilla*, *Chamaemelum nobile*, camomile: a foodplant.

2215 **lactucae** ([Denis & Schiffermüller], 1775) – *Lactuca*, the lettuce genus, which includes the foodplants.

2216 **umbratica** (Linnaeus, 1758) – *umbraticus*, belonging to the shade, retiring: Pickard *et al.* quote from Stainton 'the larva hides by day under the lower leaves of sow-thistles' and Macleod adopts the same explanation. Linnaeus described the adult as having streaked grey wings (*'alis . . . canis striatis'*) and the larva as blackish (*'nigricans'*) with three rows of red spots; he also knew the foodplant; the name probably refers to the dark shading on the forewing but possibly to the ground colour of the larva. Linnaeus makes no mention of the habits of the larva.

2217 **asteris** ([Denis & Schiffermüller], 1775) – *Aster*, the genus which includes *A. tripolium*, sea-aster, the main foodplant in Britain; at Vienna, the type locality, it must have used another member.

2218 **gnaphalii** (Hübner, 1813) – *Gnaphalium*, the cudweed genus, but not that of the larval foodplants.

subsp. **occidentalis** Boursin, 1945 – *occidentalis*, western: from the distribution of the subspecies.

2219 **lychnitis** Rambur, 1833 – *Verbascum lychnitis*, white mullein: a foodplant.

2220 **scrophulariae** ([Denis & Schiffermüller], 1775) – *Scrophularia*, the figwort genus: that of the larval foodplants.

2221 **verbasci** (Linnaeus, 1758) – *Verbascum*, the mullein genus: that of the larval foodplants.

2222 **prenanthis** Boisduval, 1840 – *Mycelis* (formerly *Prenanthes*) *muralis*, wall lettuce: the larva, however, feeds on *Scrophularia* spp.

2222a **caninae** Rambur, 1833 – *Scrophularia canina*, which does not occur in Britain: stated by Rambur to be the principal foodplant.

Calophasia Stephens, 1829 – κᾶλον (kālon), wood; φάσις (phasis), the appearance: from the resemblance between the moths and dead wood, though this is more evident in related genera. Macleod's derivation from καλός (kălos), beautiful, was an incorrect guess made without reference to Stephens.

2223 **lunula** (Hufnagel, 1766) – *lunula*, a little moon: from the lunate shape of the dorsal half of the postdiscal fascia.

2224 **platyptera** (Esper, 1788) – πλατύς (platus), broad; πτερόν (pteron), a wing: something of a misnomer, since neither wing is notably broad.

Brachylomia Hampson, 1906 – βραχύς (brakhus), short; λῶμα (lōma), a fringe: the cilia of the British representative are not noticeably short; the type species is American.

2225 **viminalis** (Fabricius, 1777) – *viminalis*, pertaining to the osier, which is a larval foodplant.

Leucochlaena Hampson, 1906 – λευκός (leukos), white; χλαῖνα (khlaina), a garment: from the pale vestiture of the species in the genus.

2226 **oditis** (Hübner, 1822) – derivation untraced; perhaps from a foodplant.

Brachionycha Hübner, 1816 – βραχύς (brakhus), short; ὄνυξ (onux), a claw: from the claw-like spur on the foretibia.

2227 **sphinx** (Hufnagel, 1766) – probably from the dorsal hump on abdominal segment 8 of the larva, reminiscent of the 'tail' on the larvae of the Linnaean family Sphinx (Sphingidae). 'From larva's appearance when resting' (Macleod); the 'sprawling' attitude with the head reflexed backwards could be regarded as enigmatic, the Sphinx having been a monster that posed riddles (see 1976, genus). The systematic placing has posed a problem, but it is unlikely that Hufnagel was bothered about that. I cannot resist mentioning the junior synonym *cassinia* ([Denis & Schiffermüller], 1775) from the great Italian astronomer John Dominic Cassini (d. 1712), bestowed because the larva contemplates the sky when it adopts its sprawling posture.

2228 **nubeculosa** (Esper, 1785) – *nubecula*, a small cloud: from the dark shading on the forewing.

Dasypolia Guenée, 1852 – δασύς (dasus), shaggy; πολιός (polios), grey: the whole body of the next species is clad in long, greyish hair-scales; a wish to express affinity with the genus 2148 *Polia* is possible but unlikely.

2229 **templi** (Thunberg, 1792) – *templum*, a temple, a church: males die in the autumn but the impregnated female overwinters, concealing itself in rocky outcrops, drystone walls or buildings; the first specimen may have been a female found in a church. This is, of course, conjecture. The type series was taken at Halland by D. or P. Osbeck and the lectotype (Karsholt & Nielsen, 1986) is a male; had those authors read what I have written, they might have selected a female!

Aporophyla Guenée, 1841 – ἄπορος (aporos), difficult, troublesome; φυλή (phulē), a tribe, loosely any taxon above the species: from the difficulty in determining the correct systematic placing of the genus.

2230 **australis** (Boisduval, 1829) – *australis*, southern: the nominate subspecies occurs on the south-western coasts of France.

subsp. **pascuea** (Humphreys & Westwood, 1843) – *pascueus*, a coined alternative to *pascuus*, belonging to pasture: the larva eats grasses.

2231 **lutulenta** ([Denis & Schiffermüller], 1775) – *lutulentus*, muddy: from the grey-brown ground colour of the forewing.

subsp. **lueneburgensis** (Freyer, 1848) – although Freyer did not give the type locality, it may be assumed to be Lüneburg, in north-western Germany.

2232 **nigra** (Haworth, 1809) – *niger*, black: from the ground colour of the forewing.

Lithomoia Hübner, 1821 – λίθος (lithos), a stone; ὅμοιος (homoios), like: from the cryptic forewing. The next species, however, prefers to rest on posts and bracken rather than rocks.

2233 **solidaginis** (Hübner, 1803) – *Virgaurea solidago* (gen. *solidaginis*), goldenrod: not, however, the larval foodplant.

Scotochrosta Lederer, 1857 – σκότος (skotos), darkness: χρῶς (khrōs), the skin, its colour, the complexion: from the dark forewing of the next species.

2234 **pulla** ([Denis & Schiffermüller], 1775) – *pullus*, dark-coloured: from the dark forewing.

Copipanolis Grote, 1874 – κόπις (kopis), a liar; the genus 2179 *Panolis*, q.v.: the next species resembles *Panolis flammea*, thus lying about its true identity.

2234a **styracis** (Guenée, 1852) – *styrax*, resinous gum: probably from a mistaken belief that the larva fed on resinous conifers, like *P. flammea*.

Lithophane Hübner, 1821 – λίθος (lithos), a stone; φαίνω, φαν- (phaino, phan-), to appear, to appear to be: with the same meaning as 2233 *Lithomoia*, q.v.

2235 **semibrunnea** (Haworth, 1809) – *semi-* half; *brunneus* (late Lat.), brown: from the dorsal half of the forewing being darker than the costal.

2236 **hepatica** (Clerck, 1759) – strictly diseased in the liver but here liver-coloured: our specimens are paler but in Clerck's figure the tegulae and costa are so coloured: '*alarum color hepaticus*' (Linnaeus, 1767).

= **socia** (Hufnagel, 1766) – *socius*, allied: allied perhaps to one of the other members of the genus named by Hufnagel in the same paper, but certainly not to the last species which was named forty-three years later. 'From moth's gregarious habits' (Macleod): the larva is a cannibal!

2237 **ornitopus** (Hufnagel, 1766) – ὄρνις, ὄρνιθος (ornis, ornithos), a bird; ποῦς (pous), the foot: from the resemblance between the black basal streak on the forewing and a bird's claw.

subsp. **lactipennis** (Dadd, 1911) – *lac*, *lactis*, milk; *pennae* (pl.), a wing: from having a paler ground colour than the nominate subspecies.

2238 **furcifera** (Hufnagel, 1766) – *furca*, a two-pronged fork; *fero*, to carry: from the shape of the black dash extending outwards from the claviform stigma.

subsp. **suffusa** Tutt, 1892 – *suffusus*, clouded, suffused: a subspecies darker than the nominate one.

2239 **lamda** (Fabricius, 1787) – bearing a mark shaped like the Greek letter 'λ': either the basal or median black streak. The letter is usually spelt 'lambda' or, by philologists, 'labda'; cf. 641 *lambdella*.

2240 **leautieri** (Boisduval, 1828) – i.h.o. M. Leautier who discovered the species in south-western France.

subsp. **hesperica** Boursin, 1957 – *hespericus*, a coined variant of *hesperius*, western: the subspecies occurring in the western part of the range.

Xylena Ochsenheimer, 1816 – ξύλον, ξύλινος (xulon, xulinos), wood, wooden: a name prompted by Linnaeus' description of 2242 *exsoleta*, q.v., and first suggested by Hübner in his *Tentamen* [1806].

2241 **vetusta** (Hübner, 1813) – *vetustus*, old: a name chosen as showing close affinity with the next species, q.v., and indirectly conveying the same sense.

2242 **exsoleta** (Linnaeus, 1758) – *exsoletus, exoletus*, mature, old: '*Ph. color ligni putrescentis admodum exsoleti*', moth exactly the colour of an old piece of rotting wood'.

Xylocampa Guenée, 1837 – ξύλον (xulon), wood; κάμπη (kampē), a caterpillar: from the cryptic larva of the next species, which closely resembles a twig.

2243 **areola** (Esper, 1789) – *areola*, a small space demarcated by lines: from the dark-ringed, pale stigmata.

Meganephria Hübner, 1820 – μέγας (megas), big; νεφροί (nephroi), the kidneys: from the large stigmata on the forewing.

2244 **bimaculosa** (Linnaeus, 1767) – *bi-*, two, twice; *maculosus*, spotted: from the large orbicular and reniform stigmata.

Allophyes Tams, 1942 – ἀλλοφυής (allophuēs), changeful in nature: from the dimorphism of the next species.

2245 **oxyacanthae** (Linnaeus, 1758) – *Crataegus oxyacantha*, hawthorn: a foodplant.

Valeria Stephens, 1829 – the Valerian gens, one of the old patrician families of Rome. Stephens attributes the name to Germar.

2246 **oleagina** ([Denis & Schiffermüller], 1775) – *oleaginus*, olive-coloured: from the forewing.

Dichonia Hübner, 1821 – διχῶς (dikhōs), doubly: from the two whitish lines on the hindwing (Spuler).

2247 **aprilina** (Linnaeus, 1758) – *Aprilis*, the month, but the meaning here is the colour of spring leaves; *aprilis* or *aperilis* is from the root of *aperio*, to open, the month being the season of opening buds. The reference is therefore to the colour of the wings, not the time of appearance. 'From time of moth's appearance in type locality, not in Britain' (Macleod); this is a guess, unchecked by reference to Linnaeus, but expressed as if it were fact. The type locality is 'Europe', and Britain is part of Europe. Cf. 2099 *praecox* and 2418 *Earias*.

Dryobotodes Warren, 1910 – the genus *Dryobota* Lederer, 1857, not represented in Britain, from δρῦς (drūs), an oak, and βόσκω, βοτ- (boskō, bot-), to feed on; εἶδος, ὠδ- (eidos, ōd-), form: a genus resembling *Dryobota*.

2248 **eremita** (Fabricius, 1775) – *eremita*, a hermit: here the larva which lives in a 'cell' consisting of a substantial spinning among leaves of oak.

Blepharita Hampson, 1907 – βλεφαρῖτις (blepharītis), of the eyelids: 'eyes large, rounded, overhung by long cilia' (Sir George Hampson).

2249 **satura** ([Denis & Schiffermüller], 1775) – *satur* (of colour), full, deep, rich: the 'beautiful arches'.

Mniotype Franclemont, 1941 – μνίον (mnion), moss; τυπή (tupē), a blow, a wound, τύπος (tupos), the mark of a blow, the stamp, character: from a fancied similarity between the variegated wing-pattern and moss. The name is probably influenced by 2254 *Antitype*, q.v.

2250 **adusta** (Esper, 1790) – *adustus*, sunburnt, swarthy: from the dark ground colour.

2250a **solieri** (Boisduval, 1840) – i.h.o. A. J. J. Solier (1792–1851), a French entomologist who lived at Marseilles.

Trigonophora Hübner, 1821 – τρίγωνον (trigōnon), a triangle; φορέω (phoreō), to carry, to wear: possibly from the shape of segment 2 of the labial palpus, which is strongly dilate with long scales. 'From markings on forewings' (Macleod); not of the next species, which was one of the eight included by Hübner.

2251 **flammea** (Esper, 1785) – *flammeus*, flame-coloured: from the reddish purple ground colour.

Polymixis Hübner, 1820 – πολύς (polus), much, many; μῖξις (mīxis) a mixing: from the intricate mingling of colour on the forewing. (πολύμιξ (polumix) means promiscuous intercourse, which you may prefer).

2252 **flavicincta** ([Denis & Schiffermüller], 1775) – *flavus*, yellow; *cinctus*, girt: from the mixture of fuscous and yellow on the forewing, the yellow being girt by the fuscous more than the other way round.

2252a **gemmea** (Treitschke, 1825) – *gemmeus*, adorned with precious stones, glittering: from the shining white markings superimposed on the variegated darker ground colour.

2253 **xanthomista** (Hübner, 1819) – ξανθός (xanthos), yellow; μεῖστος (meistos), least: from the sparse admixture of yellow scales in the pattern of the nominate subspecies.

subsp. **statices** (Gregson, 1869) – *Limonium*, formerly *Statice*, the sea-lavender genus, which includes some of the foodplants.

Antitype Hübner, 1821 – ἀντίτυπος (antitupos), that which is beaten back, an echo, a replica: from close affinity, probably with 2252 *Polymixis*, already named by Hübner.

2254 **chi** (Linnaeus, 1758) – from the black discal mark which resembles the Greek letter chi (χ).

Eumichtis Hübner, 1821 – a typographical error for *Eumictis*, εὖ (eu), well; μικτός (miktos), blended: from the mingling of green and grey scales on the forewing of the next species, the only one included by Hübner.

2255 **lichenea** (Hübner, 1813) – λειχήν, λιχήν (leikhēn, likhēn), lichen: from colour similarity between lichen and the forewing.

subsp. **scillonea** Richardson, 1958 – the subspecies found in the Isles of Scilly.

Eupsilia Hübner, 1821 – εὖ (eu), well, very; ψιλος (psilos), bald, bare: probably from the generally white reinform stigma of the next species which contrasts strongly with brownish forewing and resembles a bald patch. The genus is monotypic.

2256 **transversa** (Hufnagel, 1766) – *transversus*, lying across: from the transverse fasciae on the forewing.

Jodia Hübner, 1818 – ἰώδης (iōdēs), rust-coloured: from the orange forewing.

2257 **croceago** ([Denis & Schiffermüller], 1775) – *croceus*, orange; the termination *-ago*, see 2271: the 'orange upperwing'.

Conistra Hübner, 1821 – κόνιστρα (konistra), a place covered with dust, an arena: from the speckled forewing, especially of 2260 *C. rubiginea*.

2258 **vaccinii** (Linnaeus, 1761) – *Vaccinium*, the bilberry genus: a possible foodplant for this polyphagous species.

2259 **ligula** (Esper, 1791) – *ligula*, a little tongue or strap: from the whitish or reddish rather broad subterminal fascia of some specimens.

2260 **rubiginea** ([Denis & Schiffermüller], 1775) – *robigo, robiginis* (*rubigo*), rust, its colour: from the ground colour of the forewings.

2261 **erythrocephala** ([Denis & Schiffermüller], 1775) – ἐρυθρός (eruthros), red; κεφαλή (kephalē), the head: the 'red-headed chestnut', though, in fact, the head is more often brown than red.

Agrochola Hübner, 1821 – ἀγρός (agros), a field, the ground colour; χολή (kholē), gall, bile, its green or (as here) yellow colour: from the coloration of some of the species in the genus.

2262 **circellaris** (Hufnagel, 1766) – *circellus*, a small ring: from the stigmata, which are normally themselves of the ground colour, but darker-ringed.

= **ferruginea** ([Denis & Schiffermüller], 1775) – *ferrugineus*, the colour of iron-rust, reddish: from the ground colour of the forewing. See 2152, genus.

2263 **lota** (Clerck, 1759) – *lotus*, washed: washed colours merge softly and this applies to the pattern apart from the more sharply defined reniform stigma and subterminal fascia.

2264 **macilenta** (Hübner, 1809) – *macilentus*, lean, meagre: from the supposedly weak markings on the forewing.

2265 **helvola** (Linnaeus, 1758) – *helvola*, pale yellow: perhaps from the colour of the underside; Linnaeus describes the upperside as *rufus* (red), and the underside, which is yellow with a reddish tinge, as *rufescens* (reddish).

2266 **litura** (Linnaeus, 1761) – *litura*, a smearing on a wax writing tablet, an erasure, a blotch: from the four black costal spots on the forewing (*liturisque quatuor marginalibus nigris*) which may be said to block out what is 'written' beneath.

2267 **lychnidis** ([Denis & Schiffermüller], 1775) – *Lychnis*, a genus in the Carophyllaceae, containing possible foodplants, the larva being widely polyphagous.

2268 see below 2313

Atethmia Hübner, 1821 – ἀ, alpha privative; t, phonetic insertion; ἠθμός (ēthmos), a sieve, i.e. unspotted, the reverse of 717 *Ethmia*, q.v.: Hübner describes the pattern of the three species he placed in the genus as consisting 'only' of a central fascia and two lines. '*a*-, privative prefix, *tethmios*, fixed; from variability of colour' (Macleod); this is a guess, unsupported by any statement in Hübner's diagnosis.

2269 **centrago** (Haworth, 1809) – *centrum*, the centre; *-ago*, the conventional termination, see 2271: from the prominent median fascia.

Omphaloscelis Hampson, 1906 – ὀμφαλός (omphalos), the navel, the middle point; κηλίς (kēlis), a stain: from the discal spot on the hindwing (see next species).

2270 **lunosa** (Haworth, 1809) – *lunosus*, an adjective coined from *luna*, the moon: from the lunate discal spot on the hindwing.

Xanthia Ochsenheimer, 1816 – ξανθός (xanthos), yellow: from the predominantly yellow coloration of the forewings of the species included in the genus. Ochsenheimer took this name from Hübner's *Tentamen* [1806].

2271 **citrago** (Linnaeus, 1758) – *citrus*, the citron-tree, the orange or yellow colour of its fruit; *-ago*, a termination indicating affinity (e.g. *plumbum* and *plumbago*, lead and black lead); the use of the termination in this instance led to the convention of using it for other species with an orange or yellow ground colour. There is no reference to *citrago*, a plant yielding balm.

2272 **aurago** ([Denis & Schiffermüller], 1775) – *aurum*, gold; *-ago*, see last species: from the colour of the forewing.

2273 **togata** (Esper, 1788) – *togatus*, wearing a toga: the toga worn by senators bore the *latus clavus*, the broad purple stripe and the reference here is to the purplish median fascia on the forewing.

2274 **icteritia** (Hufnagel, 1766) – ἴκτερος (ikteros), jaundice: from the yellow forewing; see 1838 *icterata* for a fuller exposition.

2275 **gilvago** ([Denis & Schiffermüller], 1775) – *gilvus*, pale yellow; *-ago*, see 2271: from the colour of the forewing.

2276 **ocellaris** (Borkhausen, 1792) – *ocellus*, a little eye: from the white-pupilled reinform stigma.

ACRONICTINAE (2278)

Moma Hübner, 1820 – Momus, the god of mockery: suggested by *Trichosea ludifica* (Linnaeus, 1758) (*ludifico*, to mock – '*simillima* Ph. aprilinae'), a non-British species included by Hübner in the genus.

2277 **alpium** (Osbeck, 1778) – *Alpes*, gen. pl. *Alpium*, the Alps, mountains in general: the type locality is given as Göteborg, a Swedish seaport, only because the periodical in which the name appears was published there; the first specimens may have been taken in mountainous country inland.

Acronicta Ochsenheimer, 1816 – ἀκρόνυξ (akronux), nightfall: probably intended to be a name collateral with *Noctua*, since the moths are not crepuscular.

2278 **megacephala** ([Denis & Schiffermüller], 1775) – μέγας (megas), large; κεφαλή (kephalē), head: from the large head of the larva.

2279 **aceris** (Linnaeus, 1758) – *Acer*, the maple genus: one of several foodplants cited by Linnaeus.

2280 **leporina** (Linnaeus, 1758) – *lepus*, a hare, *leporinus*, of or like a hare: the similarity is with the white winter coat of the mountain hare; Linnaeus stresses this whiteness and makes a comparison with 2060 *Spilosoma lubricipeda*. 'From colour of forewings of some specimens' (Macleod): he should have written 'from colour of some hares'. Spuler is also wrong in supposing that the likeness is in the softness of the fur; Linnaeus describes the moth as *laevis* (*levis*), smooth, as he did every other species on the same page, the meaning being 'without crests'.

2281 **alni** (Linnaeus, 1767) – *Alnus*, the alder genus: alder is a foodplant.

2282 **cuspis** (Hübner, 1813) – *cuspis*, a point, a spear: from the strongly expressed 'dagger' marks on the forewing.

2283 **tridens** ([Denis & Schiffermüller], 1775) – *tri-*, three; *dens*, a tooth: from the three prongs of the black tornal streak. That the sacculus of the male genitalia is tridentate and so differs from that of the next species which has it bidentate is a happy coincidence.

2284 **psi** (Linnaeus, 1758) – from the similarity between the 'dagger' mark on the forewing and the Greek letter ψ (psi).

2285 **strigosa** ([Denis & Schiffermüller], 1775) – *strigosus*, an adjective coined from striga, a furrow, line or streak: from the well-developed subdorsal 'dagger' marks on the forewing. There is a classical Latin adjective *strigosus*, thin, and Macleod derives the name from this without explanation.

2286 **menyanthidis** (Esper, 1789) – *Menyanthes*, the bog-bean genus: bog-bean is one of the foodplants.

subsp. **scotica** Tutt, 1891 – the subspecies occurring in Scotland.

2287 **auricoma** ([Denis & Schiffermüller], 1775) – *aurum*, gold: *coma*, hair: from the reddish tufts of setae on the larva ('yellow', Macleod).

2288 **euphorbiae** ([Denis & Schiffermüller], 1775) – *Euphorbia*, the spurge genus: this includes some of the foodplants.

subsp. **myricae** Guenée, 1852 – *Myrica gale*, bog-myrtle, an important foodplant of the subspecies occurring in Scotland.

2289 **rumicis** (Linnaeus, 1758) – *Rumex*, the sorrel genus: one of several foodplants given by Linnaeus.

Simyra Ochsenheimer, 1816 – either from σιμός (simos), snub-nosed; ῥίς (rhis), a nose, or, more simply and probably, from Simyra, a Phoenician town, the name being without entomological meaning: cf. 1652, 2321 and p. 33.

2290 **albovenosa** (Goeze, 1781) – *albus*, white; *venosus*, having veins: from the veins of the forewing, which are whiter than the pale ochreous ground colour.

Craniophora Snellen, 1867 – κρανίον (kranion), a skull; φορέω (phoreō), to bear: from the markings on the forewing; the 'skull' is not at all obvious but could possibly be visualized when the wings are folded, the orbicular stigmata then forming the eyes. The genus is monotypic.

2291 **ligustri** ([Denis & Schiffermüller], 1775) – *Ligustrum vulgare*, privet: a foodplant.

Cryphia Hübner, 1818 – κρύφιος (kruphios), hidden, secret: from the 'cryptic' pattern and coloration both of the moths and larvae.

2292 **algae** (Fabricius, 1775) – *alga*, originally seaweed, then applied loosely to lichen, on which the larva feeds.

2293 **domestica** (Hufnagel, 1766) – *domesticus*, belonging to a house: from the habitat; the larva eats lichens, including those growing on houses.

2294 **raptricula** ([Denis & Schiffermüller], 1775) – *raptricula*, dim. of *raptrix*, a female robber: Denis and Schiffermüller included this species among their 'minors' to which they gave names associated with theft; see 2339 *Oligia latruncula*.

2295 **muralis** (Forster, 1771) – *muralis*, pertaining to a wall: from the larval pabulum of lichens growing on walls.

subsp. **impar** (Warren, 1884) – *impar*, unequal: here used in antithesis to ab. *par* Hübner, which is dull ochreous grey; this subspecies, which occurs at Cambridge, is by contrast well-marked.

subsp. **westroppi** Cockayne & Williams, 1956 – i.h.o. M. S. Dudley Westropp, who took the first specimens of the subspecies in Co. Cork, Republic of Ireland, in 1933.

AMPHIPYRINAE (2297)

2296 see below 2467

Amphipyra Ochsenheimer, 1816 – ἀμφί (amphi), round; πῦρ (pūr), the fire: 'flying round the light' (Pickard *et al.*, quoting Treitschke); Spuler thought the reference was to the coppery hindwing of the next species, but this is less likely.

2297 **pyramidea** (Linnaeus, 1758) – *pyramis*, a pyramid: from the conical hump on abdominal segment 8 of the larva, which Linnaeus describes as *conico-gibba*, conically humped. 'From row of triangular marks across forewings' (Macleod): Linnaeus would have known the difference between plane and solid figures.

2298 **berbera** Rungs, 1949 – *Berber*, an inhabitant of Barbary: the type locality is in North Africa.

subsp. **svenssoni** Fletcher, 1968 – i.h.o. Ingvar Svensson, a contemporary Swedish entomologist.

2299 **tragopoginis** (Clerck, 1759) – *Tragopogon pratensis*, goat's-beard, a foodplant of this polyphagous species.

Mormo Ochsenheimer, 1816 – Μορμώ, a hideous she-monster, a bugbear: cf. 235.

2300 **maura** (Linnaeus, 1758) – *Maurus*, an inhabitant of Mauritania, a Moor: Linnaeus gives Mauritania as the type locality, adding 'E. Brander', presumably the captor and the dedicatee of 1088 *Pseudosciaphila branderiana*, q.v. 'Moor, from fanciful resemblance of this dark moth, with wings extended to Moor's face' (Macleod): fanciful invention!

Dypterygia Stephens, 1829 – δι- (di-), two; πτερύξ (pterux), a wing: double-winged from the fancied image of a bird's wing superimposed on the forewing of the next species – the 'bird's wing'.

2301 **scabriuscula** (Linnaeus, 1758) – *scaber*, rough, *scabrius*, rougher, rather rough; *-culus*, dim. termination: '*cristae in dorso ipsius abdominis 4 pone thoracem gibbum*', with four crests on the dorsum of the abdomen itself, posterior to the humped thorax.

Rusina Stephens, 1829 – ῥούσιος (rhousios), Lat. *russus*, reddish: from the reddish ground colour of the next species.

2302 **ferruginea** (Esper, 1785) – *ferrugineus*, the colour of iron-rust, reddish: from the ground colour.

Polyphaenis Boisduval, 1840 – πολύς (polus), much, many; φαίνω (phaino), to shine: from the glossy forewing.

2302a **sericata** (Esper, 1787) – *sericatus*, dressed in silk: from the glossy forewing.

Thalpophila Hübner, 1820 – θάλπος (thalpos), summer heat; φιλέω (phileō), to love: a name in accord with the meaning of that of the next species.

2303 **matura** (Hufnagel, 1766) – *maturus*, ripe: from the yellow hindwing, suggestive of corn ripening in summer heat (see generic name) – the 'straw underwing'. Spuler's 'grown-up and well-developed' misses the point.

Trachea Ochsenheimer, 1816 – τραχύς (trakhus), rough: from the crested abdomen.

2304 **atriplicis** (Linnaeus, 1758) – *Atriplex, Atriplicis*, the orache genus, which contains one of the foodplants, as correctly stated by Linnaeus.

Euplexia Stephens, 1829 – εὖ (eu), well; πλέχις (plexis), weaving, πλέκω (plekō), to plait, to twist: from the moth's resting position with the wings longitudinally plicate. 'Perhaps from scalloped hind margin of wings' (Macleod).

2305 **lucipara** (Linnaeus, 1758) – *lux, lucis*, light; *pario*, to bring forth, *luciparens*, light-bearing: Linnaeus describes a black fascia with a yellow spot (the submetallic, gold-tinged reniform stigma) beyond, resembling a lamp shining out of darkness. 'From larva's shining head' (Macleod): Linnaeus did not know the life history.

Phlogophora Treitschke, 1825 – φλόξ, φλογός (phlox, phlogos), flame; φορέω (phoreō), to carry: a Greek rendering of 2305 *lucipara*, which Treitschke included in his genus. It is unfortunate that Duponchel designated 2306 *meticulosa* as type species in 1829. 'From the colour of the wings' (Macleod): Linnaeus described the wings of the next species as 'pale'.

2306 **meticulosa** (Linnaeus, 1758) – *meticulosus*, fearful in both senses (I am fearful of a fearful monster): an old puzzle. 'Timorous, from its habit of quivering when light is thrown on it' (Pickard *et al.*; Macleod), or from the moth's habit of feigning death (Spuler), or quivering when warming up for flight. Fear also makes one cringe or 'cower' which the dictionary defines as to 'crouch shrinkingly'. Linnaeus paid great attention to the way in which a moth held its wings in repose (pp. 20–22) and this species rests with plicate wings, as if 'shrinking' in the face of danger, withdrawing into itself.

Pseudenargia Boursin, 1956 – ψεῦδος (pseudos), false; the genus 2313 *Enargia*, q.v. a genus resembling *Enargia*.

2307 **ulicis** (Staudinger, 1859) – *Ulex australis*, the larval foodplant.

Callopistria Hübner, 1821 – κάλλος (kallos), beauty; ὀπίστερος (opisteros), posterior: having a beautiful backside, from the white-ringed abdomen of the next species.

2308 **juventina** (Stoll, 1782) – *juventinus*, adj. coined from *juventus*, the season of youth, the age for military service: the moth has bright colours, suggestive of young blood.

2309 **latreillei** (Duponchel, 1827) – i.h.o. P. A. Latreille (1762–1833), the distinguished French entomologist.

Eucarta Lederer, 1857 – εὖ (eu), well, good; κάρτα (karta), very, very much: Lederer is describing the next species as 'a jolly good moth'.

2310 **amethystina** (Hübner, 1803) – *amethystinus*, of the colour of amethyst, violet: from the colour of the forewing.

Ipimorpha Hübner, 1821 – ἶπος (īpos), in a mouse-trap, the piece of wood that falls and catches the mouse; μορφή (morphē), shape: clearly descriptive of the shape of the forewing in the next two species, but one would need to see Hübner's mouse-trap to know just what he means. Derivation from ἴψ, ἴπος (ips, ipos), a larva that is a pest of vine-buds, is ruled out by the sense of μορφή.

2311 **retusa** (Linnaeus, 1761) – *retusus*, blunt, retuse, defined by the dictionary as 'with the tip blunt and broadly notched': from the excavate termen of the forewing.

2312 **subtusa** ([Denis & Schiffermüller], 1775) – *sub-*, somewhat; *-tusus, retusus*, see last species: having the forewing only slightly retuse.

Enargia Hübner, 1821 – ἐνάργεια (energeia), of bright and vivid appearance: from the aspect of the next species, the only one placed by Hübner in the genus.

2313 **paleacea** (Esper, 1788) – *paleaceus*, adj. formed from *palea*, chaff: from the ochreous yellow forewing.

Parastichtis Hübner, 1821 – παρά (para), beside; στικτός (stiktos), punctured, spotted: from the subterminal series of black dots on the forewing of the next species. Macleod's derivation from στίχος (stikhos), a row, is also possible; whichever is right, Hübner misspelt the name. The unjustified emendation *Parastictus* Agassiz, 1846, favours the first explanation.

2268 **suspecta** (Hubner, 1817) – *suspectus*, mistrusted, probably over suspicion on whether it was specifically distinct from some other species. 'Of suspected classification' (Macleod); here he is ascribing foreknowledge of the problems of future systematists to Hübner himself, who placed it in *Noctua* which can hardly have occasioned him any qualms. Leraut (1980) transferred it from the Cucullinae to its present position in the Amphipyrinae.

2314 **ypsillon** ([Denis & Schiffermüller], 1775) – γ, ypsilon, the letter of the Greek alphabet: from the shape of the rather obscure claviform stigma. British authors thought it looked more like a pair of shears and accordingly named the moth the 'dingy shears'.

Dicycla Guenée, 1852 – δι- (di-), two, double; κύκλος (kuklos), a ring: a Greek rendering of the next specific name.

2315 **oo** (Linnaeus, 1758) – 'o' + 'o', from the circular orbicular and reniform stigmata: *'alis . . . oo notatis'*, with the wings marked 'oo'.

Cosmia Ochsenheimer, 1816 – κόσμιος (kosmios), well-ordered, becoming: from the attractive appearance of the species in the genus. The name was first suggested by Hübner in his *Tentamen* [1806].

2316 **affinis** (Linnaeus, 1767) – *affinis*, related, i.e. to the next species, not to 2319 *C. pyralina* as supposed by Spuler and Macleod as that species was not to be described and named for another eight years.

2317 **diffinis** (Linnaeus, 1767) – *diffinis*, a portmanteau word formed from *differens*, differing, and *affinis*, related: a species resembling and closely related to, but differing from, *C. affinis* which Linnaeus named on the same page. He used the same device when he named a danaine and its papilionine mimics *similis*, *assimilis* and *dissimilis* (Linnaeus, 1758: 479). Macleod makes the same blunder as in the last species.

2318 **trapezina** (Linnaeus, 1758) – *trapezium*, a quadrilateral with two sides parallel from the shape of the area enclosed by the ante- and postmedian fasciae.

2319 **pyralina** ([Denis & Schiffermüller], 1775) – *pyralinus*, an invented adjective meaning 'resembling a pyrale' such as 1417 *Pyralis farinalis*. 'Gr. *pur*, fire, from "fiery" red wings' (Macleod): possible but less likely.

Hyppa Duponchel, 1845 – probably an invented word of no meaning but possibly from ὑπό (hupo), under, somewhat; παός (paos), related: related, but not closely, to another genus.

2320 **rectilinea** (Esper, 1788) – *rectus*, straight; *linea*, a line: from the broad, straight, black subdorsal bar in the median area of the forewing.

Apamea Ochsenheimer, 1816 – 'Apamea, name of the town in Asia Minor where Theodore, one of the ancient Fathers of the Church, lived' (Pickard *et al.*): a name without entomological relevance; cf. 1652 *Thyatira* Ochsenheimer and 1988 *nicaea* de Prunner, both names of cities situated in Asia Minor and with ecclesiastical rather than entomological connotation. Antiochus the Great signed a peace treaty with the Romans in 188 B.C. at Apamea after his defeat at the Battle of Magnesia.

2321 **monoglypha** (Hufnagel, 1766) – μόνος (monos), one, single; γλυφίς (gluphis), a notch: from the dentate character in the subterminal line of the forewing which is shaped like a 'W', having one or two notches depending on the way in which it is regarded. 'Gr. . . . *gluphe*, carving; perhaps from crescent mark on hindwing' (Macleod); less likely, the '*mono-*' then lacking meaning.

2322 **lithoxylaea** ([Denis & Schiffermüller], 1775) – λίθος (lithos), a stone; ξύλον (xulon), wood: from the grain-like markings on the stone-coloured ground colour of the forewing.

2323 **sublustris** (Esper, 1788) – *sublustris*, glimmering faintly: from the faintly glossy texture of the forewing.

2324 **maillardi** (Geyer, 1834) – i.h.o. M. Maillard (early 19th century), Abbé at the Séminaire de St Lucien and an entomologist. See Corrigendum on p. 229.

subsp. **exulis** (Lefebvre, 1836) – *exsul*, *exul*, an exile: from the remote type locality. '*Exul*, exile, usual locality being far north' (Macleod): the species was described in a French journal from type material taken in Labrador and the name has nothing to do with the northerly British distribution.

subsp. **assimilis** (Doubleday, 1847) – *assimilis*, similar: from resemblance to subsp. *exulis*.

2325 **oblonga** (Haworth, 1809) – *oblongus*, oblong: from the black bar sometimes present between the ante- and postmedian lines. 'From shape of forewings' (Macleod); but these do not differ in shape from those of related species.

2326 **crenata** (Hufnagel, 1766) – *crenatus*, notched: from the dentate subterminal line on the forewing; cf. 2321 *monoglypha* of the same author.

2327 **epomidion** (Haworth, 1809) – ἐπωμίδιος (epōmidios), on the shoulder: from the pale blotch at the base of the costa.

= **characterea** sensu auctt. – χαρακτήρ, Lat. *character*, a characteristic mark, a letter of the alphabet: *characterea* Hübner does not occur in Britain and its markings do not concern us; *A. epomidion* has an L-shaped mark composed of the basal margin of the reniform stigma and the dorsal edge of the orbicular stigma.

2328 **lateritia** (Hufnagel, 1766) – *latericius*, *lateritius*, made of brick: from the light reddish brown forewing of the paler forms.

2329 **furva** ([Denis & Schiffermüller], 1775) – *furvus*, dark, dusky: from the dark brown forewing, from the dark brown terminal shade of the hindwing, or from both.

subsp. **britannica** Cockayne, 1950 – the subspecies found in Britain.

2330 **remissa** (Hübner, 1809) – *remissus*, sent back, relaxed, loose, of cheerful or gay temperament: I have no convincing explanation and suspect some non-entomological reason such as the figured specimen being 'sent back' to Hübner by one of his correspondents. Macleod's explanation 'gay, from appearance' is unacceptable because *remissus* does not mean 'gay-coloured' and such a description would hardly have been given by an artist to the 'dusky brocade'.

2331 **unanimis** (Hübner, 1813) – *unanimis*, of one mind, harmonious: another puzzling name, neither of the explanations offered being satisfactory. Pickard *et al.* give 'unchangeable' and it would be easy to say that this species is very little subject to variation compared with the very similar 2343 *Mesapamea secalis*; but the earlier entomologists did not recognize the varieties of the latter as such, instead thinking them distinct species (it has eight synonyms in Kloet & Hincks, one bestowed by Hübner), and when a species is first named it is unlikely that there will be a long enough series to make judgement on variability. Macleod has 'from uniform colour of wings', but they are not so and Hübner the artist would have seen plenty of pattern to depict. I think it more likely that the name means harmonious with, similar to, another species, such as *M. secalis*.

2332 **pabulatricula** (Brahm, 1791) – dim. formed from *pabulatrix*, fem. of *pabulator*, a herdsman: possibly from the type locality in grassland, but no further detail of this is known other than that it was in Germany; in Britain the species used to occur in woodland. Macleod has 'grazer, from larva's habit of feeding on grasses', but *pabulator* means the herdsman who grazes animals, not the animal that grazes grass, and it is unlikely that Brahm knew the life history. For the feminine form see 1033 *Tortrix*.

2333 **anceps** ([Denis & Schiffermüller], 1775) – *anceps*, doubtful: probably from uncertainty whether it was a distinct species.

2334 **sordens** (Hufnagel, 1766) – *sordens*, dirty: an unjustified description.

2335 **scolopacina** (Esper, 1788) – *scolopax*, a snipe, a woodcock: from a supposed colour resemblance. 'From slender body' (Macleod).

2336 **ophiogramma** (Esper, 1793) – ὄφις, ὀφιο- (ophis, ophio-), a snake; γραμμή (grammē), a marking: from the sinuous edge of the division between the dark costal area and the paler ground colour.

Oligia Hübner, 1821 – ὀλίγος (oligos), small: from the smallness of the species ('minors').

2337 **strigilis** (Linnaeus, 1758) – dim. of *striga*, a little line, a stria: from the short black projections on the veins on the distal side of the post-median line, *'fascia alba alas terminans latior, intra quam area fusca inserit 5 vel 6 strias nigras fere ad eius medium'* (a rather broad white fascia in the terminal part of the wing, inside which the fuscous [median] area thrusts five or six black striae almost reaching its middle); or again, *'denticulis setaceis intra fasciam albam terminalem'* (with little teeth like bristles inside the terminal white fascia). Pickard *et al.*, Spuler and Macleod all derive from *strigilis*, the strigil or scraper used by bathers and imagine a marking on the forewing of similar shape (whatever that may be). *Stria*, *striga* and *strigilis* are all derived from the root of *stringere*, to draw. Cf. 2488 *strigilata*.

2338 **versicolor** (Borkhausen, 1792) – *versicolor*, of changeable colour or parti-coloured: probably in the latter sense, there being a contrast between the conspicuously pale orbicular stigma and the dark ground colour, and between the darker basal area and the terminal region which is pale with a vinous red suffusion.

2339 **latruncula** ([Denis & Schiffermüller], 1775) – *latro*, a robber, *latrunculus* (dim.), a petty thief, a highwayman, also a piece or 'man' in a game of draughts. Spuler adopted the last meaning without explanation; Macleod, undoubtedly correctly, the first but added, less convincingly, that it was because the larva robbed the farmer of his grass. Denis and Schiffermüller bestowed two other names with similar meaning on 'minors', 2341 *furuncula* and 2337 *praeduncula* (= *strigilis*) from *praedo*, a robber. It is unlikely that the 'minors' were regarded as pest species in the 18th century; it is more probable that their small size suggested stealthy and furtive habits (adjectives implying robbery). Cf. also 2294 *raptricula*.

2340 **fasciuncula** (Haworth, 1809) – fascia, a band: from the median fascia of the forewing. Haworth, borrowed the dim. termination -*unculus* for a 'minor' from Denis and Schiffermüller.

Mesoligia Boursin, 1965 – μή (mē), not; 's', phonetic insertion; the genus 2337 *Oligia*, q.v.: not the same genus, but closely allied to it.

2341 **furuncula** ([Denis & Schiffermüller], 1775) – *fur*, a thief; *-uncula*, the dim. termination adopted for 'minors': see 2339.

2342 **literosa** (Haworth, 1809) – *literosus*, learned, pertaining to letters, i.e. of the alphabet: from the 'writing' round the stigmata. The moth is largely submaritime and derivation from an adjective coined from *litus*, the shore, is also possible.

Mesapamea Heinicke, 1959 – μή (mē), not; 's', phonetic insertion; the genus 2321 *Apamea*, q.v.: not the same genus, but closely allied to it.

2343 **secalis** (Linnaeus, 1758) – *Secale cereale*, rye: Linnaeus states correctly that it mines the stems.

2343a **didyma** (Esper, 1788) – δίδυμος (didumos), double (adj.), a twin (n.): from the reniform stigma which is double because it consists of a smaller pale stigma enclosed within a larger one more or less of the ground colour.

= **secalella** Remm, 1983 – dim. of 2343 *secalis*, this species on average being slightly smaller.

2343b **remmi** Rezbanyai-Reser, 1986 – i.h.o. Dr H. Remm, the contemporary Estonian entomologist who separated the previous species from *M. secalis*.

Photedes Lederer, 1857 – φῶς, φωτός (phōs, phōtos), light; ἦδος (ēdos), delight: from the habits of the next species (originally the only one in the genus), which flies by day.

2344 **captiuncula** (Treitschke, 1825) – *captio*, a taking, a robbery; *-uncula*, the conventional termination for 'minors': a name formed on the analogy of 2339 *latruncula*, q.v. '*Captio*, deception, from the way it evades capture' (Macleod).

subsp. **expolita** (Stainton, 1855) – *expolitus*, polished, neat: from the moth's attractive appearance.

subsp. **tincta** (Kane, 1895) – *tinctus*, dyed: from the pinkish suffusion of the Irish subspecies.

2345 **minima** (Haworth, 1809) – *minimus*, smallest, least: Haworth described it in the comprehensive genus *Noctua*, in which company it was indeed a very small member.

2346 **morrisii** (Dale, 1837) – i.h.o. Rev. F. O. Morris (1810–93), the British entomological author who discovered the species at Charmouth, Dorset in 1837.

subsp. **bondii** (Knaggs, 1861) – i.h.o. F. Bond (1826–89), a British entomologist.

2347 **extrema** (Hübner, 1809) – *extremus*, at either extreme end of a scale, largest or least: here in the latter sense with the same meaning as 2345 *minima*. Hübner described the species in *Noctua* and not in 2369 *Nonagria* Ochsenheimer, 1816, as stated by Macleod, another of his many anachronisms.

2348 **elymi** (Treitschke, 1825) – *Elymus arenarius* (now *Leymus arenarius*), lyme-grass, is the foodplant.

2349 **fluxa** (Hübner, 1809) – *fluxus*, pale, transient: from the weakly-expressed, indistinct markings of the forewing.

2350 **pygmina** (Haworth, 1809) – *pygminus*, an adjective coined from *Pygmaei*, the pygmies: for the same reason as 2345, q.v.

2351 **brevilinea** (Fenn, 1864) – *brevis*, short; *linea*, a line: from the short, fine basal streak on the forewing.

Eremobia Stephens, 1829 – ἐρῆμος (erēmos), solitary, deserted; βιόω (bioō), to live: Stephens gives the habitat of the next species as 'exposed heathy downs', so the reference is to its seclusion, not its aridity as thought by Macleod. The moth was formerly a rarity.

2352 **ochroleuca** ([Denis & Schiffermüller], 1775) – ὠχρός (ōkhros), pale yellow; λευκός (leukos), white: from the ground colour.

Luperina Boisduval, 1828 – possibly from *luperinus*, an adjective coined from *lupus*, a wolf to refer to the greyish colour of the moths, as suggested by Macleod. The objection is that Boisduval and his contemporaries derived generic names from Greek words and this is unlikely to be an exception. In consequence Pickard *et al.* leave the name unexplained and Spuler regards it as a typographical error for *Lyperina* from λυπηρός (lupēros), painful, causing sorrow (not sorrowful, as he supposed), with reference to the dull colours of the member species; the drawback here is that *-inus* is a Latin termination and should not have been attached to a Greek stem. In default of a better alternative, Macleod's Latin source seems the least unsatisfactory.

2353 **testacea** ([Denis & Schiffermüller], 1775) – *testaceus*, brick-coloured: this is a variable species and Viennese specimens may have been redder than British ones; alternatively, their bricks may have been less red.

2354 **nickerlii** (Freyer, 1845) – i.h.o. F. A. Nickerl (1813–71), who lived at Prague and probably discovered the species since the type locality is given as Czechoslovakia.

subsp. **gueneei** Doubleday, 1864 – i.h.o. A. Guenée (1809–80), the French entomologist and entomological writer, and personal friend of Doubleday.

subsp. **leechi** Goater, 1976 – i.h.o. M. J. Leech, who collaborated with B. Goater in the research leading up to the bestowal of subspecific status on the Cornish population.

subsp. **knilli** Boursin, 1964 – i.h.o. S. A. Knill-Jones, who discovered the subspecies in the south-west of Ireland.

2355 **dumerilii** (Duponchel, 1827) – i.h.o. A. M. C. Dumeril (1774–1860), a French naturalist.

2356 **zollikoferi** (Freyer, 1836) – i.h.o. C. T. Zollikofer (1774–1843), a Swiss naturalist.

Amphipoea Billberg, 1820 – ἀμφί (amphi), round; πόα, ποία (poa, poia), grass: from the habitat.

2357 **lucens** (Freyer, 1845) – *lucens*, shining: from the gloss (Spuler); 'from bright mark on forewings' [i.e. the reniform stigma, the 'ear'] (Macleod). Either of these may be correct, while a third possibility is the gleam of the resting moth by the light of a lantern, as it is often seen by collectors.

2358 **fucosa** (Freyer, 1830) – *fucosus*, painted, either in the sense beautiful or copied, counterfeit, spurious: probably in the second sense, a copy of 2360 *A. oculea*, and perhaps not distinct from it.

subsp. **paludis** (Tutt, 1888) – *palus*, *paludis*, a marsh: from the habitat.

2359 **crinanensis** (Burrows, 1908) – from the Crinan Canal in Scotland, the type locality.

2360 **oculea** (Linnaeus, 1761) – *oculeus*, eyed (*oculus*, an eye): from the reniform stigma, redesignated an 'ear' by British entomologists.

Hydraecia Guenée, 1841 – ὕδωρ (hudōr), water; οἰκέω (oikeō), to dwell: from the rather damp habitat. Later, Guenée (1852) altered the spelling to *Hydroecia*.

2361 **micacea** (Esper, 1789) – *micaceus*, an adjective coined from *micare*, to quiver or glitter: seems unapt, but most moths quiver when they are warming up prior to flight, and most glitter when seen by the light of a lantern as they rest on herbage. Esper may have seen one of these.

2362 **petasitis** Doubleday, 1847 – *Petasitis hybridus*, butterbur, is the foodplant.

2363 **osseola** Staudinger, 1882 – *osseus*, of bone, bone-coloured, *osseolus*, dim., somewhat bone-like: from the ground colour.

subsp. **hucherardi** Mabille, 1907 – i.h.o. M. Hucherard who had recently discovered the subspecies at Royan, Charente-Maritime (France).

Gortyna Ochsenheimer, 1816 – Gortyna, a town in Crete: no entomological application. Ochsenheimer quite often turned to towns for generic names; cf. 2154 *Mamestra*, etc. and p. 33.

2364 **flavago** ([Denis & Schiffermüller], 1775) – *flavus*, yellow; the termination *-ago*, see 2271.

2365 **borelii** Pierret, 1837 – i.h.o. M. Borel (early 19th century), who took the first specimen in the woods of Sainte-Geneviève, near Paris.

subsp. **lunata** Freyer, 1839 – *lunatus*, crescent-shaped: from the dark, crescent-shaped strigula in the reniform stigma.

Calamia Hübner, 1821 – κάλαμος (kalamos), a reed: from the habitat.

2366 **tridens** (Hufnagel, 1766) – *tridens*, three-toothed: Hufnagel describes three sometimes imperceptible specks in the reniform stigma.

subsp. **occidentalis** Cockayne, 1954 – western, the subspecies occurring in the west of Ireland.

Celaena Stephens, 1829 – κελαινός (kelainos), dark, swarthy: from the dark coloration of some specimens of the next species.

2367 **haworthii** (Curtis, 1829) – i.h.o. A. W. Haworth (1767–1833), who had recently published the fourth volume of his *Lepidoptera Britannica*.

2368 **leucostigma** (Hübner, 1808) – λευκός (leukos), white; στίγμα (stigma), a spot: from the small pale reniform stigma.

subsp. **scotica** Cockayne, 1944 – the subspecies occurring in Scotland.

Nonagria Ochsenheimer, 1816 – *Nonagria*, an old name for the island of Andros in the Aegean Sea: Treitschke, Ochsenheimer's collaborator, gives this explanation and it is in keeping with the latter's practice of using places for his source of generic names. Spuler suggests *non*, not, and ἄγριος (agrios), wild: not in the wild, i.e. not in the open, because the larvae are internal feeders; it is unlikely that Ochsenheimer, a man of letters, would have coined a hybrid name from Latin and Greek. Macleod imputes to him a typographical error, suggesting 'Nonacria, epithet of Greek princess Atalanta'.

2369 **typhae** (Thunberg, 1784) – *Typha latifolia*, bulrush, reedmace: the larval foodplant.

Archanara Walker, 1866 – Walker's names are often meaningless, and this one seems to belong to that category.

2370 **geminipuncta** (Haworth, 1809) – *gemini*, twins; *punctum*, a spot: from the two white dots in the reniform stigma, the 'twin-spotted wainscot'.

2371 **dissoluta** (Treitschke, 1825) – *dissolutus*, loose, dissolved: according to Spuler, because the black dots present in the reniform stigma of the next species are absent or dissolved away in this one.

2372 **neurica** (Hübner, 1808) – νεῦρον (neuron), a sinew, a vein of an insect's wing: the veins of the forewing are obscurely darker and the postdiscal line is represented by a series of fuscous spots on the veins.

2373 **sparganii** (Esper, 1790) – *Sparganium erectum*, bur-reed: a foodplant.

2374 **algae** (Esper, 1789) – evidently from a mistaken belief that the larva fed on algae; Ochsenheimer (1816) renamed the species *cannae* (*canna*, a reed).

Rhizedra Warren, 1911 – ῥίζα (rhiza), a root; ἕδρα (hedra), a seat, an abode: from the habits of the larva which lives and feeds in the rootstock of its foodplant.

2375 **lutosa** (Hübner, 1803) – *lŭtosus*, muddy: probably from the damp habitat but possibly from clouding sometimes present on the fore- and/or hindwing. Derivation from *lūtosus*, an adjective coined from *lūteus* (itself an adjective), yellow, would imply a richer shade than the artist Hübner would be likely to ascribe to this species.

Sedina Urbahn, 1933 – *Sedyn*, the old name of Stettin (now Szczecin) in Poland, a town with entomological associations; it was Zeller's home and the *Stettiner Entomologische Zeitung* was published there from 1840–1944. The type locality of the next species is in Poland, but not at Stettin.

2376 **buettneri** (Hering, E., 1858) – i.h.o. J. G. Büttner (19th century), a Latvian entomologist.

Arenostola Hampson, 1908 – ἀρῆνα (arēna), sand; στολή (stolē), a garment: from the sandy-coloured vestiture of the next species.

2377 **phragmitidis** (Hübner, 1803) – *Phragmites australis*, formerly *Arundo phragmites*, the common reed: the larval foodplant.

Oria Hübner, 1821 – perhaps ὡρεῖον (hōreīon), a granary: from the larval foodplants of the next species, the only one placed in the genus by Hübner, these being various cereal crops. '*Oros*, whey; from colour of wings' (Macleod); this is possible though the wings are too deep an ochreous yellow. Hübner's diagnosis gives no clue.

2378 **musculosa** (Hübner, 1808) – There are two possible derivations. (1) *musculosus*, an adjective formed from *musculus* a little mouse: from the behaviour of the female, which runs down the cornstalks (MBGBI 10: 265), though Hübner may not have known this; or from the dark irroration between the veins (Spuler), but that is hardly mouse-coloured. (2) *musculosus*, fleshy: Spuler, who gives both derivations, suggests that it refers to the pale sinews or veins of the forewing. Macleod says 'fleshy' without explanation.

Coenobia Stephens, 1850 – κοινόβιος (koinobios), living in community with others: from the larval habits of the next species, the larvae being gregarious in autumn and winter. 'Lat. *coenum*, mud, Gr. *bioö*, to live; frequents marshes' (Macleod): a hybrid derivation, unlikely but not impossible from Stephens.

2379 **rufa** (Haworth, 1809) – *rufus*, reddish: the 'small rufous'.

Charanycha Billberg, 1820 – χαρά (khara), delight; νύξ, νυκτός (nux, nuctos), night, νύχιος (nukhios), nightly: from nocturnal habits, a Greek rendering of Noctua.

2380 **trigrammica** (Hufnagel, 1766) – *tri-*, three; *gramma*, a line: the 'treble lines'.

Hoplodrina Boursin, 1937 – ὅπλον (hoplon), a weapon; *-drina*, the termination of 2387 *Caradrina*: a genus related to *Caradrina* but differing in the genital armature.

2381 **alsines** (Brahm, 1791) – *Stellaria alsine*, bog-stitchwort: one of the foodplants of this polyphagous species.

2382 **blanda** ([Denis & Schiffermüller], 1775) – *blandus*, smooth: from the smooth-scaled, rather glossy forewing; Heslop (1964) called it the 'smooth rustic'.

2383 **superstes** (Ochsenheimer, 1816) – *superstes*, surviving, outliving: the moths in this and related genera are notoriously hard to determine and there has long been controversy over the number of valid species; this one has survived the threat of reduction to synonymy. 'Surviving, from rarity' (Macleod); had he taken the authorship of names into account, he would have known that this one had been given by a German, then working in Austria where he was taking leading roles at the Royal Opera, and that the species is common in central Europe; Ochsenheimer was not concerned with its British distribution, where it is a rare migrant.

2384 **ambigua** ([Denis & Schiffermüller], 1775) – *ambiguus*, doubtful: from uncertainty over specific distinction from, for example, 2382 *H. blanda*; not from 2383 *H. superstes*, as supposed by Spuler, as that is an anachronism.

Spodoptera Guenée, 1852 – σποδός (spodos), wood ashes, dust; πτερόν (pteron), a wing: from the irroration of fuscous on the forewing of the next species.

2385 **exigua** (Hübner, 1808) – *exiguus*, small: from its wingspan being less than that of most related species.

2386 **littoralis** (Boisduval, 1833) – *littoralis*, of the shore: one of the type localities cited by Boisduval is the Mascarine region on the coast of Algeria.

2386a **litura** (Fabricius, 1775) – *litura*, a smearing, an erasure, a blotch: presumably from a character in the wing pattern; cf. 2266 *litura*.

2386b **eridania** (Stoll, 1780) – Ἠριδανός (Ēridanos), the Eridanus, a river of western Europe mentioned by the early Greek writers and variously identified as the Po, the Rhone and especially the Rhine. If there is any entomological significance it is that the river (to the Greeks) was western and this is a Nearctic species.

2386c **cilium** Guenée, 1852 – *cilium*, an eyelash, one of the hairs in a fringe or fascicle: from the antenna which Guenée describes as having rather long fascicles of cilia. The concept is plural but Guenée uses the singular as he did for two other *Spodoptera* spp. which he named in the same work, *S. pecten* (*pecten*, a comb) from the pectinated antenna and *S. filum* (*filum*, a thread) from the simple antenna.

Caradrina Ochsenheimer, 1816 – stated by Ochsenheimer himself to be the name of a river in Albania.

2387 **morpheus** (Hufnagel, 1766) – Morpheus, the god of dreams: no entomological significance is necessary, but Hufnagel, who describes the moth as 'dirty', may have seen an affinity between the dingy, dusky forewing and one of the divinities of the night. 'From larva's sluggishness' (Macleod); a baseless guess. There is no evidence that Hufnagel knew the life history and, indeed, it is unlikely that he did so.

2388 **flavirena** Guenée, 1852 – *flavus*, yellow; *renes*, the kidneys: from the reniform stigma, which is finely outlined with yellowish white.

2389 **clavipalpis** (Scopoli, 1763) – *clavus*, a club; *palpus*, the labial palpus: segment 2 of the labial palpus is clad in dense scales, almost concealing segment 3, and the apex of segment 2 and the whole of segment 3 are whitish and conspicuous. This character is not variable as supposed by Macleod who wrote 'from blunt-headed palps of some moths'.

Perigea Guenée, 1852 – περίγειος (perigeios), around the earth: either from the wide distribution of this tropical genus or from a migratory tendency of its members. Guenée did not include the next species.

2390 **capensis** (Guenée, 1852) – a Latinized adjective from the French 'cap', the type locality being Cape Province, South Africa.

= **conducta** Walker, 1856 – *conduco*, to bring together, to assemble, to hire, especially to hire men as mercenaries: it is possible that the type specimen was obtained by purchase.

Chilodes Herrich-Schäffer, 1849 – the genus 1290 *Chilo*, q.v.; εἶδος, ὠδ- (eidos, ōd-), shape, form: from resemblance to the pyralid genus. 'From moth's long palps' (Macleod): he cannot have looked at a specimen, for the palps of the next species are very short; the resemblance lies in the forewing and the habitat.

2391 **maritimus** (Tauscher, 1806) – *maritimus*, pertaining to the sea, maritime: the type locality is on the shores of the Baltic Sea, but otherwise it is not maritime.

Athetis Hübner, 1821 – ἄθετος (athetos), without position or place: an anomalous genus hard to place in the systematic list.

2392 **pallustris** (Hübner, 1808) – an error for *palustris*, *paluster*, fem. *palustris* pertaining to a fen: from the habitat.

2392a **hospes** (Freyer, 1831) – *hospes*, a guest or host: perhaps from a migratory tendency; at any rate, the sole British specimen was a migrant.

Acosmetia Stephens, 1829 – ἀ, alpha privative; κόσμος (kosmos), adornment, κοσμητική (kosmēticē), the art of dress and adornment, cosmetics: from the dull coloration of the next species and suggested by its name.

2393 **caliginosa** (Hübner, 1813) – *caliginosus*, obscure, gloomy: from the dull coloration.

Stilbia Stephens, 1829 – στίλβω (stilbō), to glisten: from the glossy forewing of the next species – 'wings . . . very shining'.

2394 **anomala** (Haworth, 1812) – ἀνώμαλος (anōmalos), irregular, anomalous: the moth hardly looks like a noctuid, its ample hindwings suggesting a pyrale, the 'anomalous'.

Synthymia Hübner, 1823 – σύνθημα (sunthēma), anything agreed upon: this name appears to be Graecized form of the name of the next species, the only one Hübner included in the genus; however, in 1808 he had named it *monogramma* and he retained that name when fifteen years later he established this genus. The law of priority had not yet been invented and it is reasonable to assume that he knew that *fixa* and *monogramma* were synonyms but preferred to use his own name.

2395 **fixa** (Fabricius, 1787) – *fixus*, settled, fixed: a difficult, anomalous species; when its systematic position was at last settled, the achievement merited commemoration in the specific name. See also genus.

Elaphria Hübner, 1818 – ἐλαφρία (elaphria), lightness, agility: from the lively behaviour of the small moths.

2396 **venustula** (Hübner, 1790) – *venustulus*, lovely little, charming: a tribute from an artist to a very pretty species.

Panemeria Hübner, 1823 – πανημέρειος (panēmereios), all day long: from the diurnal behaviour of the next species.

2397 **tenebrata** (Scopoli, 1763) – *tenebrae* (pl.), darkness, *tenebratus*, shaded: from the dark shading on the forewing and the dark base and border of the hindwing. Scopoli described the moth in Phalaena, not Noctua, and the termination *-ata* indicates that he thought it was a geometrid.

HELIOTHINAE (2400)

Periphanes Hübner, 1821 – περιφανής (periphanēs), seen all round, manifest, famous: from the two species Hübner included in the genus, which stand out from others for their conspicuous beauty.

2398 **delphinii** (Linnaeus, 1758) – *Delphinium*, larkspur, correctly given by Linnaeus as the foodplant.

Pyrrhia Hübner, 1821 – πυῤῥός (purrhos), flame-coloured: from the orange-brown ground colour of the species Hübner included in the genus.

2399 **umbra** (Hufnagel, 1766) – *umbra*, a shadow: from the terminal shade of the hindwing.

Heliothis Ochsenheimer, 1816 – ἡλιώτης (hēliōtēs), of the sun: from the diurnal habits of the members of the genus.

2400 **armigera** (Hübner, 1803) – *armiger*, a weapon-bearer, a shield-bearer: to express affinity with 2403 *H. peltigera* and 2405 *Protoschinia scutosa*, q.v. 'Warlike, from devastating attacks made by larva on crops in some countries, particularly U.S.A.' (Macleod). Hübner was naming, from the adult, a species new to science, its type locality Europe; how could he possibly have known that in years to come the very similar *H. zea* (Boddie, 1850), which was at one stage misidentified as *H. armigera*, would be regarded as a pest species in the New World? *H. armigera* itself does not occur in the U.S.A.

2401 **viriplaca** (Hufnagel, 1766) – Viriplaca, the name of the Roman goddess who presided over the peace of families, placating the troublesome men (*viri*): there is no entomological significance.

2402 **maritima** (Graslin, 1855) – *maritimus*, belonging to the sea, maritime: from the type locality at Vendée, on the Atlantic coast of France, where it was first found on salt-marshes.

subsp. **warneckei** Boursin, 1964 – i.h.o. G. Warnecke (1884–1962), a German entomologist.

subsp. **bulgarica** Draudt, 1938 – the subspecies occurring in Bulgaria and Jugoslavia which has been doubtfully recorded once in Britain.

2403 **peltigera** ([Denis & Schiffermüller], 1755) – *pelta*, a small, light shield, usually crescent-shaped; *gero*, to carry: from the shape of the reniform stigma.

2404 **nubigera** Herrich-Schäffer, 1851 – *nubes*, a cloud: *gero*, to carry: from the large reniform stigma and the suffusion extending from it. The termination *-gera* marks affinity with the other members of the genus but the name abandons the theme of weaponry.

Protoschinia Hardwick, 1970 – πρῶτος (prōtos), foremost, front; the genus 2406 *Schinia*, q.v.: from the setae on the foreleg, which are situated apically.

2405 **scutosa** ([Denis & Schiffermüller], 1775) – *scutosus*, an adj. coined from *scutum*, a shield: from the reniform stigma; cf. 2400 and 2403.

Schinia Hübner, 1818 – σχοίνος (skhoinos), a rush, an arrow: from the sharp spurs on the foreleg.

2406 **rivulosa** (Guenée, 1852) – *rivulosus*, an adj. coined from *rivulus*, a rivulet: presumably from wavy, 'rivulet' markings.

ACONTIINAE (2415)

Eublemma Hübner, 1821 – εὖ (eu), well; βλεμμα (blemma), a look, a glance: from the moths being good to look at.

2407 **ostrina** (Hübner, 1803) – *ostrinus*, purple: from the purplish markings.

2408 **parva** (Hübner, 1803) – *parvus*, small: one of the smallest noctuids.

2409 **minutata** (Fabricius, 1794) – *minutus*, very small: the termination *-ata* shows that Fabricius regarded it as a geometer.

= **noctualis** (Hübner, 1796) – either from *noctu*, by night: a nocturnal species, in contrast to many of the other small noctuids which fly or are easily disturbed by day; or *Noctua*, a member of the genus, the original designation *Pyralis noctualis*, signifying 'the pyralid that resembles a noctuid'.

Protodeltote Ueda, 1984 – πρῶτος (prōtos), foremost, front; the genus 2411 *Deltote* q.v.: a more primitive genus than *Deltote* or perhaps just the one that precedes it.

2410 **pygarga** (Hufnagel, 1766) – πύγαργος (pugargos), a white-rumped Libyan antelope (πυγή (pugē), the rump; ἀργός (argos), bright, shining): from the white tornal spot which covers the moth's anus when the wings are folded.

Deltote R. L., 1817 – δελτωτός (deltōtos), shaped like the Greek capital delta, Δ: from the triangular shape of the forewing, though this is not as noticeable as in the 'deltoids', an old name for the 'snouts' (Hypeninae). For the possible identity of 'R.L.', the nomenclator, see 1572 *Aricia*.

= **Lithacodia** Hübner, 1818 – λίθαξ (lithax), stony, stone; εἶδος, ὠδ- (eidos, od-), form, appearance: from the marbling on the forewing, the 'marbled white-spot'.

= **Eustrotia** Hübner, 1821 – εὔστρωτος (eustrōtos), well-spread with clothes, well-dressed:

from the smart wing pattern. 'Perhaps from hairy antennae' (Macleod); very unlikely, the antenna being only weakly ciliate.

2411 **deceptoria** (Scopoli, 1763) – *deceptor*, a deceiver: perhaps because the moth is deceptively like a tortricid.

2412 **uncula** (Clerck, 1759) – *uncula*, a little hook: from the somewhat hook-shaped reniform stigma projecting from the costal streak; the 'silver hook'.

2413 **bankiana** (Fabricius, 1775) – i.h.o. Sir Joseph Banks (1743–1820), who sailed with Captain Cook and was a friend of Fabricius; *-ana*, because described as a tortricid (Pyralis *usu* Fabricius).

Emmelia Hübner, 1821 – ἐμμέλεια (emmeleia), harmony in music, gracefulness in general: from the beauty of the next species.

2414 **trabealis** (Scopoli, 1763) – *trabealis*, pertaining to, or clad in, the *trabea*, a robe of state: from the striking black and yellow pattern.

Acontia Ochsenheimer, 1816 – ἀκοντίας (akontias), a species of snake: from the quick movements of the moths (Treitschke).

2415 **lucida** (Hufnagel, 1766) – *lucidus*, bright, shining: from the white basal patch.

2416 **aprica** (Hübner, 1803) – *apricus*, sunny, bright: from the wing pattern.

2417 **nitidula** (Fabricius, 1787) – dim. of *nitidus*, shining: from the wing pattern.

CHLOEPHORINAE (2422)

Earias Hübner, 1825 – ἔαρ (eär), the spring: from the green ground colour, suggesting spring foliage; cf. 2099 *praecox* and 2247 *aprilina*. 'From time of moth's appearance' (Macleod): very unlikely, since this is late May to July.

2418 **clorana** (Linnaeus, 1761) – for *chlorana*, χλωρός (khlōros), bright green: from the ground colour; *-ana*, because Linnaeus described it as a tortricid.

2419 **biplaga** Walker, 1866 – *bi-*, two; *plaga*, a blow, a stripe, a weal: from the purplish dorsal spots, one on each forewing, which suggest weals.

2420 **insulana** (Boisduval, 1833) – *insula*, an island: from the type localities which are islands in the Indian Ocean; *-ana*, because Boisduval described it as a tortricid.

2420a **vittella** (Fabricius, 1794) – *vitta*, a chaplet, in entomology a longitudinal stripe: from the forewing which is green with broad cream-coloured costal and dorsal vittae; *-ella*, because Fabricius described it as a tineid.

Bena Billberg, 1820 – apparently a meaningless neologism.

2421 **prasinana** (Linnaeus, 1758) – *prasinus*, leek-green: from the ground colour; *-ana*, because Linnaeus described it as a tortricid.

Pseudoips Hübner, 1822 – ψεῦδος (pseudos), false; ἴψ (ips), a worm that eats vine-buds: perhaps to be taken literally, from the destructive behaviour of the larvae. Hübner first proposed this name in his *Tentamen* [1806], giving as the example [1033 *Tortrix*] *viridana* which is often extremely damaging to the foliage of trees, especially the oak. '*Ips*, perhaps short for old generic name *Sarrothripus*' (Macleod). *Sarrothripus* Curtis, 1824 is the junior name by two years, or, more precisely, by eighteen years from time when *Pseudoips* had first been proposed.

= **Chloephora** Stephens, 1827 – χλόη (khloē), the green of new leaves; φορέω (phoreō), to bear: from the bright green ground colour of the species in the genus.

2422 **fagana** (Fabricius, 1781) – *Fagus sylvatica*, the beech: a foodplant; *-ana*, because Fabricius placed it in Tortrix (Pyralis *usu* Fabricius).

subsp. **britannica** (Warren, 1913) – the subspecies occurring in Britain.

SARROTHRIPINAE (2423)

Nycteola Hübner, 1822 – νύξ, νύκτος (nux, nuktos), night; possibly also ἠώς (ēōs), dawn: species that fly in the twilight of dawn; however, the termination *-eola* may be simply an unorthodox diminutive.

= **Sarrothripus** Curtis, 1824 – σάρωθρον (sarōthron), a sweeping-broom; ποῦς (pous), the foot: from the brushes of hair-scales attached to the foreleg.

2423 **revayana** (Scopoli, 1772) – i.h.o. M. Revay (18th century), a French entomologist.

2424 **degenerana** (Hübner, 1796) – *degener*, departing from race: Hübner described the species as a tortricid and may have had well-founded misgivings after observing characters untypical of that family.

PANTHEINAE

From the genus *Panthea* Hübner, 1820 – Πάνθειον, the Pantheon, a temple consecrated to all the gods, especially that at Rome: there is no entomological application. The genus, which is not represented in Britain, is placed in the Acronictinae by Leraut (1980).

Colocasia Ochsenheimer, 1816 – κολοκασία (kolokasia), a beautiful plant resembling the water-lily which is found in marshy parts of Egypt: there appears to be no entomological application.

2425 **coryli** (Linnaeus, 1758) – *Corylus avellana*, the hazel: a foodplant.

Charadra Walker, 1865 – χαράδρα (kharadra), a mountain stream: no entomological significance.

2426 **deridens** (Guenée, 1852) – *deridens*, laughing at, aping: perhaps from resemblance to another species. It is American.

Raphia Hübner, 1821 – possibly from ῥάφη (rhaphē), a large kind of radish, perhaps akin to Lat. *rapum*, *rapa*, rape: possibly a foodplant of one of the three species included by Hübner. *Rhaphia* Agassiz, 1846, is an unjustified emendation.

2427 **frater** Grote, 1864 – *frater*, a brother: probably from affinity with or similarity to another American species.

PLUSIINAE (2439)

Chrysodeixis Hübner, 1821 – χρυσός (khrusos), gold; δεῖξις (deixis), a display: from the metallic-marked forewings.

2428 **chalcites** (Esper, 1789) – χαλκίτης, Lat. *chalcitēs*, a precious stone of copper colour mentioned by Pliny: from the metallic markings of the forewing.

2429 **acuta** (Walker, 1858) – *acutus*, pointed: from the apex of the forewing, which is supposedly more pointed than in related species.

Ctenoplusia Dufay, 1970 – κτείς, κτενός (kteis, ktenos), a comb; the genus 2439 *Plusia*: from the strong spines on the ventral margin of the valva in the male genitalia.

2430 **limbirena** (Guenée, 1852) – *limbus*, a border; *renes*, the kidneys: from the complex white spot in the disc of the forewing, which partly encircles the reniform stigma.

2431 **accentifera** (Lefebvre, 1827) – *accentus*, the accentuation of a word, hence the mark by which it is indicated; *fero*, to carry: from the dark costal and dorsal streaks on the forewing.

Trichoplusia McDunnough, 1944 – θρίξ, τρίχος (thrix, trikhos), hair; the genus 2439 *Plusia*: from the strong dorsal and lateral abdominal crests.

2432 **ni** (Hübner, 1803) – the letters ν + ι (nu + iota): from the formation of the reniform stigma which resembles the combination of these Greek characters.

Diachrysia Hübner, 1821 – διάχρυσος (diakhrusos), interwoven with gold: from the extensive metallic markings.

2433 **orichalcea** (Fabricius, 1775) – *orichalceus*, adj. coined from *orichalcum*, brass: from the metallic markings.

2434 **chrysitis** (Linnaeus, 1758) – χρυσίτης, fem. χρυσῖτις (khrusitēs, khrusītis), like gold: from the metallic forewing.

2435 **chryson** (Esper, 1789) – Byzantine Gr. χρυσών (khrusōn), gold treasure: from the metallic markings.

Pseudoplusia McDunnough, 1944 – ψεῦδος (pseudos), false; the genus 2439 *Plusia*: a genus resembling *Plusia*. For *Pseudo-* see p. 35.

2435a **includens** (Walker, 1858) – *includens*, shutting in: the 'silver Y' character is surrounded or 'shut in' by pale lilac.

Macdunnoughia Kostrowicki, 1961 – i.h.o. J. H. McDunnough (1877–1962), the Canadian entomologist who revised the subfamily.

2436 **confusa** Stephens, 1850 – *confusus*, confused: possibly with 2448 *Syngrapha circumflexa*.

Polychrysia Hübner, 1821 – πολύς (polus), much; χρυσός (khrusos), gold: from the forewing, which has the whole ground colour pale golden.

2437 **moneta** (Fabricius, 1787) – *Moneta*, a title of Juno, in whose temple on the Capitoline Hill at Rome the mint was situated; hence our word 'money': from the metallic markings on the forewing.

Euchalcia Hübner, 1821 – εὖ (eu), well, intensive prefix; χαλκίον (khalkion), copper: from the metallic markings.

2438 **variabilis** (Piller & Mitterpacher, 1783) – *variabilis*, changeable, variable: from the pattern of contrasting colours rather than variability between specimens.

Plusia Ochsenheimer, 1816 – πλούσιος (plousios), rich: from the gold or silver markings on the forewing. This is one of the names Ochsenheimer took from Hübner's privately circulated *Tentamen* [1806].

2439 **festucae** (Linnaeus, 1758) – *Festuca fluitans*, a species of fescue, is given by Linnaeus as the foodplant.

2440 **putnami** Grote, 1873 – i.h.o. G. P. Putnam, a member of the Publications Committee of the Buffalo Society of Natural Sciences, in whose *Bulletin* the first description appeared.

subsp. **gracilis** Lempke, 1966 – *gracilis*, slender, meagre, poor, unadorned, but here probably supposed to mean graceful. cf. 280 *Gracillaria* and 2186 *gracilis*.

Autographa Hübner, 1821 – αὐτόγραφος (autographos), written in one's own hand: from the characters (Y, etc.) 'written by the moths themselves' on their forewings.

2441 **gamma** (Linnaeus, 1758) – from the metallic mark shaped like the Greek gamma (γ) on the forewing. Linnaeus calls it 'golden', not silver.

2442 **pulchrina** (Haworth, 1809) – an adjective formed from the adjective *pulcher*, beautiful: the 'beautiful golden Y'. Haworth (1802) had proposed that all the names of the Noctuidae should end in *-ina* (see p. 22).

2443 **jota** (Linnaeus, 1758) – *iota*, the Greek letter 'ι': Linnaeus describes the golden mark on the forewing as resembling an iota or a question mark, but he had already used the latter for his preceding species, 2447 *interrogationis*.

2444 **bractea** ([Denis & Schiffermüller], 1775) – *bractea*, a thin plate of metal, gold-leaf: from the presence of a more extensive area of gold than on most other plusiines.

2445 **biloba** (Stephens, 1830) – *bi-*, two; λοβός (lobos), a lobe, as of the ear: from the metallic streak on the forewing, which is deeply excavate on its costal side so as to resemble two lobes.

2446 **bimaculata** (Stephens, 1830) – *bi-*, two; *maculatus*, spotted: the silver marking on the forewing consists of two spots.

Syngrapha Hübner, 1821 – probably a syncopated word from σύν (sun), with, and the genus 2441 *Autographa*, q.v.: another genus of species with metallic characters 'written' on their wings. cf. the formation of 986 *Syndemis*.

2447 **interrogationis** (Linnaeus, 1758) – *interrogationis signum*, a question mark: Linnaeus describes such a white mark on the forewing.

2448 **circumflexa** (Linnaeus, 1767) – *circumflexus*, bent round: from the metallic mark on the forewing which is arched like a circumflex accent.

Abrostola Ochsenheimer, 1816, an error for *Habrostola*, an emendation proposed by Sodoffsky (1837) and adopted by some authors – ἁβρός (habros), graceful, beautiful; στολή (stolē), a robe: from the attractive species.

2449 **trigemina** (Werneburg, 1864) – *trigeminus*, one of a set of triplets: one of three very similar moths, one of the others being the next species.

2450 **triplasia** (Linnaeus, 1758) – *triplasius*, triple: Linnaeus describes the ante- and postmedian fasciae, which divide the forewing into three areas, and also three blue-grey spots (the stigmata) in the median section; the synonym *tripartita* Hufnagel, 1766, is also descriptive of the triple division of the forewing.

CATOCALINAE (2451)

Catocala Schrank, 1802 – κάτω (katō), below; καλός (kalos), beautiful: from the brightly coloured hind- or underwing; Schrank translates his name 'Prachteule', splendid owls.

2451 **fraxini** (Linnaeus, 1758) – *Fraxinus excelsior*, the ash-tree: stated wrongly by Linnaeus to be the foodplant.

2452 **nupta** (Linnaeus, 1767) – *nupta*, a bride: one wonders why Linnaeus started the convention of naming species with brightly coloured underwings after the fairer sex and brides and fiancées in particular. Did 18th century brides in Sweden wear gaudy underwear (red flannel petticoats) to stimulate the groom, or did Linnaeus think they ought to do so? cf. 446, 985, 1661, 2057, 2058, 2068, 2071, 2107, 2109 and other members of this genus.

2453 **electa** (Vieweg, 1790) – *electus*, chosen, *electa*, a fiancée: see 2452.

2453a **elocata** (Esper, 1788) – *elocatus*, hired out: a prostitute, assuming Esper was playing the game described under 2452; cf. his name 2455a *nymphagoga*.

2454 **promissa** ([Denis & Schiffermüller], 1775) – *promissus*, promised, pledged in marriage: see 2452.

2455 **sponsa** (Linnaeus, 1767) – *sponsus*, promised in marriage; *sponsa*, a fiancée or bride: see 2452.

2455a **nymphagoga** (Esper, 1788) – νυμφαγωγός (numphagōgos), the person who leads the bride from her home to the bridegroom's house: see 2452.

Minucia Moore, 1885 – The name of a Roman gens, its most famous member having been the half-legendary Lucius Minucius Esquilinus Augurinus (5th century B.C), a public benefactor in time of famine. A commemorative statue stood near the Porta Minucia. No entomological significance. 'Vestal virgin' (Macleod).

2456 **lunaris** ([Denis & Schiffermüller], 1775) – *lunaris*, pertaining to the moon: from the somewhat lunate reniform stigma, the 'lunar double stripe'.

Clytie Hübner, 1823 – κλύτος (klutos), famous, glorious.

2457 **illunaris** (Hübner, 1813) – *in-*, not; *lunaris*, see 2456: a similar species but lacking the conspicuous lunate reniform stigma.

Caenurgina McDunnough, 1937 – καινουργός (kainourgos), new-made, a novelty; *-ina*, a Latin adjectival termination: a 'new-made' genus. Affinity with *Caenurgia* Walker, 1858, is implied.

2458 **crassiuscula** (Haworth, 1809) – *crassius*, rather stout; *-culus*, dim. termination: rather a stout little moth.

Mocis Hübner, 1823 – μῶκος (mōkos), mockery: Hübner's diagnosis offers no clue; perhaps the species are similar and so ape each other.

2459 **trifasciata** (Stephens, 1830) – *tri-*, three; *fascia*, a band: from the pattern of the forewing, the 'triple-barred'.

Dysgonia Hübner, 1823 – δυσ- (dus-), a prefix with the sense ill, hard, with difficulty, as in dyslexia; γωνία (gōnia), an angle: Hübner's diagnosis describes an irregular postmedian line; the prefix may have no precise meaning.

2460 **algira** (Linnaeus, 1767) – Algeria: the type locality.

Grammodes Guenée, 1852 – γραμμώδης, γραμμοειδής (grammōdes, grammoeides), in the form of lines: from the sharply-defined white fasciae on both fore- and hindwing of the next species.

2461 **stolida** (Fabricius, 1775) – *stolidus*, dull, stupid: Fabricius named a related species *geometrica* at the same time and geometry may have reminded him of the '*pons asinorum*'.

Callistege Hübner, 1823 – κάλλος (kallos), beauty; στέγη (stegē), a roof, here perhaps just a covering: from the beautiful vestiture of the wings.

2462 **mi** (Clerck, 1759) – mi, a Latinism for the Greek letter mu, our 'm': from the hindwing underside where a large 'M' may be seen in the discal area (Linnaeus, 1767). Macleod wrongly supposed the marking was on the forewing upperside.

Euclidia Ochsenheimer, 1816 – i.h.o. Euclid (*fl.* 300 B.C.), the Greek geometrician: from the geometrical patterns to be seen on the wings. Ochsenheimer took the name from Hübner's *Tentamen* [1806].

2463 **glyphica** (Linnaeus, 1758) – γλυφή (gluphē), an emblem, a device: *'alis . . . maculis hieroglyphicis nigris'*, the wings with black hieroglyphic markings.

OPHIDERINAE

From the genus *Ophideres* Boisduval, 1832, now reduced to synonymy: perhaps from ὄφις (ophis), a snake, with reference to the undulating gait of the larvae in which one or more pairs of prolegs are non-functional or absent.

Catephia Ochsenheimer, 1816 – κατήφεια (katēpheia), a casting of the eye downwards in a gesture implying dejection, sorrow or shame: from the black and white pattern of the next species, colours traditionally associated with mourning.

2464 **alchymista** ([Denis & Schiffermüller], 1775) – Latinized from the Arabic *al-kimia*, an alchemist, who traditionally wore a dark robe, like this moth.

Tyta Billberg, 1820 – apparently meaningless; cf. 2421 *Bena*.

2465 **luctuosa** ([Denis & Schiffermüller], 1775) – *luctuosus*, mournful: from the black and white coloration suggestive of mourning.

Diphthera Hübner, 1809 – διφθέρα (diphthera), a prepared hide, a piece of leather: Treitschke explained this as the Fell (i.e. hide) of Amalthea, the goat that suckled Zeus and was placed among the stars as a reward (Pickard *et al.*). Hübner first proposed the name in his *Tentamen* [1806], giving as an example [2247 *Dichonia*] *aprilina*, another brightly-coloured species.

2465a **festiva** (Fabricius, 1775) – *festivus*, gay, festive: from the brilliant coloration; cf. the Danai festivi of Linnaeus (p. 16).

Lygephila Billberg, 1820 – λύγη (lugē), darkness; φιλέω (phileo), to love: the moths fly as soon as it gets dark.

2466 **pastinum** (Treitschke, 1826) – *pastinum*, a two-pronged dibber used especially for planting vines, or the ground that has been dug and trenched ready for planting: probably the second meaning, from the numerous parallel transverse strigulae on the forewing which resemble fine furrows for planting seeds. There is no marking on the forewing resembling a two-pronged dibber as was assumed by Macleod.

2467 **craccae** ([Denis & Schiffermüller], 1775) – *Vicia cracca*, the tufted vetch: a foodplant.

Tathorhynchus Hampson, 1894 – τείνω, ταθ- (teino, tath-), to stretch, extend; ῥύγχος (rhugkhos), a snout: from the labial palpus which has a long, porrect tuft of hair-scales on segment 2.

2296 **exsiccata** (Lederer, 1855) – *exsiccatus*, dried out: either because the moth is the colour of a dead leaf, or because Lebanon, the type locality, is a dry part of the world. The species was formerly included in the Amphipyrinae. 'Dried up, from shrunken body' (Macleod); an imaginary character.

Synedoida Edwards, 1878 – the genus *Syneda* Guenée, 1852, now a junior synonym of *Drasteria* Hübner, 1818; εἶδος (eidos), form: from similarity between the genera; *Syneda* may be derived from συνήδομαι (sunēdomai), to rejoice.

2468 **grandirena** (Haworth, 1809) – *grandis*, large; *renes*, the kidneys: from the large reniform stigma, the 'great kidney'.

Scoliopteryx Germar, 1811 – σκολιός (skolios), crooked; πτέρυξ (pterux), a wing: from the crenate termen of the forewing of the next species.

2469 **libatrix** (Linnaeus, 1758) – *libatrix* (fem.), one who makes a libation to the gods. Linnaeus supplies no clue to the meaning. The moth's vestiture may have suggested to him a stately robe, suitable for the ceremonial duty; cf. the vernacular name 'herald' and also 1716 *sacraria*, q.v. Alternatively he may have thought that the naming of this fine moth called for a celebratory drink!

Phytometra Haworth, 1809 – φυτόν (phuton), a plant; μετρέω (metreō), to measure: a name formed on the analogy of 1666 *Geometra*, but it is the plant and not the earth that the larva measures. The caterpillar of the next species is a looper in its early instars and a half looper later. In 1809, *Noctua* was still an enormous genus, or rather family, and its break-down into smaller groups had only just begun. Haworth was innovating boldly when he set up two new genera to accommodate the noctuids that had a reduced number of prolegs. The first of these, *Hemigeometra* (half-earth-measurers) comprised 1661 *Archiearis* and 2451 *Catocala* (to us a strange association, but both had coloured underwings), and the second, *Phytometra*, thirty-six species including plusiines such as 2441 *Autographa gamma*. *Hemigeometra* has now gone entirely and *Phytometra* is reduced to a single species. It is sad that Haworth's important contribution to the study of taxonomy has received such scant recognition.* 'From larva's habit of lying full length on plants' (Macleod).

2470 **viridaria** (Clerck, 1759) – *viridis*, green: from the ground colour. The termination '*-aria*' is puzzling; Clerck did not assign it to a family and can hardly supposed it to belong to the Geometrae pectinicornes (p. 21) which have that ending since the male antenna is not bipectinate. The noun *viridarium* means a pleasure garden.

Anomis Hübner, 1821 – ἄνομος (anomos), without law, here probably meaning anomalous. Hübner included one species, not the next.

2471 **sabulifera** (Guenée, 1852) – *sabulo*, coarse sand, gravel; *fero*, to carry: possibly from the type locality in Ethiopia. The single British specimen may have been accidentally introduced.

Colobochyla Hübner, 1825 – κολοβός (kolobos), docked, stunted; χεῖλος (kheilos), a lip: from the short labial palpus of the next species, the only one included in the genus. Macleod wrongly derives the name from χηλή (khēlē), a claw, saying that it refers to the degenerate prolegs of the larva.

2472 **salicalis** ([Denis & Schiffermüller], 1775) – *Salix*, the willow genus, which includes the larval foodplants; the termination *-alis* because the authors described it as a pyrale (*Pyralis salicalis*).

Laspeyria Germar, 1811 – i.h.o. J. H. Laspeyres (1769–1809), a prominent German entomologist who had recently died; he had also experienced a distinguished public career, having been burgomaster of Berlin.

2473 **flexula** ([Denis & Schiffermüller], 1775) – *flexulus*, dim. of *flexus*, bent: from the undulate termen of the forewing.

Rivula Guenée, 1845 – *rivulus*, a rivulet: probably from the oblique postmedian line on the forewing of the next species, resembling the 'rivulets' of the Geometridae. *R. sericealis* was described by Scopoli in Phalaena, the name used by the successors of Linnaeus for his Geometrae + Pyrales (see p. 29).

2474 **sericealis** (Scopoli, 1763) – *sericeus*, silky: from the glossy forewing: the termination *-alis* because Scopoli thought it was a pyrale (see genus).

Parascotia Hübner, 1825 – παρά (para), beside, other than (as in paradox); the genus *Scotia* Hübner, 1821, a junior synonym of 2084 *Agrotis*, from σκοτία (skotia), darkness. Hübner was not expressing affinity; he just wanted to use the word again for a genus of dark-coloured moths.

2475 **fuliginaria** (Linnaeus, 1761) – *fuligo, fuliginis*, soot: from the dark ground colour; *-aria* because Linnaeus described it as a geometrid.

Orodesma Herrich-Schäffer, 1868 – ὀρός (oros), whey, its pale colour; δέσμα (desma), a binding, a fetter: Herrich-Schäffer compares the larval spinning with that of 2028 *Calliteara pudibunda* and the name is apparently descriptive of the colour of the cocoon.

2475a **apicina** Herrich-Schäffer, 1868 – *apex, apicis*, the tip: from the rusty-brown apical area of the forewing ('rostbraune Spitze').

HYPENINAE (2476)

Many of the moths in this subfamily were at first considered to belong to the Pyralidae because of the way in which they held their wings in repose (see p. 21) and in consequence their names have the termination *-alis*. Later they were placed in a distinct family under the name Deltoidae, again from the deltoid or triangular shape they assume when the wings are folded. Recent research by Kitching (1984) revives the early family name Herminia

*Nye (1975) accepts *Hemigeometra* Haworth, 1809, as a valid genus with *Phalaena fraxini* Linnaeus, 1758, as the type species, but this has not been adopted by Bradley & Fletcher, 1979; 1983; 1986.

Latreille, 1802, (2489) as the Herminiidae, but retains 2482 *Schrankia*, 2485 *Hypenodes* and 2476 *Hypena* in the Noctuidae. Here the arrangement and nomenclature of Kloet & Hincks (1972) and Bradley & Fletcher (1986) is adopted.

Hypena Schrank, 1802 – ὑπήνη (hypēnē), a moustache or beard: from the setose labial palpus and perhaps also from the scale-tufts present on the legs of some species. The name at first held family rank; see p. 30.

2476 **crassalis** (Fabricius, 1787) – *crassus*, thick: probably from the densely pilose and consequently thickened labial palpus.

2477 **proboscidalis** (Linnaeus, 1758) – προβοσκίς (proboskis), an elephant's trunk: from the long, porrect labial palpi. It is more logical to call the two palpi a proboscis than it is the haustellum; an elephant's proboscis is a nose modified to serve also as an arm, whereas the haustellum is a modified mouth. The nose and the mouth are different organs.

2478 **obsitalis** (Hübner, 1813) – *obsitus*, sown all over, covered, filled: from the forewing being covered with scales or a pattern (Spuler; Macleod). This applies to almost every moth and seems pointless. The pattern is formed by tufts of *raised* scales and the word *obsitus* was used to describe the barnacles, etc., adhering to the backs of marine creatures; this is probably what Hübner had in mind.

2479 **obesalis** Treitschke, 1828 – *obesus*, fat: a large species, but probably to express affinity with 2476 *crassalis* and with the same meaning.

2480 **rostralis** (Linnaeus, 1758) – *rostrum*, the beak of a bird or the 'beak' of a trireme: from the long labial palpus.

Plathypena Grote, 1873 – πλατύς (platus), broad; the genus 2476 *Hypena*.

2481 **scabra** (Fabricius, 1798) – *scaber*, rough: from the raised scales on the forewing.

Schrankia Hübner, 1825 – i.h.o. F. von P. von Schrank (1747–1835), a German entomologist (see p. 29). He was Professor of Theology and later of Botany at Ingoldstadt in Bavaria and the author of *Fauna boica*, one of the most important works in the history of taxonomy, in which the name Hypena, source of the subfamily name Hypeninae, first appeared.

2482 **taenialis** (Hübner, 1809) – *taenia*, a band: from the postmedian fascia which is strongly expressed and outlined distally with white, the 'white-lined snout'.

2483 **intermedialis** Reid, 1972 – *intermedius*, intermediate: from possessing some of the characters of the preceding species and some of the next species; it is regarded by some entomologists as a cross between them.

2484 **costaestrigalis** (Stephens, 1834) – *costa*, the anterior margin of the wing; *striga*, a streak: from the subcostal streak on the forewing.

Hypenodes Doubleday, 1850 – the genus 2476 *Hypena*; εἶδος, ὠδ- (eidos, ōd-), form, appearance: a genus resembling *Hypena*, q.v.

2485 **humidalis** Doubleday, 1850 – *humidus*, damp: from the habitat, acid heaths and mosses. For the introduction of this name instead of the next see p. 37.

= **turfosalis** (Wocke, 1850) – Latinized from the German torf, O.E. turf, peat: from the habitat on peat moors. Peat is called 'turf' in Ireland.

Idia Hübner, 1813 – ἴδιος (idios), private, distinct: a genus which is distinct from those related to it. The word 'idiot' comes from ἰδιώτης (idiōtes), a person who detaches himself from public affairs.

= **Epizeuxis** Hübner, 1818 – ἐπίζευξις (epizeuxis), a fastening together, a linking: a genus linked to related genera.

2486 **aemula** (Hübner, 1813) – *aemulus*, a rival.

2487 **lubricalis** (Geyer, 1832) – *lubricus*, slippery, oily.

Pechipogo Hübner, 1825 – πῆχυς (pēkhus), the forearm; πώγων (pōgōn), a beard: from the expansible pencil of hair-scales on the male foreleg.

2488 **strigilata** (Linnaeus, 1758) – *strigilis*, dim. of *striga*, a streak: '*postice striga transversa recta pallida*', with a pale straight subterminal streak. It is in fact darker than the ground colour, though it is sometimes obscurely pale-edged. The termination *-ata* is because Linnaeus described it in Geometra.

= **barbalis** (Clerck, 1759) – *barba*, a beard: from the expansible pencil of hair-scales on the male foreleg. Clerck figures the leg with its tuft of scales beside the complete insect and this may be the earliest structural drawing, appearing the year after *Systema Naturae*. On his title page Clerck states that the names he uses are those of Linnaeus, which means that the latter was probably the originator of this one. The pyrale termination as opposed to the geometrid termination of *strigilata* exemplifies the difficulty experienced by the early entomologists in deciding whether the abdomen was exposed or covered by the wings when the moth was at rest (see p. 21).

Herminia Latreille, 1802 – Latinized from the French 'herminé', adorned with ermine: from the tufts of hair-scales on legs, comparing them to the ermine trimming on robes; cf. 180 *herminata* and 481 *Eperminia*. The name originally held family rank and it does so once again (see subfamily introduction). The derivation from the name of a Roman gens (Pickard *et al.*; Macleod) is clearly wrong because all Latreille's family names, though fancifully conceived, are based on structural characters. A description of the tuft on the leg of the last species (see *barbalis* above) follows the generic diagnosis.

2489 **tarsipennalis** (Treitschke, 1835) – *tarsus* the distal part of the leg; *penna*, a feather: from the expansible scale-tuft on the foreleg of the male; '. . . of some males' (Macleod)!

2490 **lunalis** (Scopoli, 1763) – *luna*, the moon: from the thick and conspicuous crescentic reniform stigma on the forewing.

2491 **tarsicrinalis** (Knock, 1782) – *tarsus*, the distal part of the leg; *crinis*, hair: from the scale-tuft on the foreleg of the male.

2491a **zelleralis** Wocke, 1850 – i.h.o. P. C. Zeller (1808–83), the German entomologist.

2492 **grisealis** ([Denis & Schiffermüller], 1775) – *griseus* (late Lat.), grey: from the ground colour.
= **nemoralis** (Fabricius, 1775) – *nemus*, *nemoris*, a grove, a glade: from the habitat.

Macrochilo Hübner, 1825 – μακρός (makros), large; χεῖλος (kheilos), a lip: from the long, porrect labial palpus.

2493 **cribrumalis** (Hübner, 1793) – *cribrum*, a sieve: from the sprinkling of fuscous dots forming the postmedian and subterminal fasciae on the forewing. Cf. 1458 and 2053.

Paracolax Hübner, 1825 – παρά (para), alongside, often denoting similarity; κόλαξ (kolax), a flatterer, who may imitate as part of his technique: probably from similarity between the four species included by Hübner (2488, 2490, 2492 and 2494).

2494 **tristalis** (Fabricius, 1794) – *tristis*, sad, sombre: from the sober colour.
= **derivalis** (Hübner, 1796) – *derivo*, to divert a stream: possibly from the median fascia which is angled near the costa but otherwise pursues a straight course; or possibly the '*de-*' means 'un-' and the meaning is that the transverse lines are not wavy like 'rivulets'. Cf. 1747 *derivata*.

Trisateles Tams, 1939 – τρισ- (tris-), thrice; ἀτελής (atelēs), not brought to an end, fruitless: there had been three previous attempts to fit the next species into the systematic list – after its birth in Pyralis, Hübner had placed it in 939 *Aethes*, Duponchel in 841 *Sophronia* and Spuler in *Standfussia*; Tams' name expresses the hope that his attempt to supply an acceptable generic name, the fourth, will be more successful.

2495 **emortualis** ([Denis & Schiffermüller], 1775) – *emortualis*, pertaining to death: 'olivâtre', olive-coloured, was Duponchel's explanation, but this describes the moth rather than the name; from the ground colour which resembles dead leaves (Spuler), perhaps the best interpretation; 'from larva's having only rudimentary false legs' (Macleod) is certainly wrong because the life history was unknown until well after the name was bestowed and rudimentary legs are not dead legs; for the same reason, the name cannot refer to the diet of dead leaves. It might conceivably have reference to the 'death' of the fugitive greenish tinge of fresh specimens; this may have been what Duponchel was implying by 'olivâtre'.

The following species is not in the British check list but was described from Britain and appears in MBGBI (10: 414).

AGARISTIDAE

From *Agarista* Leach, 1814. Agarista was the wife of Pericles, the Athenian leader during the Peloponnesian War (5th century B.C.).

Eudryas Boisduval, 1836 – εὖ (eu), well; δρυάς (druas), a dryad or wood-nymph: from the actual or supposed woodland habitat.

[2496] **staejohannis** Walker, 1856 – St John: the type specimen was found resting on the door of the Church of St John, Horselydown, east London.

CORRIGENDUM

Mikkola & Goater (1988) have shown that neither *Apamea maillardi* nor *A. exulis*, the subspecies formerly ascribed to it, occur in Britain. The entry for 2324 **maillardi** (p. 213) should therefore be amended to read as follows:

2324 **zeta** (Treitschke, 1825) – ζῆτα (zēta), zed, the sixth letter of the Greek alphabet. Many early entomologists when confronted by a series of forms or varieties apparently of a single species used to list them under the Greek letters and this was probably the sixth in such a complex. When Treitschke, working from notes left him by Ochsenheimer, decided that it merited specific status, he used the label it had held as a form for the name.

= **maillardi** sensu auctt. – i.h.o. M. Maillard (early 19th century), Abbé at the Séminaire de St Lucien and an entomologist. *A. maillardi* (Geyer, 1834) occurs in the mountains of southern central Europe but not in Britain.

subsp. **marmorata** (Zetterstedt, 1840) – *marmoratus*, marbled: from the variegated forewing pattern. This is the subspecies occurring in the Shetlands.

= subsp. **exulis** sensu auctt. – *exsul*, *exul*, an exile: from the remote type locality. '*Exul*, exile, usual locality being far north' (Macleod): Lefebvre described it in a French journal in 1836 from material taken in Labrador and the name has nothing to do with a northerly distribution in Britain or elswhere. Its range extends from Labrador to Iceland or possibly the Faroes and it does not occur in Britain at all.

subsp. **assimilis** (Doubleday, 1847) – *assimilis*, similar: from resemblance to subsp. *exulis*, then supposed to occur in Britain. This is the subspecies found mainly in the Scottish Highlands.

Additional Reference

Mikkola, K. & Goater, B., 1988. The taxonomic status of *Apamea exulis* (Duponchel [*sic*]) and *A. assimilis* (Doubleday) in relation to *A. maillardi* (Geyer) and *A. zeta* (Treitschke) (Lepidoptera: Noctuidae). *Entomologist's Gaz.* **39**: 249–257.

Appendix 1

People commemorated in the scientific names of Lepidoptera

Many of the names in the Systematic Section were bestowed in honour of the collector who discovered the species, a well-known naturalist past or present, the wife, a relative or friend of the nomenclator, or even a historical figure unconnected with entomology but in some cases, like Croesus and Euclid, suggested by a character shown by the insect. Names without entomological significance taken from classical mythology or legend are excluded. About 260 names are listed below, followed by the Log Book numbers of the associated species.

Adkin, R. (1849–1935): 1517
A British entomologist who lived in Sussex.

Albers, J. A. (1772–1821): 1217
A German entomologist who lived at Bremen.

Allis, T. H. (1817–70): 687
A British entomologist who lived in Yorkshire.

Alströmer, C. (d. 1792): 695
A Swedish pupil of Linnaeus, best known as a botanist.

Aristotle (384–332 B.C.): 751
The Greek philosopher whose *Historia animalium* attempted the first scientific study of insects.

Ashworth, J. H. (19th century): 2129
The British collector who discovered the species that bears his name.

Bacot, A. W. (1866–1922): 183
A British collector.

Bankes, E. R. (1861–1929): 178, 1456[18]
A British collector who lived in Dorset.

Banks, Sir Joseph (1743–1820): 2413
A distinguished British naturalist who sailed with Captain Cook in the *Endeavour*.

Barrett, C. G. (1836–1904): 2169
A British entomologist and author of *The Lepidoptera of the British Islands*.

Baynes, E. S. A. (1890–1972): 1532
An Irish entomologist and author of *A revised catalogue of Irish Lepidoptera*.

Bechstein, J. M. (1757–1810): 275
A German entomologist.

Bedell, G. (1805–77): 264, 616
A British microlepidopterist.

Beirne, B. P.: 47
A contemporary Irish, English and Canadian entomologist and author of *The male genitalia of the British Stigmellidae* and *A list of the Microlepidoptera of Ireland*.

Benander, P. (1885–1976): 559, 565
A Swedish entomologist.

Bennet, E. (early 19th century): 1488
The British collector who discovered the species that bears his name.

Bergmann, T. C. (d. 1784): 1035
A Swedish entomologist.

Binder von Kriegelstein, C. F. (late 18th–early 19th century): 512
An Austrian entomologist.

Bjerkander, C. (1735–95): 386
A Swedish coleopterist.

Blankaart, S. (late 18th century): 326
A Dutch entomologist.

Blomer, C. (d. 1835): 1872
A British collector.

Bohemann, C. H. (1796–1868): 19
A Swedish entomologist.

Boisduval, J. A. (1799–1879): 1453[22]
The French entomologist and entomological author.

Bond, F. (1826–89): 2346
A British collector.

Bonnet, C. (1720–93): 421
A Swiss entomologist.

Borel, –. (early 19th century): 2365
The French collector who discovered the species that bears his name.

Borkhausen, M. B. (1732–1807): 644
A German entomologist.

Bosc, L. A. G. (1759–1828): 1050
A French entomologist.
Bowes, A. J. L. (1913–42): 1521, 1939
A British collector.
Brahm, N. J. (late 18th–early 19th century): 866
A German collector who lived at Mainz.
Bradford, E. S.: 33
A contemporary British collector and entomological artist.
Brander, E. (18th century): 1088
A Swedish pupil of Linnaeus.
Bretherton, Mrs J.: 171
The wife of R. F. Bretherton, contemporary British entomologist.
Brock, J. K. (early 19th century): 410
A German collector.
Brongniart, A. (1770–1847): 313
A French collector.
Brünnich, M. T. (18th century): 592, 1155
A Danish collector.
Buckler, W. (1814–84): 1493
A British entomologist and author of *The larvae of British butterflies and moths*.
Buol, Baron von (18th century): 1210
A Viennese collector.
Büttner, J. G. (19th century): 2376
A Latvian entomologist.
Capper, S. J. (1825–1912): 1494
A British collector who lived in the north-west of England.
Caradja, A. von (1861–1955): 871a
A German entomologist.
Carpenter, the Hon. Mrs B. (19th century): 951
A friend or relative of Lord Walsingham, formerly the Hon. Thomas de Grey, q.v.; she was the first to rear the species which bears her name.
Cassini, J. D. (d. 1712): 2227
A famous Italian astronomer.
Chambers. V. T. (1831–83): 638
An American entomologist.
Christy, W. M. (1863–1939): 1796
The British collector who took the type material of the species which bears his name.
Clemens, J. B. (1829–67): 901
An American entomologist.
Clerck, K. A. (1710–65): 263
A Swedish pupil of Linnaeus and an entomological artist.
Colquhoun, H. (19th century): 1031
A Scottish medical practitioner and collector.
Conway, –. (18th century): 1011
A British collector and friend of Fabricius.
Croesus, (6th century B.C.): 151, 1035
A king of Lydia, famed for his wealth.
Crombrugghe de Picquendaele, G. E. M. de (late 19th–early 20th century): 1489
A Belgian entomologist.
Curtis, W. (1746–99): 449
A British naturalist, author of *Flora Londiniensis* and *A short history of the brown-tail moth*.
Curzon, E. R. (19th century): 1828
A British collector who first took the subspecies that bears his name.
Dahl, G. (early 19th century): 2121
The Viennese collector who first reared the species that bears his name.
DeGeer, Baron K. (1720–78): 148
A Swedish naturalist who provided material for Linnaeus to name.
de Grey, the Hon. Thomas, later 6th Baron Walsingham (1843–1919): 961
The distinguished British microlepidopterist.
Demarné, –. (19th century): 1135
A German collector who lived at Neustrelitz.
Demary, M. (early 19th century): 276
A French entomologist and the first secretary of the Entomological Society of France.
Denis, M. (1729–1800): 638
A Viennese entomologist and part author of . . . *Schmetterlinge der Wienergegend*.
Doubleday, H. (1809–75): 1078
A British entomologist and author of *A synomymic list of British Lepidoptera*.
Douglas, J. W. (1814–1905): 178, 398, 677
A British microlepidopterist and authority on the Gelechiidae.
Drury, D. (1725–1803): 746, 896
A British apothecary and entomologist.
Dumeril, A. M. C. (1774–1860): 2355
A French naturalist.
Ekeblad, Count C. (18th century): 123
A member of the Swedish court who supplied material for Linnaeus to name.
Erxleben, J. C. P. (1744–77): 447
A German naturalist.
Esper, E. J. C. (1742–1810): 649
A German entomological author.
Euclid (*fl.* 300 B.C.): 2463
The Greek mathematician.
Fabricius, J. C. (1745–1808): 385
A Danish entomologist and systematist, the most distinguished of the pupils of Linnaeus.
Farren, W. (1836–87): 479
A professional entomological dealer and

collector who lived at Cambridge and discovered the species that bears his name.

Fischer von Röslerstamm, J. E. (1783–1866): 446
A German entomologist and entomological author.

Fischer von Waldheim, G. (1770–1853): 391
A German entomologist.

Fletcher, W. H. B. (1853–1941) or
Fletcher, T. Bainbrigge (1878–1950): 914
It is uncertain which was the dedicatee, the former, author of *Lepidoptera* in *The Victoria County History of Sussex*, a county in which the moth occurs, being the more likely. The latter lived in Gloucestershire where the species had not then been recorded.

Forsskahl (or Forsskål), P. (1732–63): 1036
A Swedish pupil of Linnaeus and mainly a botanist, though his university dissertation dealt with Lepidoptera.

Forster, J. R. (1729–98): 394, 1002
A Polish naturalist of British descent who settled in England and accompanied Captain Cook on one of his voyages; he was a close friend of Fabricius. Later he became Professor of Natural History at Halle University.

Francillon, J. (1744–1816): 950
A British entomologist.

Freyer, C. F. (1794–1885): 631
A German entomologist.

Friese, G.: 445
A contemporary German entomologist.

Frisch, J. L. (1660–1743): 517
A German entomologist.

Frölich, F. A. G. (early 19th century): 358
A German entomologist.

Geoffroy, E. L. (1727–1810): 652
A French entomologist.

Gerning, J. C. (18th century): 1008
A German entomologist who lived at Frankfurt.

Gleichen, W. F. von (1717–83): 594
A Viennese naturalist.

Glitz, C. T. (1818–89): 507
A German entomologist.

Gödart, J. (1620–68): 411
A Dutch entomologist.

Goossens, T. (1827–89): 1831
A French naturalist.

Graves, P. P. (1876–1953): 1546
A British and Irish entomologist.

Greville, R. K. (early 19th century): 1096
A Scottish medical practitioner who discovered the subspecies named in his honour.

Grotius, H. (1583–1645): 1006
A Dutch jurist and national hero.

Guenée, A. (1809–80): 1017, 1284, 2354
A French entomologist and entomological author.

Gysselin, J. V. G. (19th century): 442
An Austrian collector.

Haase, E. (1857–94): 465a
A German entomologist. J. A. Haas (late 18th century) is also possible.

Harris, M. (1731–88): 315
An English entomologist and author of *The Aurelian*.

Hartmann, P. I. (1727–91): 941
A German Professor of Physics and botanist.

Hartwieg, F. (1877–1962): 1658
A German doctor and friend of the nomenclator.

Hast, R. (mid-18th century): 1053
A brilliant Finnish pupil of Linnaeus who died while still a student.

Hauder, F. (1860–1923): 295
An Austrian collector who discovered the species named after him.

Haworth, A. H. (1767–1833): 11, 395, 1813, 2367
The author of *Lepidoptera Britannica*.

Hebenstreit, J. E. (1703–51): 983
A German entomologist.

Heeger, E. (d. 1866): 317
An Austrian entomologist.

Heinemann, H. von (1812–71): 125a
A German entomological author.

Heller, J. F. (b. 1813): 905
A Viennese professor and entomologist.

Hering, E. M. (1893–1967): 39, 1191
A German professor and authority on leaf-mining insects.

Hermann, J. (1738–1800): 746
A German naturalist.

Herrich-Schäffer, G. A. W. (1799–1874): 1236a
A German entomological author.

Heyden, C. H. G. von (1793–1866): 216a, 957
Senator of Frankfurt-on-Main and German microlepidopterist.

Hofmann, O. (1835–1900): 311, 647
A German entomological author.

Hohenwarth, S. von (1745–1822): 1200
A Viennese professor and bishop and authority on Alpine Lepidoptera.

Hornig, J. von (1819–86): 509, 740
A Russian entomologist.

Hübner, J. (1761–1826): 99, 837
The distinguished German entomological artist and author.

Hucherard, –. (late 19th–early 20th century): 2363
A French collector who discovered the species that bears his name.

Huggins, H. G. (1891–1977): 1894
A British entomologist who also collected in Ireland.

Illiger, J. C. W. (1775–1825): 481
An Austrian entomologist and editor.

Isert, P. E. (1756–87): 1165
A German physician who became a naturalized Dane and worked and collected insects in Guinea.

Janiszewska, J.: 910
A contemporary lady professor at Wrocklau, Poland.

Jung, J. C. (18th century): 1251
A court official at Uffenheim and a collector (ascription uncertain).

Kaekeritz, –. (18th century): 698
Presumed to have been a pupil of Linnaeus.

Kessler, H. F. (1816–97): 433
A German entomologist.

Kleemann, C. F. K. (1735–89): 360
A German entomologist.

Klimesch, J.: 814a
Contemporary Austrian microlepidopterist.

Klimesch, Frau M. (d. *c.* 1974): 104
The late wife of Dr J. Klimesch.

Knaggs, H. G. (1832–1908): 837
A British doctor of medicine and entomologist.

Knill-Jones, S. A.: 2354
A contemporary British collector who discovered the subspecies named after him.

Kollar, V. (1797–1860): 298
A Viennese entomologist.

Krössmann, D. W. (mid-19th century): 836
A schoolmaster and collector who lived at Hanover.

Kühn, –. (19th century): 1475[55]
The Professor of Agriculture at Halle who sent the species named after him to Zeller for identification.

Kuznetzov, V. I. (contemporary) or
Kuznetzov, N. Y. (1873–1948): 697a
It is uncertain which of these Russian entomologists is the dedicatee.

Lafaury, –. (19th century): 984
A French collector who discovered the species named after him.

Lang, H. G. (late 18th century): 880
A German jewel-cutter and amateur entomologist.

Larsen, C. C. R. (1846–1920): 844
A Danish entomologist.

Laspeyres, J. H. (1769–1809): 1240, 2473
Burgomaster of Berlin and an amateur entomologist.

Latham, J. (early 19th century): 902
A British doctor of medicine, ornithologist and entomologist who discovered the species named after him.

Latreille, P. A. (1762–1833): 900, 2309
The distinguished French entomologist and taxonomist.

Leautier, –. (early 19th century): 2240
A French collector who discovered the species named after him.

Leche, J. (mid-18th century): 1000
A Finnish entomologist.

Leech, M. J.: 2354
A contemporary British collector who jointly discovered the subspecies named after him.

Leplastrier (Le Plastrier), 'Mr' (early 19th century): 1218
A British professional collector who lived at Dover.

Leeuwenhoek, A. van (1632–1723): 899
A Dutch microscopist and entomologist.

Lienig, Madam (d. 1855): 897, 1416, 1518
A Latvian entomologist who collaborated with Zeller.

Linder, J. (1830–69): 210
A German coleopterist.

Linné, C. von (Linnaeus) (1707–78): 903
The Swedish Professor of Natural History at Uppsala and author of *Systema Naturae*.

Ljung (Ljungh or Liung), S. I. (1757–1828): 974
A Swedish entomologist.

Löfling, P. (1727–56): 1032
A Swedish botanist and pupil of Linnaeus.

Logan, R. F. (1827–87): 300
A Scottish entomologist.

Lorey, –. (early 19th century): 2208
A French doctor of medicine and collector who lived at Dijon.

Lorquin, P. J. M. (1797–1873): 1058
A French collector who took the type material of the species that bears his name.

'Louis' (early 19th century): 22
The presumed dedicatee of a species named by Sircom (19th century), a British

microlepidopterist who lived at Bristol.

Lüders, L. (19th century): 1224
Probably the collector who discovered the species in question.

Luff, W. A. (1851–1910): 184
A collector who lived in the Channel Islands.

Luz, J. F. (early 19th century): 135
An Austrian soldier and amateur entomologist.

Lyonet, P. (1706–89): 262
A distinguished French naturalist who later became a naturalized Dutchman.

McDunnough, J. H. (1877–1962): 2436
A Canadian entomologist.

Machin, W. (1822–94): 557
The British entomologist who discovered the species.

Maesting, –. (18th century): 341
The presumed name of an entomologist after whom the species was named.

Maillard, –. (early 19th century): 2324
A French abbé and entomologist.

Mann, J. J. (1804–89): 926
A Viennese entomologist.

Massey, –. (late 19th–early 20th century): 1571
A British collector who supplied Tutt with the type material of the subspecies that bears his name.

Mayer, U. (early 19th century): 518
A Viennese timber inspector who was also an entomologist.

Mees, A. (d. 1915): 202
A German entomologist.

Megerle von Mühlfeld, J. C. (1765–1832): 617
A German entomologist.

Mendes, M. C. (1874–1944): 591
A French entomologist.

Messing, –. (19th century): 469
A German musician and amateur entomologist who lived at Neustrelitz.

Metaxa, L.; Metaxa, T. (19th century): 143
Two Italian brothers who were naturalists and collaborated in their work. They are probably joint dedicatees of the species that bears their name.

Metzner, –. (d. 1861): 723, 726, 1196
A German senior civil servant and collector who lived at Frankfurt-on-Oder.

Milhauser, J. A. (mid-18th century): 2004
A (?) German friend of Fabricius.

Millière, P. (1811–87): 1827
A French entomologist.

Mitterpacher, L. (d. 1814): 1120
Professor of Natural History at Pesth and an entomological author.

Morris, F. O. (1810–93): 2346
A British parson, naturalist and entomological author who discovered the species named after him.

Möschler, B. H. (early 20th century): 2066
Presumed dedicatee of the species cited.

Moses (*c.* 1600 B.C): 744a
The Jewish leader who was found as an infant amongst the flags growing by the R. Nile.

Mouffet, T. (1553–1604): 762
One of the earliest British entomologists and author of *Theatrum Insectorum*.

Mühlig, G. C. (19th century): 574
A German collector who lived at Frankfurt-on-Main.

Müller, –. (19th century): 322
There are several candidates for this dedicatee. The one named below is possible but one of Zeller's contemporaries is more likely.

Müller, O. F. (1730–84): 388
A Danish entomologist and author of *Fauna Insectorum Fridrichsdalina*.

Mutuura, A.: 1379
A contemporary Japanese entomologist.

Mygind, –. (d. 1787): 1070
A Danish entomologist who settled in Vienna.

Nicelli, Graf von (19th century): 359
A German entomologist.

Nikerl, F. A. (1813–71): 2354
A Czechoslovakian entomologist.

Nylander, W. (1822–99): 103
A Finnish entomologist.

Ochsenheimer, F. (1767–1822): 251, 1238
A German actor who settled in Vienna and was the author of the first four volumes of *Die Schmetterlinge von Europa*.

Oehlmann, G. (d. *c.* 1815): 131
A professional German insect dealer who lived at Liepzig.

Olivier, G. A. (1756–1814): 650
A French entomologist.

Orstadius, E. T. (1861–1939): 604
A Swedish entomologist.

Osthelder, L. (1877–1954): 1311
An Austrian entomologist.

Panzer, G. W. F. (1755–1829): 141
A German entomologist, author and illustrator.

Pelham-Clinton, E. C., later 10th Duke of Newcastle (1920–88): 817
A British entomologist who discovered the species which bears his name.

Pentz, D. (18th century): 1031
A Swedish entomologist.

Petiver, J. (1660–1718): 1273
A British apothecary and naturalist; the 'father' of British entomology.

Petry, A. (1858–1932): 159
A German schoolmaster and microlepidopterist who lived at Nordhausen.

Pfeiffer, J. B. (early 19th century): 158
A German naturalist who lived at Augsburg.

Phillips, R. A. (1866–1945): 1511
An Irish naturalist who lived in Cork.

Picard, J.: 1508b
A contemporary French entomologist.

Pierce, F. N. (1861–1943): 238, 927, 942
A British entomologist who specialized in the description of the genitalia.

Piller, M. (18th century): 1012
A priest, professor and entomologist who lived in Vienna.

Poda von Neuhaus, N. (1723–98): 977
An Austrian entomologist who was Professor of Physics at Graz.

Prout, L. B. (1864–1943): 188
A British entomologist.

Purdey, W. (1844–1922): 1207
A British soldier and entomologist.

Putnam, G. P. (19th century): 2440
An American entomologist who lived at Buffalo.

Rambur, J. P. (1801–70): 1403
A French entomologist.

Raschke, J. G. (1763–1815): 883
A German entomologist.

Ratzeburg, J. T. C. (1801–71): 1163
A German entomologist and entomological author.

Ray, J. (1627–1705): 345
The distinguished British botanist and author of *Historia Insectorum*.

Réaumur, R. A. Ferchault de (1683–1757): 150, 185
The distinguished French scientist and entomologist.

Remm, H.: 2343b
A contemporary Estonian entomologist.

Reutti, C. (1830–94): 840
A German entomologist.

Revay, –. (18th century): 2423
A French entomologist.

Rheede tot Draakenstein, H. A. van (late 17th–early 18th century): 1239
A Dutch naturalist, author and Governor of Malabar.

Richardson, N. M. (1855–1925): 202
The British collector who discovered the species named after him.

Rösel von Rosenhof, A. J. (1705–59): 400
The editor of the earliest magazine, published in Germany, devoted to the study of insects.

Sang, J. (1828–87): 12, 845
A British entomologist who lived in the north-east of England.

Saporta, le Comte de (19th century): 319
A French entomologist who collected in the south of France.

Schaller, J. G. (1734–1813): 1047
A German entomologist.

Schiffermüller, I. (1727–1809): 634
A Viennese entomologist and principal author of . . . *Schmetterlinge der Wienergegend*.

Schmidt, A. (d. 1899): 861, 895
The possible dedicatee of one or both of the species given, but there are other German candidates of the same name.

Schrank, F. von Paula von (1747–1835): 2482
A German naturalist, professor and author of *Fauna boica*.

Schreber, J. C. D. von (1739–1810): 352
A German entomologist and professor at Leipzig.

Schreckenstein, R. von (d. 1808): 485
A German entomologist.

Schreibers, K. F. A. von (1775–1852): 922
A German entomologist.

Schulz, J. D. (18th century): 1073
A German entomologist who lived at Hamburg.

Schumacher, C. F. (1757–1830): 1013
A German entomologist.

Schütze, K. J. (late 19th century): 1454a[20]
The German entomologist who reared the type material of the species named after him.

Schwarz, C. (d. 1810): 141
A German entomologist.

Sehestedt, Graf O. R. (1757–1838): 387
A Norwegian entomologist and pupil of Fabricius.

Serville, J. G. Audinet de (1775–1858): 1256
A French entomologist specializing in Coleoptera and Hymenoptera.

Shepherd, E. (d. *c.* 1883): 1046
A British entomologist who discovered the species named after him.

Smeathman, H. (1750–87): 947
A British entomologist who studied termites.

Solander, D. C. (1738–82): 1156
A Swedish naturalist who settled in England and became curator of the Natural History Department of the British Museum.

Solier, A. J. J. (1792–1851): 2250a
A French entomologist who lived at Marseilles.

Sorhagen, L. F. (1836–1914): 908
A German entomologist.

Sparrmann, A. (1748–1820): 9
A Swedish entomologist.

Sparshall, J. (early 19th century): 2023
A collector who reputedly took the type specimen of an Australian species in Norfolk.

Spinola, Marchese M. (1780–1857): 337
An Italian entomologist.

Spuler, A. (1869–1937): 904
A German entomologist and author of *Die Schmetterlinge Europas*.

Stainton, H. T. (1822–92): 340a
The distinguished British microlepidopterist and author of *The natural history of the Tineina*.

Standfuss, M. R. (1854–1917): 2104
A German entomologist.

Stange, G. (19th century): 822a
A German schoolmaster who discovered the species named after him.

Steinkellner, –. (18th century): 667
A Viennese professor and entomologist.

Stephens, J. F. (1792–1852): 169, 592, 907, 1020
A British entomologist and author of *Illustrations of British Entomology*.

Ström, H. (1726–98): 1151
A Norwegian entomologist.

Svensson, I.: 87, 2298
A contemporary Swedish entomologist.

Suomalainen, Esko: 598a
A contemporary Finnish professor and entomologist.

Swammerdam, J. J. (1637–80): 140, 436, 437, 440
A Dutch entomologist.

Sylvestre, I. (late 18th century): 1206, 1701
The possible dedicatee of one or both of the species cited.

Tengström, J. M. J. von (1821–90): 60
A Swedish entomologist.

Thomson, G.: 1551
A contemporary Scottish entomologist.

Thunberg, K. P. (1743–1828): 1
A Swedish naturalist and successor of Linnaeus as Professor of Botany at Uppsala.

Tischer, C. F. A. von (1777–1849): 123
A German entomologist. 'von' may not be correct; see p. 48.

Toll, Gräfin J. von: 660
The widow of Graf S. von Toll, (1893–1961), a Polish entomologist, who named the species cited in her honour.

Traun, –. (d. 1748): 1235
An Austrian field marshal unconnected with entomology.

Treitschke, F. (1776–1842): 159
A German entomologist who collaborated with Ochsenheimer and completed *Die Schmetterlinge von Europa* after the latter's death.

Uddmann, I. (18th century): 1175
A Finnish entomologist and entomological author.

Verhuell, Q. M. R. (mid-19th century): 199
A Dutch entomologist.

Vine, A. C. (1844–1917): 846
A British collector who lived at Brighton and collected the type material of the species named after him.

Waga, A. (1799–1890): 769
A French entomologist.

Wailes, G. (1802–82): 255
A British collector who lived in Co. Durham.

Warnecke, G. (1884–1962): 2402
A German entomologist.

Weaver, R. (19th century): 43, 228
A British collector who lived in Birmingham.

Weir, J. J. (1822–94): 678, 1221
A British entomologist.

Westropp, M. S. Dudley (20th century): 2295
An Irish collector or visitor to Ireland who took the type material of the subspecies that bears his name in Co. Cork.

Whittle, F. G. (1854–1921): 190
A British entomologist who lived at Southend.

Wilkes, B. (d. 1749): 733
A British entomological artist.

'Willi' (18th century): 944
A presumed dedicatee. Latin and Greek have no letter W and all lepidopterous names starting with it are derived from people or places except for *w-album* and *w-latinum*.

Wocke, M. F. (1820–1906): 527
A German entomologist and author.

Wolff, N. L. (1900–78): 217
A Danish entomologist.

Wood, J. H. (1841–1914): 1066
A Herefordshire medical practitioner and entomologist who discovered the species named in his honour.

Wood, R. (early 19th century): 901
A Manchester collector who forwarded the species cited to Curtis to be named and figured, but did not discover it.

Yeates, T. P. (d. 1782): 714
A British collector.

'Yildiza' (20th century): 222
The species cited is possibly named in honour of a person.

Zeller, P. C. (1808–83): 71, 435, 1429, 2491a
A distinguished German professor and microlepidopterist.

Ziegler, –. (late 18th–early 19th century): 894
A Viennese medical practitioner and entomologist.

Zincken, J. L. (early 19th century): 129, 1451a[9a]
A German medical practitioner and entomologist who lived at Brunswick.

Zoega, J. (18th century): 938
A Swiss pupil of Linnaeus who discovered the species that bears his name.

Zollikofer, C. T. (1774–1843): 2356
A Swiss naturalist.

Appendix 2

Geographical names

The use of geographical names with no direct association with the insect or group of insects on which they were bestowed has never been popular. Those on the British list are given below. It will be seen that Ochsenheimer (9), his collaborator and successor Treitschke (3) and Stephens (4) were the only authors who favoured this source. The names given by Fabricius (3) almost certainly have a double meaning. The other authors gave only one or two each and in some cases the derivation is doubtful.

2321 **Apamea** Ochsenheimer, 1816 – a town in Asia Minor with early historical and ecclesiastical connections.

1572 **Aricia** R. L., 1817 – a town in Latium where Diana had a temple.

1110 **Bactra** Stephens, 1834 – a town in Afghanistan with early historical associations.

1007 **Capua** Stephens, 1834 – a town in Campania.

2387 **Caradrina** Ochsenheimer, 1816 – a river in Albania.

1864 **Chesias** Treitschke, 1825 – from Chesium on the island of Samos, where there was a temple of Diana.

1542 **Colias** Fabricius, 1807 – a promontory on the coast of Attica where there was a temple of Aphrodite. The name may be a pun.

1427 **Corcyra** Ragonot, 1885 – the old name for the island of Corfu, off the west coast of Greece. The moth for which the genus was erected has associations with Greece, but not directly with this island.

2181 **Egira** Duponchel, 1845 – Aegira, a city in the Peloponnesus.

188 **eppingella** Tutt, 1900 – from Epping Forest, a large woodland locality in south Essex, England.

2386b **eridania** Stoll, 1780 – from the semi-mythical River Eridanus, variously identified with the Po, the Rhone or the Rhine.

1539 **gorganus** Fruhstorfer, 1922 – perhaps from Gorganus, a town near the Caspian Sea.

2364 **Gortyna** Ochsenheimer, 1816 – a town in Crete.

1037 **holmiana** Linnaeus, 1758 – Holmia, the Latin for Stockholm; the type locality is Sweden, but the moth has no direct connection with Stockholm.

1696 **Idaea** Treitschke, 1827 – from Mt. Ida in Asia Minor.

1584 **Ladoga** Moore, 1898 – possibly from Lake Ladoga in north-western U.S.S.R.

1990 **livornica** Esper, 1780 – from Livorno or Leghorn in Italy.

1927 **Lycia** Hübner, 1825 – doubtfully from a country in Asia Minor.

2154 **Mamestra** Ochsenheimer, 1816 – a town in Lesser Armenia.

1611 **Melitaea** Fabricius, 1807 – possibly from Melita, the old name for Malta, but a double meaning is likely.

1878 **Minoa** Treitschke, 1825 – *Minous*, Cretan.

1466 **Mussidia** Ragonot, 1888 – from Mussidan, a town in the Dordogne, unassociated with the species Ragonot placed in the genus.

2191 **Mythimna** Ochsenheimer, 1816 – from Mithimna, a town in the island of Mytelene.

175 **Narycia** Stephens, 1836 – a town in Greece.

1988 **nicaea** de Prunner, 1798 – a city in Asia Minor with early ecclesiastical associations.

2077 **Nola** Leach, 1815 – a town in Campania.

2369 **Nonagria** Ochsenheimer, 1816 – the old name for the island of Andros, in the Aegean Sea.

1013 **Olindia** Guenée, 1845 – perhaps from Olinda, a town in South America.
1530 **phyleus** Drury, 1775 – of or belonging to Phyle, a town in Attica.
1552 **Pontia** Fabricius, 1807 – possibly from Pontia, an island off the coast of Latium, but a pun is likely.
2376 **Sedina** Urbahn, 1913 – from Sedyn, the old name of Stettin, where the *Stettiner Entomologische Zeitung* used to be published.
2290 **Simyra** Ochsenheimer, 1816 – a town in Phoenicia.
1966 **Siona** Duponchel, 1829 – perhaps from Sion, a town and canton in Switzerland; possibly from Mt. Sion.
1558 **Strymonidia** Tutt, 1908 – via the genus *Strymon* Hübner, from the River Strymon, now the Struma, in northern Greece.
1767 **Thera** Stephens, 1831 – an island in the Aegean Sea.
1652 **Thyatira** Ochsenheimer, 1816 – a city of Asia Minor with biblical connections.
1538 **Zerynthia** Ochsenheimer, 1816 – from Zerinthus in Thrace, where Apollo had a temple.

Appendix 3

Unresolved names

The explanations given in this work for the scientific names of Lepidoptera range from complete certainty as when the author himself gives the reason, through probability and possibility to a small minority where I have no worthwhile solution to offer. The names in the last class, kept to the minimum, are listed below in the hope that readers will be able to suggest suitable interpretations. Some obviously have meanings that have eluded me; others may be malformed; others again may be anagrams or meaningless neologisms.

Billberg
- 2421 **Bena**
- 2465 **Tyta**

Bode
- 1650 **Sabra**

Boisduval
- 2113 **Spaelotis**

Busck
- 796 **Aroga**

Clemens
- 201 **Tenaga**
- 202 **Eudarcia**
- 248 **pomiliella**

Duponchel
- 2320 **Hyppa**

Fabricius
- 1586 **villida**
- 2180 **crini**

Fruhstorfer
- 1581 **eutyphron**

Godart
- 1614 **tircis**

Grote
- 1451 **Pyla**

Guenée
- 1360 **Hellula**
- 1418 **manihotalis**
- 1479 **Plodia**
- 1888 **Ligdia**

Hübner
- 1570 **Everes**
- 2120 **Diarsia**
- 2226 **oditis**

Hulst
- 1453 **Pima**

Koçak
- 222 **yildizae**

Linnaeus
- 373 **salmachus**

Moore
- 1623 **Chazara**
- 1560 **Rapala**

Ragonot
- 1438 **Numonia**

Walker
- 975 **Homona**
- 975 **menciana**
- 1351b **crisonalis**
- 1401 **Maruca**
- 1412 **Daraba**
- 1461 **Assara**
- 2035 **Thumatha**
- 2095 **Feltia**

Appendix 4

Apparent errors in R. D. Macleod's *Key to the names of British butterflies and moths*

Below are listed the names which are certainly or in all probability wrongly explained in the *Key*, or where there are factual errors in the explanation. In some cases the derivation is incorrectly given; in others there are mistakes such as anachronisms; others involve false assumptions regarding the biology of the insect concerned or the knowledge of it available when the name was bestowed; yet others are at variance with the original diagnosis. Only names in current use or those that I have listed in synonymy are included; not all the errors are referred to in my text. The systematic order of Bradley & Fletcher (1986) is followed.

The purpose of the list is to correct popular misconceptions.

36 quinquella
64 continuella
66 sorbi
67 plagicolella
117 confusella
119 salaciella
122 spatulella
123 Tischeria
129 Incurvaria
144 Nemophora
158 pfeifferella
160 castaneae
166 Zygaena
173 Apoda
181 Taleporia
184 lapidella
192 Pachythelia
195 Sterrhopteryx
205 borreonella
209 rutella
210 Lindera
211 Haplotinea
212 insectella
237 Niditinea
239 columbariella
265 Bucculatrix
278 Opogona
279 antistacta
286 alchimiella
309 torquillella
311 Acrocercops
332 corylifoliella
350 insignitella
367 saligna
370 Sesia
382 scopigera
388 myllerana
389 Choreutis
393 equitella
403 glabratella
409 quadriella
414 cornella
424 Yponomeuta
444 Ocnerostoma
451 Ypsolopha
457 lucella
458 alpella
464 Plutella
476 Acrolepia
479 Cataplectica
480 profugella
481 Epermenia
487 Metriotes
487 modestella
488 Goniodoma
490 lutipennella
491 gryphipennella
528 chalcogrammella
547 discordella
568 versurella
584 alticolella
589 clypeiferella
590 Perittia
591 Mendesia
637 tinctella
638 augustella
640 Batia
648 Endrosis
648 sarcitrella
658 Carcina
668 lobella
683 depressana
686 Exaeretia
686 ciniflonella
688 applana
693 putridella
706 costosa
707 prostratella
716 rotundella
750 Psamathocrita
755 gemmella
756 Parachronistis
757 Recurvaria
771 alburnella
776 diffinis
791 distinctella
808 Platyedra
809 malvella
816 obsoletella
824 streliciella
839 congressariella
852 Anacampsis
876 normalis
881 terminella
901 woodiella
925 Phtheochroa
940 rutilana
955 ambiguella
985 Cacoecimorpha
986 Syndemis
991 Clepsis
999 Adoxophyes
999 orana
1000 Ptycholoma
1006 Epagoge
1006 grotiana
1034 Spatalistis
1048 variegana

1049 permutana
1067 cespitana
1068 rivulana
1069 aurofasciana
1070 mygindiana
1071 arbutella
1072 metallicana
1081 penthinana
1088 branderiana
1096 sauciana
1098 sellana
1102 nigricostana
1121 upupana
1131 subsequana
1134 ramella
1149 crenana
1157 Crocidosema
1158 ustomaculana
1159 naevana
1160 Acroclita
1167 Gypsonoma
1168 sociana
1172 nitidulana
1181 grandaevana
1183 foenella
1188 caecimaculana
1190 aspidiscana
1192 conterminana
1194 aemulana
1205 Spilonota
1213 logaea
1215 leucotreta
1237 germmana
1239 rhediella
1242 internana
1277 senectana
1294 Crambus
1313 pinella
1316 falsella
1320 craterella
1321 chrysonuchella
1327 Ancylolomia
1327 tentaculella
1329 Donacaula
1329 forficella
1330 mucronellus
1332 Scoparia
1334 ambigualis
1336 Eudonia
1338 Dipleurina
1356 Evergestis
1356 forficalis
1359 Cynaeda
1361 Pyrausta
1369 polygonalis
1371 verticalis
1373 pandalis
1381 Anania
1402 Diasemia
1405 ruralis
1408 unionalis
1416 Pyralis
1421 pinguinalis
1422 dimidiata
1427 cephalonica
1428 Aphomia
1449 similella
1444 obductella
1447 hostilis
1435 tumidana
1436 tumidella
1437 consociella
1460 ceratoniae
1470 pinguis
1468 Nyctegretis
1481 sinuella
1482 nimbella
1476 cautella
1477 figulilella
1487 Agdistis
1489 Oxyptilus
1494 Capperia
1495 Maraschmarcha
1497 acanthadactyla
1523 Oidaematophorus
1532 Erynnis
1534 Pyrgus
1553 Anthocharis
1555 Callophrys
1562 dispar
1613 athalia
1614 Pararge
1631 Poecilocampa
1642 Gastropacha
1644 versicolora
1667 Comibaena
1673 Hemistola
1674 Jodis
1683 immorata
1692 immutata
1694 ternata
1706 humiliata
1713 aversata
1738 obscurata
1742 bilineata
1749 comitata
1753 latentaria
1758 pyraliata
1759 Ecliptopera
1760 Chloroclysta
1760 siterata
1761 miata
1765 Cidaria
1769 variata
1773 corylata
1774 Colostygia
1780 Coenocalpe
1781 Horisme
1784 procellata
1790 dubitata
1811 Eupithecia
1812 inturbata
1827 intricata
1828 satyrata
1838 cognata
1839 succenturiata
1843 distinctaria
1852 abbreviata
1858 Chloroclystis
1860 rectangulata
1864 legatella
1870 Odezia
1872 blomeri
1876 Hydrelia
1880 Trichopteryx
1883 viretata
1889 Semiothisa
1890 alternaria
1899 Isturgia
1903 Plagodis
1904 dolabraria
1923 Colotois
1936 Menophra
1940 Deileptenia
1951 Aethalura
1953 cremiaria
1956 exanthemata
1960 Theria
1961 Campaea
1970 Perconia
1993 celerio
1995 vinula
2000 dromedarius
2005 anceps
2006 Pheosia
2011 Pterostoma
2017 pigra
2019 curtula
2025 Orgyia
2026 antiqua
2031 Leucoma
2036 Setina
2036 irrorella
2040 mesomella
2043 Eilema

2043 sororcula
2047 complana
2049 deplana
2054 Utetheisa
2059 sannio
2063 mendica
2064 Phragmatobia
2070 Syntomis
2070 phegea
2074 caca
2078 confusalis
2083 cursoria
2097 polyodon
2099 praecox
2112 interjecta
2113 Spaelotis
2114 augur
2116 sobrina
2117 glareosa
2118 Lycophotia
2120 mendica
2135 agathina
2136 typica
2137 occulta
2139 Cerastis
2141 Mesogona
2142 Anarta
2144 melanopa
2148 bombycina
2156 contigua
2161 blenna
2181 conspicillaris
2182 cruda
2186 gracilis
2191 turca
2204 obsoleta
2223 Calophasia
2236 socia
2247 aprilina
2251 Trigonophora
2269 Atethmia
2280 leporina
2285 strigosa
2297 pyramidea
2300 maura
2305 Euplexia
2305 lucipara
2306 Phlogophora
2268 suspecta
2316 affinis
2317 diffinis
2324 exulis
2325 oblonga
2330 remissa
2331 unanimis
2332 pabulatricula
2337 strigilis
2339 latruncula
2344 captiuncula
2347 extrema
2352 Eremobia
2369 Nonagria
2379 Coenobia
2383 superstes
2387 morpheus
2389 clavipalpis
2391 Chilodes
2400 armigera
2411 Eustrotia
2418 Earias
2422 Pseudoips
2432 ni
2453 electa
2462 mi
2466 pastinum
2296 exsiccata
2470 Phytometra
2472 Colobochyla
2489 Herminia
2489 tarsipennalis
2495 emortualis

REFERENCES

Every name that is explained bears a reference and to list them all in full would almost double the length of the book. For species with log book numbers 1–400, 1525–1660 and 1991–2496 the references are given in fuller form in the species headings in the volumes of MBGBI that have already been published. The remainder can be traced from card indexes held in the Department of Entomology at the British Museum (Natural History). The references given below are to authors and works mentioned for reasons other than that they give the first citation of a name.

Three works, Pickard *et al.* (1858), Spuler (1903–10) and Macleod (1959), are referred to so frequently that I have deemed it unnecessary to give the date on each occasion. The date is also omitted after an author's name when only one of his works is cited here and that consists of a series of volumes published over a number of years.

Agassiz, D. J. L., 1987. *A recorder's log book and label list of British butterflies and moths: addenda & corrigenda*, 16 pp. Colchester.

Albin, E., 1720. *The natural history of English insects*, 100 pls. London.

Aldrovandi, U., 1602. *De animalibus insecti libri septem*, [x], 767, [43] pp., ill. Bologna.

Allan, P. B. M., 1947. *A moth-hunter's gossip* (Edn 2), 269 pp. London.

———, 1980. *Leaves from a moth-hunter's notebook*, 281 pp. Faringdon.

Aristotle, *c.* 330 B.C. *Historia animalium.*

Beavis, I. C., 1988. *Insects and other invertebrates in classical antiquity*, xv, 269 pp. Exeter.

Bible, The, 1611. Authorized Version.

Billberg, C. J., 1820. *Enumeratio insectorum in museo Christ. Joh. Billberg*, ii, 138 pp. Stockholm.

Boisduval, J. A., 1834. *Icones historiques des Lépidoptères nouveaux ou peu connus* **2**, 208 pp., 37 pls. Paris.

———, 1836–57. *Histoire naturelle des Insectes* **1**, **5–10**, Paris.

Bradley, J. D. & Fletcher, D. S., 1974. Addenda & corrigenda to the Lepidoptera part of Kloet & Hincks check list of British insects (Edn 2) 1972. *Entomologist's Gaz.* **25**: 219–223.

———, 1979. *A recorder's log book or label list of British butterflies and moths*, 136 pp. London.

———, 1983. Addenda & corrigenda *in* Hall-Smith, D. H., *A recorder's log book or label list of British butterflies and moths, Index*, 59 pp. Leicester.

———, 1986. *An indexed list of British butterflies and moths*, 119 pp. Orpington.

Bradley, J. D. & Martin, E. L., 1956. An illustrated list of the British Tortricidae part 1: Tortricinae and Sparganothinae. *Entomologist's Gaz.* **7**: 151–156, 10 pls.

Bradley, J. D., Tremewan, W. G. & Smith, A., 1973. *British tortricoid moths. Cochylidae and Tortricidae: Tortricinae*, viii, 251 pp., 47 pls (26 col.), 52 text figs. London.

Butler, A. G., 1879. Descriptions of new species of Lepidoptera from Japan. *Ann. Mag. nat. Hist.* (5) **4**: 437–457.

Clemens, B., 1860. Contributions to American Lepidopterology. Nos 3, 5. *Proc. Acad. nat. Sci. Philad.* **12**: 4–15; 203–221.

Clerck, K., 1759. *Icones insectorum rariorum*, [xii], [iii] pp., 55 pls. Stockholm.

Cooper, B. A., & O'Farrell, A. F., 1946. A check list of British Macrolepidoptera. *Amateur Entomologist's Society Pamphlet* **5**, 32 pp. London.

Cowan, C. F., 1971. About the *Accentuated List*. *Entomologist's Rec. J. Var.* **83**: 388–390.

Curtis, J., 1824–39. *British entomology*, **1–16**. London.

Darwin, C. R., 1859. *The origin of species by means of natural selection* **1**, **2**. London.

Davies, M. & Kathirithamby, J., 1986. *Greek insects*, xvi, 211 pp., 37 text figs. London.

[Denis, M. & Schiffermüller, I.], 1775. *Ankündung eines systematischen Werkes von den Schmetterlinge der Wienergegend*, 324 pp. Vienna.

Diakonoff, A. & Hinton, H. E., 1956. Observations on species of Lepidoptera infesting stored products, XV: on a new genus of Nemapogoninae (Tineidae). *Entomologist* **89**: 31–36.

Donovan, E., 1792–1813. *Natural history of British insects*, **1–16**. London.

Doubleday, H., 1849. *A synonymic list of British Lepidoptera*, 27 pp. London.

Douglas, J. D., 1850–51. On the British species of the genus *Gelechia* of Zeller. *Trans. ent. Soc. London* **1** (NS): 14–21.

Duponchel, P. A. J., 1826–42. *Histoire naturelle des Lépidoptères de France*, **6–11**, Suppl. **1–4**. Paris.

Durrant, J. H., 1897. *Aproaerema*, n. n. *Entomologist's mon. Mag.* **33**: 221.

———, 1911. Descriptions of two new British species of *Rhyacionia* Hb. [Lep. Tin.] [*sic*]. *Ibid.* **47**: 251–253.

Emmet, A. M., 1987a. Addenda and corrigenda to the British list of Lepidoptera. *Entomologist's Gaz.* **38**: 31–52.

———, 1987b. *Endothenia ericetana* (Humphreys & Westwood, 1845) (Lepidoptera: Tortricidae). *Entomologist's Gaz.* **38**: 66.

Fabricius, J. C., 1775. *Systema Entomologiae*, . . . [xxvii], 832 pp. Leipzig.

———, 1781. *Species insectorum exhibentes eorum differentias specificas synonyma auctorum, loca natalia, metamorphosin adjectis observationibus, descriptionibus*, 518 pp. Hamburg.

———, 1787. *Mantissa insectorum sistens species nuper detectas adiectis synonymis, observationibus, descriptionibus, emendationibus*, **1**, **2**. Hafnia.

———, 1793–94. *Entomologia systematica emendata et aucta*, **1**, **2**. Hafnia.

———, 1798. *Supplementum Entomologiae systematicae* . . ., 572 pp. Hafnia.

———, 1807. Systema Glossatorum. In *Magazin Insecktenk.* (Illiger) **6**: 277–289.

Fischer von Röslerstamm, J. E., 1834–43. *Abbildungen zur Berichtigung und Ergänzung der Schmetterlingeskunde, besonders der Mikrolepidopterologie, als Suppl. zu Treitschke's und Hübner's Europ. Schmett.*, 304 pp., 100 col. pls. Leipzig.

Forster, J. R., 1770. *A catalogue of English insects*, 16 pp. London.

———, 1771. *Novae species insectorum, centuria*, viii, 100 pp., London.

Geoffroy, E. L., 1762. *Histoire abrégée des insectes qui se trouvent aux environs de Paris*, . . ., **1**, **2**. Paris.

———, 1785. *In* Fourcroy, A. F. de, *Entomologia Pariensis*, vii, 544 pp. Paris.

Goater, B., 1986. *British pyralid moths*, 175 pp., 8 col. pls. Colchester.

Gilbert, P., 1977. *A compendium of the biographical literature on deceased entomologists*, [6], 455 pp. London.

Goeze, J. E. A., 1777–83. *Entomologische Beyträge zu der Ritter Linne*, 1–4. Leipzig.

Gregson, C. S., 1871. Description of an *Ephestia* new to science. *Entomologist* **5**: 385.

Guenée, A., 1845. *Europae Micro-Lepidopterorum index methodicus*, vi, 106 pp. Paris.

———, 1852–57. *Histoire naturelle des insectes*, 5–10. Paris.

Hampson, G. F., 1907. Descriptions of new genera and species of Syntomidae, Arctiadae, Agaristidae and Noctuidae. *Ann. Mag. nat. Hist.* **19**(7): 221–257.

Harris, M., 1766. *The Aurelian, a natural history of British moths and butterflies*, 92 pp., 44 col. pls. London.

———, 1775a. *The Aurelian*, . . . (Edn 2), xvii, 90 pp., 45 col. pls. London.

———, 1775b. *The English Lepidoptera; or, the Aurelian's pocket companion*, x, 66 pp., 1 col. pl. London.

Haworth, A. H., 1802. *Prodromus Lepidopterorum Britannicorum*, vi, 39, 6 pp. Holt.

———, 1803–28. *Lepidoptera Britannica*, **1–4**. London.

Heath, J. & Emmet, A. M., 1976–91. *The moths and butterflies of Great Britain and Ireland*, **1**, **2**, **7**(1), **7**(2), **9**, **10**. London and Colchester.

Hemming, F., 1937. *A bibliographical and systematic account of the entomological works of Jacob Hübner*, **1**, **2**. London.

Herrich-Schäffer, G. A. W., 1843–56. *Systematische Bearbeitung der Schmetterlinge von Europa*, **1–6**. Regensburg.

Heslop, I. R. P., 1947. *Indexed check-list of the British Lepidoptera, with the English name of each of the* 2313 *species*, 85 pp. London.

———, 1964. *Revised indexed check-list of the British Lepidoptera* (Library edition), 145 pp. [No place of publication.]

Homer, *c.* 750 B.C. The Iliad.

———, *c.* 750 B.C. The Odyssey.

Howarth, T. G., 1973. *South's British butterflies*, xiii, 210 pp., 48 col. pls, 57 maps. London.

Hübner, J., 1786–90. *Beiträge zur Gesichte der Schmetterlinge*, **1**, **2**. Augsburg.

———, 1796–1838*. *Sammlung europäischer Schmetterlinge*, **1–9**. Augsburg.

———, [1806]. *Tentamen determinationis, digestionis atque denominationis singularum stirpium Lepidopterorum, peritis ad inspiciendum et dijudicandum communicatum, a Jacobo Hübner* [An attempt to establish the composition, arrangement and nomenclature of the individual genera of the Lepidoptera, set before experts for their consideration and critical appraisal, by Jacob Hübner], 2pp. See facsimile reprint in Hemming, 1937, **1**: 599–600.

———, 1806–38*. *Sammlung exotischer Schmetterlinge*, **1–3**. Augsburg.

———, 1816–26. *Verzeichniss bekannter Schmetterlinge*, 432 pp. Augsburg.

*Completed by Geyer after Hübner's death.

Hufnagel, –., 1766–67. Tabelle . . . 1–6. *Berlinischer Magazin* **2**: 54–90, 174–195, 391–473; **3**: 202–215, 280–309, 393–426. [Only the author's surname is given.]

Humphreys, H. N. & Westwood, J. O., 1843–45. *British moths and their transformations*, **1**, **2**. London.

International code of zoological nomenclature, third edition, adopted by the XX General Assembly of the International Union of Biological Sciences, 1985. Eds Ride, W. D. L. *et al.*, xx, 338 pp. London, Berkeley & Los Angeles.

Illiger, J. K. W., 1798. *Verzeichniss der Käfer Preussend* . . ., xlii, 510 pp. Halle. [There is an appendix (pp. 487–510) on insect classification.]

Karsholt, O. & Nielsen, E. S., 1986. The Lepidoptera described by C. P. Thunberg. *Ent. scand.* **16**: 433–463.

Kitching, I. J., 1984. An historical review of the higher classification of the Noctuidae (Lepidoptera). *Bull. Br. Mus. nat. Hist.* (Ent.) **49**: 153–234.

Kloet, G. S. & Hincks, W. D., 1972. A check list of British insects: Lepidoptera (Edn 2). *Handbk ident. Br. insects* 11(2), viii, 153 pp. London.

Kluk, K., 1801–02. *Zwierzat domowych i dzikich osobliwie kraiowych*, etc., 79–111. Warsaw.

Knoch, A. W., 1783. *Beiträge zur Insektengeschichte* **3**, 138 pp., 5 pls. Leipzig.

Latreille, P. A., 1796. *Précis des caractères génériques des insectes*, xiii, 201, [7] pp. Bordeaux.

———, 1804. *Lépidoptères; lepidoptera* in *Histoire naturelle générale et particulière des Crustacés et des Insectes*, **14**: 69–141. Paris.

———, 1806–09. *Genera Crustaceorum et Insectorum, secundum ordinem naturalem in familiis disposita*, **1–4**. Paris.

Lederer, J., 1859. Classification der europäischer Tortricinen (part). *Wien. ent. Monatschr.* **3**: 328–352.

Leraut, P., 1980. *Liste systematique et synonymique des Lépidoptères de France, Belgique et Corse*, 334 pp. Paris.

Lewin, W., 1795. *The Papilios of Great Britain*, 97 pp., 46 col. pls. London.

Lhoste, J., 1987. *Les entomologistes français*, 1750–1950, 351 pp., ill.

Liddell, H. G. & Scott, R., 1869. *A Greek-English lexicon* (Edn 6), xvi, 1865 pp. Oxford.

——— & ———, revised and augmented by Jones, H. S., 1968. *A Greek-English lexicon*, xlv, 2042 pp., Supplement, xi, 153 pp. Oxford.

Lienig, F. & Zeller, P. C., 1846. Lepidopterologische Fauna von Lievland und Curland. *Isis, Leipzig* **1846**: 175–302.

Linnaeus, C., 1746. *Fauna Suecica* . . ., xxviii, 412 pp., 2 pls. Stockholm.

Linnaeus, C., 1756. *Amoenitates Academicae* **3**, 464 pp. Wetstenium.
———, 1758. *Systema Naturae* (Edn 10) **1**, 824 pp. Stockholm.
———, 1761. *Fauna Suecica . . .* (Edn 2), xlvi, 578 pp., 2 pls. Stockholm.
———, 1764. *Museum Ludovicae Ulricae Reginae* **1**, viii, 722 pp. Stockholm.
———, 1767. *Systema Naturae* (Edn 12). Lepidoptera, pp. 774–900. Stockholm.
Mac Cana, P., 1983. *Celtic mythology* (Edn 2), 144 pp. Feltham, Middlesex.
Macleod, R. D., 1959. *Key to the names of British butterflies and moths*, 86 pp. London.
MBGBI, see Heath & Emmet.
Meyrick, 1886. On the classification of the Pterophoridae. *Trans. ent. Soc. Lond.* **1886**: 1–21.
———, 1895. *A handbook of British Lepidoptera*, viii, 843 pp. London.
———, 1925. *In* Wytsman, M. P., Lepidoptera Heterocera Fam. Gelechiadae. *Genera Insectorum* **184**: 1–290, 5 pls.
———, 1928. *A revised handbook of British Lepidoptera*, vi, 914 pp., text figs. London.
Mouffet, T., 1634. *Insectorum sive minimorum animalium Theatrum*, xx, 362, iv pp., text ills. London.
Müller, O. F., 1764. *Fauna insectorum Fridrichsdalina . . .*, 96 pp. Hafnia & Leipzig.
Nye, I. W. B., 1975. *The generic names of moths of the world* 1 Noctuoidea (part): Noctuidae, Agaristidae and Nolidae, 568 pp., 1 pl. London.
Obraztsov, N. S., 1960. Die Gattung der Palaearktischen Tortricidae II. Die Unterfamilie Olethreutinae (part). *Tijdschr. Ent.* **103**: 111–143, text figs, pls.
Ochsenheimer, F., 1807–16. *Die Schmetterlinge von Europa*, **1–4**. Leipzig.
Ó Corráin, D. & Maguire, F., 1981. *Gaelic personal names*, 188 pp. Dublin.
Ovid (Publius Ovidius Naso), *c.* 6 A.D. *Metamorphoses*, **1–15**.
Petiver, J., 1695–1703a. *Musei Petiveriani centuria* **1–10**. London.
———, 1702b–1706. *Gazophylacium naturae et artis* **1–10**, 12 pp., 100 pls. London.
———, 1717. *Papilionum Britanniae icones*, 2 pp., 6 pls. London.
Pickard, H. A. *et al.* (Council members of Oxford University Entomological Society and Cambridge Entomological Society), 1858. *An accentuated list of the British Lepidoptera*, xlvi, 118 pp. London.
Pierce, F. N. & Metcalfe, J. W., 1922. *The genitalia of the group Tortricidae of the Lepidoptera of the British Isles*, xxii, 101 pp., 34 pls. Oundle.
Pliny the Elder (Caius Plinius Secundus), *c.* 60 A.D. *Naturalis Historia*, **8–11**.
Ray, J., 1710. *Historia Insectorum*, xv, 400 pp. London.
———, 1724. *Synopsis methodica Stirpium Britannicarum* (Edn 3), xii, 482, xxx pp. London.
Réaumur, R. A. F., 1734–43. *Mémoires pour servir à l'histoire des insectes*, **1–6**. Paris.
Rennie, J., 1832. *A conspectus of butterflies and moths found in Britain*, xxxvii, 287 pp. London.
Retzius, A. I., 1783. *Genera et species insectorum . . .*, 220 pp. Leipzig.
Schäffer, J. C., 1766. *Icones insectorum circa Ratisbonam indigenorum . . .* **1**, 100 pls. Ratisbon.
Schnack, K., (Ed.), 1985. *Catalogue of the Lepidoptera of Denmark*, 163 pp. Copenhagen.
Schrank, F. von Paula von, 1798–1803. *Fauna Boica, durchgedachte Geschichte der in Baiern einheimischen und sahmen Thiere* **1–3**. Nuremburg.
Scopoli, J. A., 1763. *Entomologia carniolica exhibens Insecta Carnoliae indigena et distributa in ordines, genera, species, varietates. Methodo linnaeana.* xxxvi, 422 pp., 43 pls. Vienna.
———, 1777. *Introductio ad Historiam naturalem, sistens genera lapidum. plantarum et animalium . . .*, [vi], 506, [34] pp. Prague.
Skinner, B., 1984. *Colour identification guide to moths of the British Isles*, x, 267 pp., 42 col. pls, 57 text figs. Harmondsworth.
Smith, W., 1855. *A Latin-English dictionary*, vi, 1214 pp. London.
———, 1877. *A smaller classical dictionary* (Edn 17), viii, 464 pp. London.
Sodoffsky, C. H. G., 1837. Etymologische Untersuchungen die Gattungsnamen der Schmetterlinge. *Bull. Soc. Nat. Moscou* **1837**(6): 76–99.
South, R., 1907–08. *The moths of the British Isles* **1, 2**. London.
———, 1961. *The moths of the British Isles* (Edn 4) **1, 2**. London.

Spuler, A., 1903–10. *Die Schmetterlinge Europas* **1–4**. Stuttgart.

Stainton, H. T., 1848. A monograph on the British Argyromyges. *Zoologist* **6**: 2079–2097, 2152–2164.

———, 1849a. Descriptions of new British Micro-Lepidoptera. *Zoologist* **7**: lxi–lxiv.

———, 1849b. On the species of *Depressaria*, a group of Tineina, and the allied genera *Orthotaelia* and *Exaeretia*. *Trans. ent. Soc. Lond.* **5**: 151–173.

———, 1851. *A supplementary catalogue of the British Tineidae & Pterophoridae*, 28 pp. London.

———, 1853. An introduction to the study of Nepticulae. *Zoologist* **11**: 2952–2960.

———, 1854. *Insecta Britannica*. Lepidoptera: Tineina, viii, 313 pp., 10 pls. London.

———, 1855–73. *The natural history of the Tineina* **1–13**. London. [References to this work are indicated by an asterisk.]

———, [1857]. Lepidoptera. New species in 1857. *Entomologist's Annu.* **1858**: 85–98.

———, 1859. A new Tinea. *Ent. wkly Intell.* **6**: 183.

———, [1865]. *In* Knaggs, H. G., Notes on British Lepidoptera (excepting Tineina) for 1865. *Entomologist's Annu.* **1866**: 147–149.

———, 1887. Note to Ragonot, E. L., *Coleophora Mühligiella*. *Entomologist's mon. Mag.* **24**: 41–42.

Staudinger, O., 1859. Diagnosen nebst kurzen Beschreibungen neuer andalusischer Lepidopteren. *Stettin. ent. Ztg* **1859**: 211–259.

Stephens, J. F., 1827–35. *Illustrations of British entomology, . . .* Haustellata **1–4**. London.

Tams, W. H. T., 1939. Further notes on the generic names of British moths. *Entomologist* **72**: 133–141.

Treitschke, F., 1825–35. *Die Schmetterlinge von Europa* **5–10**. Dresden & Leipzig.

Virgil (Virgilius Maro), *c*. 40 B.C. *Eclogues*.

———, *c*. 20 B.C. *Aeneid* **1–12**.

Walker, F., 1854–66. *List of specimens of lepidopterous insects in the collection of the British Museum* **1–35**. London.

Westwood, J. O., 1838. *The entomologist's text-book*, x, 432 pp., 4 pls, text figs. London.

Westwood, J. O., 1855. *The butterflies of Great Britain with their transformations delineated and described*, xl, 140 pp., 19 col. pls. London.

Wilkes, B., 1741–42. *Twelve new designs of British butterflies*, 12 col. pls. London.

———, 1747–49. *The English moths and butterflies . . .* , 8, [22], 64, [4] pp. 120 col. pls. London.

Zeller, P. C., 1839. Versuch einer naturgemässen Eintheilung der Schaben. *Isis, Leipzig* **1839**: 167–220.

———, 1846. Die Arten der Blattminierergattung Lithocolletis. *Linn. ent.* **1**: 166–261.

———, 1848. Die Gattungen der mit Augendeckeln versehenen blattminierenden Schaben. *Ibid.* **3**: 248–344.

———, 1849. Beitrag zur Kenntnis der Coleophoren. *Ibid.* **4**: 191–416.

———, 1850. Verzeichniss der von Herrn Jos. Mann beobachteten Toscanischen Microlepidoptera (part). *Stettin. ent. Ztg* **11**: 195–212.

———, 1854. Die *Depressaria* und einige ihnen nahe stehende Gattungen. *Linn. ent.* **9**: 189–403.

Zincken, J. L., 1818. Monographie der Gattung *Phycis*. *Magazin Ent. (Germar)* **3**: 116–176.

(See Additional Reference on p. 229).

Index

Conventions

(i) Generic and specific names in current use are in Roman type. Suprageneric names are in capitals.
(ii) Synonyms are in italics.
(iii) Names in current use and their synonyms are followed by their Log Book number (see p. 9), supraspecific names by the first Log Book number to follow.
(iv) Non-British species are listed under the Log Book number of the British species under which they are cited or that of the first species to follow.
(v) Homonyms are followed by the generic name or, if both occur in the same genus, by the author's name. Subspecific homonyms are followed by the specific name. Names consisting of the same adjective are treated as homonyms, even if they differ in gender termination, e.g. *formosana* and *formosanus* (see p. 36).
(vi) Principal page numbers are given in bold type.
(vii) 'Historical' names, i.e. those once used with a different meaning from that which they hold today, have no Log Book number and are followed by *usu* (as used by) and an author's name. For instance, *Noctua* Linnaeus is the current genus, NOCTUA *usu* Linnaeus is the second of his seven divisions of the Phalaenae; PYRALIS *usu* Fabricius is a synonym of the present Tortricidae. Early authors inflected these names, so the plural form is given where appropriate.
(viii) Taxonomic revision has resulted in certain species and genera being moved so that they now appear out of Log Book numerical sequence. In such cases an asterisk precedes the Log Book number and the name should be located by reference to the page number. The most extensive revision is in the Phycitinae (Log Book nos 1432–1486) and the subfamily is therefore given a special index on p. 288.
(ix) A name may be explained by reference to the synonym of another genus or species. The synonym is then given the Log Book number of the genus or species under which it is cited; for example, *Scotia* is a synonym of 2084 *Agrotis*, but is indexed as 2475 *Scotia*, having been cited in the explanation of 2475 *Parascotia*.

Index

Special Index to the Phycitinae (see p. 135)

List of Monochrome Plates